Business Statistics for Competitive Advantage with Excel 2016

Cynthia Fraser

Business Statistics for Competitive Advantage with Excel 2016

Basics, Model Building, Simulation and Cases

Springer

Cynthia Fraser
McIntire School of Commerce
University of Virginia
Charlottesville, VA, USA

ISBN 978-3-319-32184-4 ISBN 978-3-319-32185-1 (eBook)
DOI 10.1007/978-3-319-32185-1

Library of Congress Control Number: 2016939396

Printed on acid-free paper

This Springer imprint is published by Springer Nature
The registered company is Springer International Publishing AG Switzerland

Table of Contents

Preface

Exceptional managers know that they can create competitive advantages by basing decisions on performance response under alternative scenarios. To create these advantages, managers need to understand how to use statistics to provide information on performance response under alternative scenarios. Statistics are created to make better decisions. Statistics are essential and relevant. Statistics must be easily and quickly produced using widely available software, Excel. Then results must be translated into general business language and illustrated with compelling graphics to make them understandable and usable by decision makers. This book helps students master this process of using statistics to create competitive advantages as decision makers.

Statistics are essential, relevant, easy to produce, easy to understand, valuable, and a powerful source of competitive advantage.

The examples, assignments, and cases used to illustrate statistics for decision making come from business problems

McIntire Corporate Sponsors and Partners, such as Alcoa, Rolls-Royce, Procter & Gamble, and Dell, and the industries that they do business in, provide many realistic examples. The book also features a number of examples of global business problems, including those from important emerging markets in China, India, and Chile. Students are excited when statistics are used to study real and important business problems. This makes it easy to see how they will use statistics to create competitive advantages in their internships and careers.

Learning is hands on with Excel and shortcuts

Each type of analysis is introduced with one or more examples. Following is an example of how to create the statistics in Excel, and what the numbers mean in English.

Included in Excel sections are screenshots which allow students to easily master Excel. Featured are a number of popular Excel shortcuts, which are, themselves, a competitive advantage.

Powerful PivotTables and PivotCharts are introduced early and used throughout the book. Results are illustrated with graphics from Excel.

In each chapter, assignments or cases are included to allow students to practice using statistics for decision making and competitive advantage. Beginning in Chapter 11, Harvard Business School cases are suggested which provide additional opportunities to use statistics to advantage.

Focus is on what statistics mean to decision makers and how to communicate results

From the beginning, results are translated into English. In Chapter 7, results are condensed and summarized in PowerPoints and memos, the standards of communication in businesses. Later chapters include example memos for students to use as templates, making communication of statistics for decision making an easy skill to master.

Instructors, give your students the powerful skills that they will use to create competitive advantages as decision makers. Students, be prepared to discover that statistics are a powerful competitive advantage. Your mastery of the essential skills of creating and communicating statistics for improved decision making will enhance your career and make numbers fun.

New in the Fourth Edition

The financial and economic events of 2008–2010 changed business dramatically. Examples have been updated to illustrate how the impacts of recent changes can be acknowledged to build powerful, valid models.

Global examples include analysis of several emerging markets multinationals in Chile. New emerging markets create unique opportunities for global business, and the Fourth Edition moves beyond the BRICs to explore these.

Acknowledgements

First, Second and Third editions of *Business Statistics for Competitive Advantage* were used in the Integrated Core Curriculum at The McIntire School, University of Virginia, and I thank the many bright, motivated and enthusiastic students who provided comments and suggestions.

Cynthia Fraser
Charlottesville, VA

Chapter 1
Statistics for Decision Making and Competitive Advantage

In the increasingly competitive global arena of business in the Twenty First century, the select few business graduates distinguish themselves by enhanced decision making backed by statistics. Statistics are useful when they are applied to improve decision making. No longer is the production of statistics confined to quantitative analysis and market research divisions in firms. Managers in each of the functional areas of business use statistics daily to improve decision making. Excel and other statistical software live in our laptops, providing immediate access to statistical tools which can be used to improve decision making.

1.1 Statistical Competences Translate into Competitive Advantages

The majority of business graduates can create descriptive statistics and use Excel. Fewer have mastered the ability to frame a decision problem so that information needs can be identified and satisfied with statistical analysis. Fewer can build powerful and valid models to identify performance drivers, compare decision alternative scenarios, and forecast future performance. Fewer can translate statistical results into general business English that is easily understood by everyone in a decision making team. Fewer have the ability to illustrate memos with compelling and informative graphics. Each of these competences provides competitive advantage to those few who have mastery. This text will help you to attain these competences and the competitive advantages which they promise.

Most examples in the text are taken from real businesses and concern real decision problems. A number of examples focus on decision making in global markets. By reading about how executives and managers successfully use statistics to increase information and improve decision making in a variety of mini-case applications, you will be able to frame a variety of decision problems in your firm, whether small or multi-national. The end-of-chapter assignments will give you practice framing diverse problems, practicing statistical analyses, and translating results into easily understood reports or presentations.

Many examples in the text feature bottom line conclusions. From the statistical results, you read what managers would conclude with those results. These conclusions and implications are written in general business English, rather than statistical jargon, so that anyone on a decision team will understand. Assignments ask you to feature bottom line conclusions and general business English.

Translation of statistical results into general business English is necessary to insure their effective use. If decision makers, our audience for statistical results, don't understand the conclusions and implications from statistical analysis, the information created by analysis will not be used. An appendix is devoted to writing memos that your audience will read and understand, and to effective PowerPoint slide designs for effective presentation of results. Memos and PowerPoints are predominant forms of communication in businesses. Decision making is compressed and information must be distilled, well written and illustrated. Decision makers read memos. Use memos to make the most of your analyses, conclusions and recommendations.

C. Fraser, *Business Statistics for Competitive Advantage with Excel 2016*,
DOI 10.1007/978-3-319-32185-1_1

In the majority of examples, analysis includes graphics. Seeing data provides an information dimension beyond numbers in tables. To understand well a market or population, you need to see it, and its shape and dispersion. To become a master modeler, you need to be able to see how change in one variable is driving a change in another. Graphics are essential to solid model building and analysis. Graphics are also essential to effective translation of results. Effective memos and PowerPoint slides feature key graphics which help your audience digest and remember results. PivotTables and PivotCharts are featured in Chapters 2 and 9. These are routinely used in business to efficiently organize and effectively display data. When you are at home in the language of PivotTables and PivotCharts, you will have a competitive advantage. Practice using PivotTables and PivotCharts to organize financial analyses and market data. Form the habit of looking at data and results whenever you are considering decision alternatives.

1.2 The Path Toward Statistical Competence and Competitive Advantage

This text assumes basic statistical knowledge, and reviews basics quickly. Basics form the foundation for essential model building. Chapters 2 and 3 present a concentrated introduction to data and their descriptive statistics, samples and inference. Learn how to efficiently describe data and how to infer population characteristics from samples.

Inference from Monte Carlo simulation based on a decision maker's assumptions is introduced in Chapter 4, and revisited in Chapters 6 and 13. Model building with simple regression begins in Chapter 5 and occupies the focus of much of the remaining chapters. To be competitive, business graduates must have competence in model building and forecasting. A model building mentality, focused on performance drivers and their synergies is a competitive advantage. Practice thinking of decision variables as drivers of performance. Practice thinking that performance is driven by decision variables. Performance will improve if this linkage becomes second-nature.

The approach to model building is steeped in logic and begins with logic and experience. Models must make sense in order to be useful. When you understand how decision variables drive performance under alternate scenarios, you can make better decisions, enhancing performance. Model building is an art that begins with logic.

Model building chapters include nonlinear regression. Nearly all aspects of business performance behave in nonlinear ways. We see diminishing or increasing changes in performance in response to changes in drivers. It is useful to begin model building with the simplifying assumption of constant response, but it is essential to be able to grow beyond simple linear models to realistic models which reflect nonconstant response. Visualize the changing pattern of response when you consider decision alternatives and the ways they drive performance.

1.3 Use Excel for Competitive Advantage

This text features widely available Excel software, including many commonly used shortcuts. Excel is powerful, comprehensive, and user friendly. Appendices with screenshots follow each chapter to make software interactions simple. Recreate the chapter examples by following the steps in the Excel sections. This will give you confidence using the software. Then forge ahead and generalize your analyses by working through end of chapter assignments. The more often you use the statistical tools and software, the easier analysis becomes.

1.4 Statistical Competence Is Powerful and Yours

Statistics and their potential to alter decisions and improve performance are important to you. With more and better information from statistical analysis, you will be equipped to make superior decisions and outperform the competition. You will find that the competitive advantages from statistical competence are powerful and yours.

Chapter 2
Describing Your Data

This chapter introduces *descriptive* statistics, center, spread, and distribution shape, which are almost always included with any statistical analysis to characterize a dataset. The particular descriptive statistics used depend on the *scale* that has been used to assign numbers to represent the characteristics of entities being studied. When the distribution of continuous data is bell shaped, we have convenient properties that make description easier. Chapter 2 looks at dataset types and their description.

2.1 Describe Data with Summary Statistics and Histograms

We use numbers to measure aspects of businesses, customers and competitors. These measured aspects are *data*. Data become meaningful when we use statistics to describe patterns within particular *samples* or collections of businesses, customers, competitors, or other entities.

Example 2.1 Yankees' Salaries: Is It a Winning Offer? Suppose that the Yankees want to sign a promising rookie. They expect to offer $1M, and they want to be sure they are neither paying too much nor too little. What would the General Manager need to know to decide whether or not this is the right offer?

He might first look at how much the other Yankees earn. Their 2005 salaries are in Table 2.1:

Table 2.1 Yankees' salaries (in $MM) in alphabetical order

Crosby	$.3	Johnson	$16.0	Posada	$11.0	Sierra	$1.5
Flaherty	.8	Martinez	2.8	Rivera	10.5	Sturtze	.9
Giambi	1.34	Matsui	8.0	Rodriguez	21.7	Williams	12.4
Gordon	3.8	Mussina	19.0	Rodriguez F	3.2	Womack	2.0
Jeter	19.6	Phillips	.3	Sheffield	13.0		

What should he do with this data?

Data are more useful if they are ordered by the aspect of interest. In this case, the Manager would re-sort the data by salary (Table 2.2):

Table 2.2 Yankees sorted by salary (in $MM)

Rodriguez	$21.7	Williams	$12.4	Rodriguez F	$3.2	Sturtze	$.9
Jeter	19.6	Posada	11.0	Martinez	2.8	Flaherty	.8
Mussina	19.0	Rivera	10.5	Womack	2.0	Crosby	.3
Johnson	16.0	Matsui	8.0	Sierra	1.5	Phillips	.3
Sheffield	13.0	Gordon	3.8	Giambi	1.3		

C. Fraser, *Business Statistics for Competitive Advantage with Excel 2016*,
DOI 10.1007/978-3-319-32185-1_2

Now he can see that the lowest Yankee salary, the *minimum*, is \$300,000, and the highest salary, the *maximum*, is \$21.7M. The difference between the maximum and the minimum is the *range* in salaries, which is \$21.4M, in this example. From these statistics, we know that the salary offer of \$1M falls in the lower portion of this range. Additionally, however, he needs to know just how unusual the extreme salaries are to better assess the offer.

He'd like to know whether or not the rookie would be in the better paid half of the Team. This could affect morale of other players with lower salaries. The *median,* or middle, salary is \$3.8M. The lower paid half of the team earns between \$300,000 and \$3.8M, and the higher paid half of the team earns between \$3.8M and \$21.7M. Thus, the rookie would be in the bottom half. The Manager needs to know more to fully assess the offer.

Often, a *histogram* and a *cumulative distribution plot* are used to visually assess data, as shown in Figures 2.1 and 2.2. A histogram illustrates central tendency, dispersion, and symmetry.

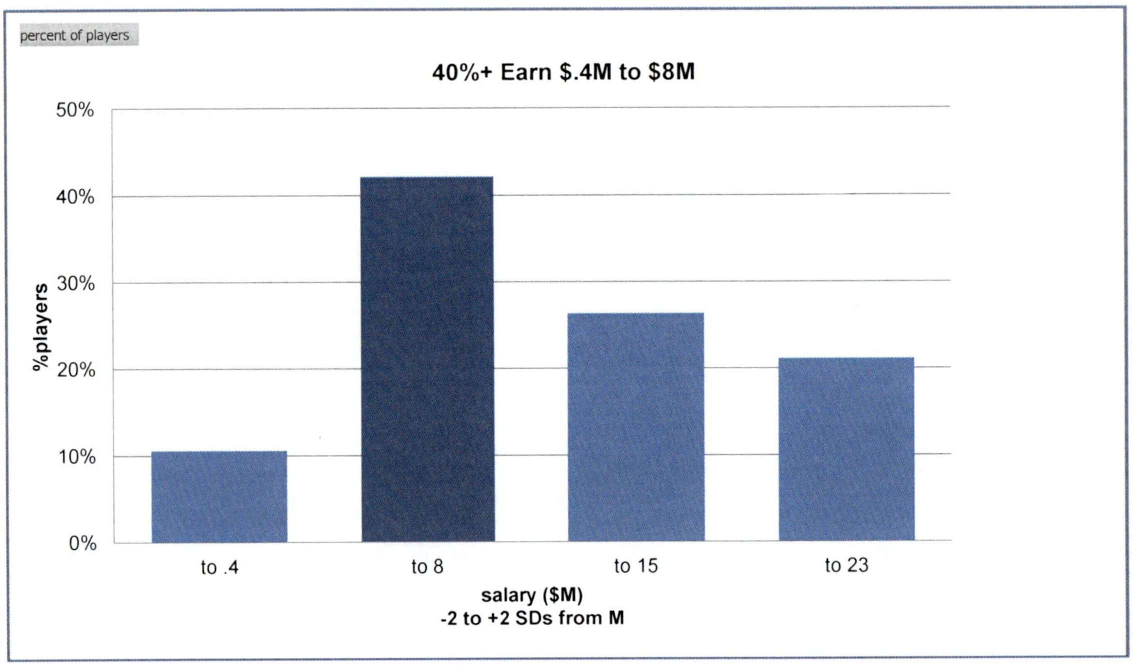

Figure 2.1 Histogram of Yankee salaries

The histogram of team salaries shows us that a large proportion, more than 40%, earn more than \$400,000, but less than the average, or *mean*, salary of \$8M.

The cumulative distribution makes it easy to see the median, or 50th percentile, which is one measure of central tendency. It is also easy to find the *interquartile range*, the range of values that the middle 50% of the datapoints occupy, providing a measure of the data dispersion.

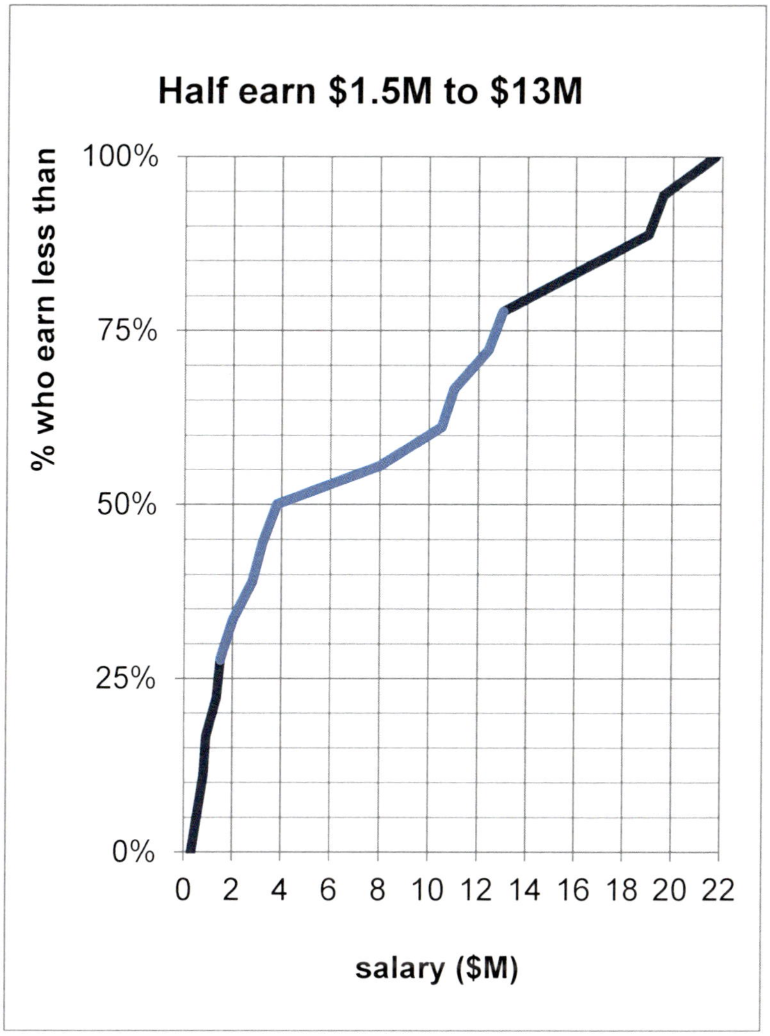

salary ($MM)	
25%	1.4
median	4
75%	13

Figure 2.2 Cumulative distribution of salaries

The cumulative distribution reveals that the *Interquartile Range,* between the 25th percentile and the 75th percentile, is more than $10M. A quarter earns less than $1.4M, the 25th percentile, about half earn between $1.5 and $13M, and a quarter earns more than $13M, the 75th percentile. Half of the players have salaries below the *median* of $4M and half have salaries above $4M.

Example 2.2 Executive Compensation: Is the Board's Offer on Target? The Board of a large corporation is pondering the total compensation package of the CEO, which includes salary, stock ownership, and fringe benefits. Last year, the CEO earned $2,000,000. For comparison, The Board consulted Forbes' summary of the total compensation of the 500 largest corporations. The histogram, cumulative frequency distribution and descriptive statistics are shown in Figures 2.3 and 2.4.

Total Compensation ($M)	*% Executives*
< 1.0	43
1.1 to 2.0	27
2.1 to 3.0	13
3.1 to 4.0	7
4.1 to 5.0	2
5.1 to 6.0	2
6.1 to 7.0	1
7.1 to 8.0	1
8.1 to 9.0	1
9.1 to 10.0	0
10.1 to 11.0	0
>11.0	3

Figure 2.3 Histogram of executive compensation

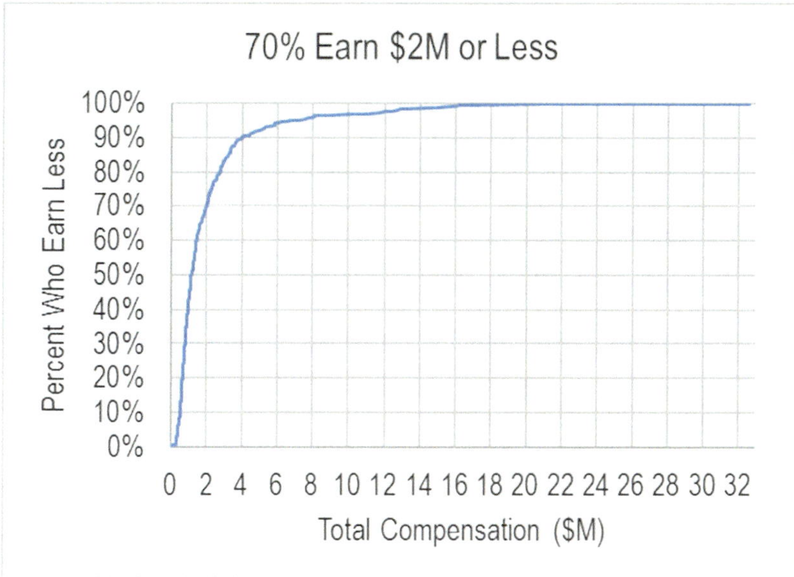

	Total Compensation ($M)
M	2.1
SD	3.0
75th percentile	2.3
median	1.1
25th percentile	.7

Figure 2.4 Cumulative distribution of total compensation

The average executive compensation in this sample of large corporations is $2.1M. Half the sample of 447 executives earns $1.1M (the median) or less. One quarter earns less than $.7M, the middle half, or *interquartile range,* earns between $.7M and $2.3M, and one quarter earns more than $2.3M.

2.2 Round Descriptive Statistics

In the examples above, statistics in the output from statistical packages are presented with many decimal points of accuracy. The Yankee manager in Example 2.1 and The Board considering executive compensation in Example 2.2 will most likely be negotiating in hundred thousands. It would be distracting and unnecessary to report descriptive statistics with significant digits more than two or three. In the **Yankees** example, the average salary is $8,000,000 (*not* $7,797,000). In the **Executive Compensation** example, average total compensation is $2,200,000 (not $2,215,262.66). It is deceptive to present results with many significant digits, creating an illusion of precision. In addition to being honest, statistics in two or three significant digits are much easier for decision makers to process and remember. If more significant digits don't affect a decision, round to fewer and make your statistics easier to process and remember.

2.3 Share the Story That Your Graphics Illustrate

Use your graphics to support the conclusion you have reached from your analysis. Choose a "bottom line" title that shares with your audience what it is that they should be able to see. Often this title should relate specifically to your reasons for analyzing data. In the executive compensation example, The Board is considering a $2M offer. The chart titles capture Board interest by highlighting this critical value. The "bottom line," that a $2M offer is relatively high, when compared with similar firms, makes the illustrations relevant.

Many have the unfortunate and unimaginative habit of choosing chart titles which name the type of chart. "Histogram of executive salaries" tells the audience little, beyond the obvious realization that they must form their own, independent conclusions from the analysis. Choose a "bottom line" title so that decision makers can take away your conclusion from the analysis. Develop the good habit of titling your graphics to enhance their relevance and interest.

2.4 Data Is Measured with Quantitative or Categorical Scales

If the numbers in a dataset represent amount, or magnitude of an aspect, **and** if differences between adjacent numbers are equivalent, the data are *quantitative* or *continuous.* Data measured in dollars (i.e., revenues, costs, prices and profits) or percents (i.e., market share, rate of return, and exam scores) are continuous. Quantitative numbers can be added, subtracted, divided or multiplied to produce meaningful results.

With quantitative data, report central tendency with the *mean, M:*

$$\mu = \frac{\sum x_i}{N} \quad \text{for describing a } \textit{population} \text{ and}$$

$$\overline{X} = \frac{\sum x_i}{N} \quad \text{for describing a } \textit{sample} \text{ from a population,}$$

where x_i are data point values, and

N is the number of data points that we are describing.

The *median* can also be used to assess central tendency, and the *range, variance,* and *standard deviation* can be used to assess dispersion.

The *variance* is the average squared difference between each of the data points and the mean:

$$\sigma^2 = \frac{\sum (x_i - \mu)^2}{N} \quad \text{for a population and}$$

$$s^2 = \frac{\sum (x_i - \overline{X})^2}{(N-1)} \quad \text{for a sample from a population.}$$

The *standard deviation SD,* σ for a population and s for a sample, is the square root of the variance, which gives us a measure of dispersion in the more easily interpreted, original units, rather than squared units.

To assess distribution symmetry, assess its skewness:

$$\text{Skewness} = \frac{n}{(n-1)(n-2)} \sum \left(\frac{x_i - \bar{x}}{s} \right)^3$$

Skewness of zero indicates a symmetric distribution, and skewness between −1 and +1 is evidence of an approximately symmetric distribution.

If numbers in a dataset are arbitrary and used to distinguish categories, the data are *nominal,* or *categorical.* Football jersey numbers and your student ID are nominal. A larger number doesn't mean that a player is better or a student is older or smarter. Categorical numbers can be tabulated to identify the most popular number, occurring most frequently, the *mode,* to report central tendency. Categorical numbers cannot be added, subtracted, divided or multiplied.

Quantitative measures convey the more information, including direction and magnitude, while categorical measures convey the less, sometimes direction, and sometimes, merely category membership. One, more informative type of categorical data are *ordinal* scales that used to rank order data, or to convey direction, but not magnitude. With ordinal data, an element (which could be a business, a person, a country) with the most or best is coded as '1', second place as '2', etc. With ordinal numbers, or rankings, data can sorted, but not added, subtracted, divided or multiplied. As with other categorical data, the mode represents the central tendency of ordinal data.

When focus is on membership in a particular category, the *proportion* of sample elements in the category is a continuous measure of central tendency. Proportions are quantitative and can be added, subtracted, divided or multiplied, though they are bounded by zero, below, and by one, above.

2.5 Continuous Data Are Sometimes Normal

Continuous variables are often *Normally distributed*, and their histograms resemble symmetric, bell shaped curves, with the majority of data points clustered around the mean. Most elements are "average" with values near the mean; fewer elements are unusual and far from the mean.

Skewness reflects lack of symmetry. Normally distributed data have skewness of zero, and approximately Normal data have skewness between −1 and +1.

If continuous data are Normally distributed, we need only the mean and standard deviation to describe this data and description is simplified.

Example 2.3 Normal SAT Scores. Standardized tests, such as SAT, capitalize on Normality. Math and verbal SATs are both specifically constructed to produce Normally distributed scores with *mean* M = 500 and *standard deviation* SD = 100 over the population of students (Figure 2.5):

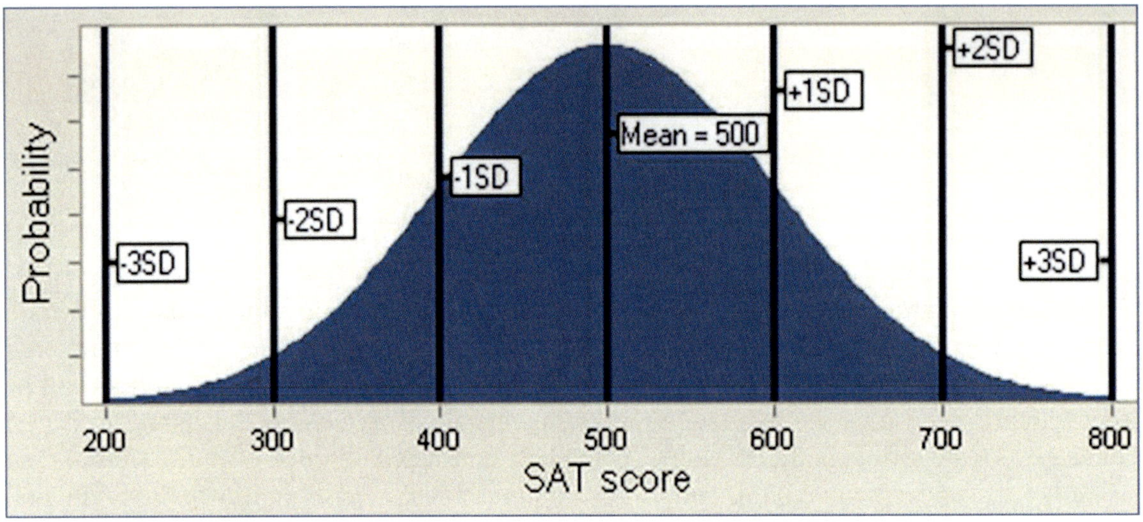

Figure 2.5 Normally distributed SAT scores

2.6 The Empirical Rule Simplifies Description

Normally distributed data have a very useful property described by the *Empirical Rule:*

- 2/3 of the data lie within one standard deviation of the mean
- 95% of the data lie within two standard deviations of the mean

This is a powerful rule! *If data are Normally distributed, data can described with just two statistics: the mean and the standard deviation.*

Returning to SAT scores, if we know that the average score is 500 and the standard deviation is 100, we also know that

- 2/3 of SAT scores will fall within 100 points of the mean of 500, or between 400 and 600,
- 95% of SAT scores will fall within 200 points of the mean of 500, or between 300 and 700.

Example 2.4 Class of Business Students' SATs: Normal & Exceptional. Descriptive statistics and a histograms of Math SATs of a third year class of business students reveal an interquartile range from 640 to 730, with mean of 690 and standard deviation of 70, as shown in Figure 2.6. Skewness is −.5, indicating approximate symmetry, an approximately Normal distribution.

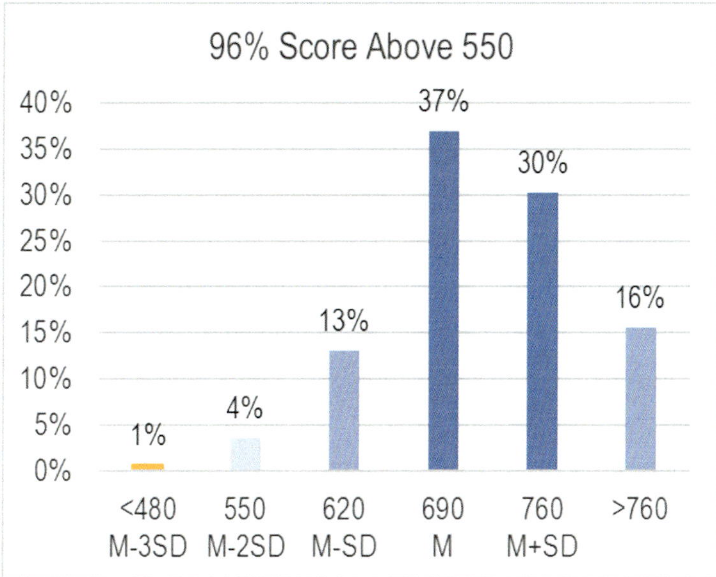

95%	800
90%	790
75%	730
median	690
25%	640
10%	600
5%	560
mean	690
sd	70
skewness	−.4

Figure 2.6 Histograms and descriptive statistics of class of business students' math SATs

These scores are bell shaped. However, there are "too many" perfect scores of 800.

The Empirical Rule would predict that 2/3 of the class would have scores within one standard deviation, 70 points, of the mean of 690, or within the interval 620 to 760. There actually 67% (=37%+30%).

The Empirical Rule would also predict that only 2–1/2% of the class would have scores more than two standard deviations below or above the mean of 690: scores below 550 and above 830. We find that 5% actually do have scores below 550, though none score above 830 (since a perfect SAT score is 800). This class of business students has Math SATs that are nearly Normal, but not exactly Normal.

To summarize students' SAT scores, report:

- Business students' Math SAT scores are approximately Normally distributed with *mean* of 690 and *standard deviation* of 70.
- Relative to the larger population of all SAT takers, the smaller *standard deviation* in business students' Math SAT scores, 70 versus 100, indicates that this class of business students is a more homogeneous group than the more varied population.

2.7 Outliers Can Distort the Picture

Outliers are extreme elements, considered unusual when compared with other sample elements. Because they are extraordinary, they can distort descriptive statistics.

Revisiting the **Executive Compensation** example, why is the *mean*, $2.2M, so much larger than the *median*, $1.1M? There is a group of *outliers,* shown as *MORE* than three standard deviations above the mean in Figure 2.7, who are compensated extraordinarily well. Each collects a compensation package of more than $11.1M, a compensation level that is more than three standard deviations greater than the mean.

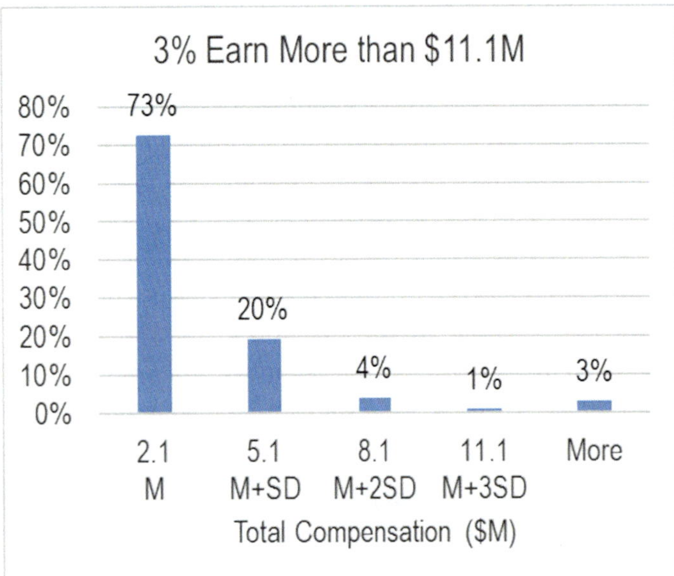

total compensation ($M) 0 to +3 SDs from M	*% Executives*
≤ 2.1	73
2.2 to 5.1	20
5.1 to 8.1	4
8.1 to 11.1	1
>11.1	3

Figure 2.7 Histogram and descriptive statistics by SDs from M

Because extraordinary executives exist, the distribution of compensation is *skewed,* with relatively few exceptional executives being exceptionally well compensated.

2.8 Central Tendency, Dispersion and Skewness Describe Data

The baseball salaries and executive compensation examples focused on two measures of *central tendency*: the *mean*, or average, and the *median*, or middle. Both examples also refer to a measure of *dispersion* or variability: the *range* separating the minimum and maximum. *Skewness* reflects distribution symmetry. SATs are approximately symmetric and Normal; Executive compensation values are skewed. To describe data, we need statistics to assess central tendency, dispersion, and skewness. The statistics we choose depends on the *scale* which has been used to code the data we are analyzing.

2.9 Describe Categorical Variables Graphically

Numbers representing category membership in nominal, or categorical, data are described by tabulating their frequencies. The most popular category is the *mode.* Visually, we show our tabulations with a *Pareto* chart, which orders categories by their popularity.

Example 2.5 Who Is Honest & Ethical? Figure 2.8 shows a column chart of results of a survey of 1014 adults by Gallup:

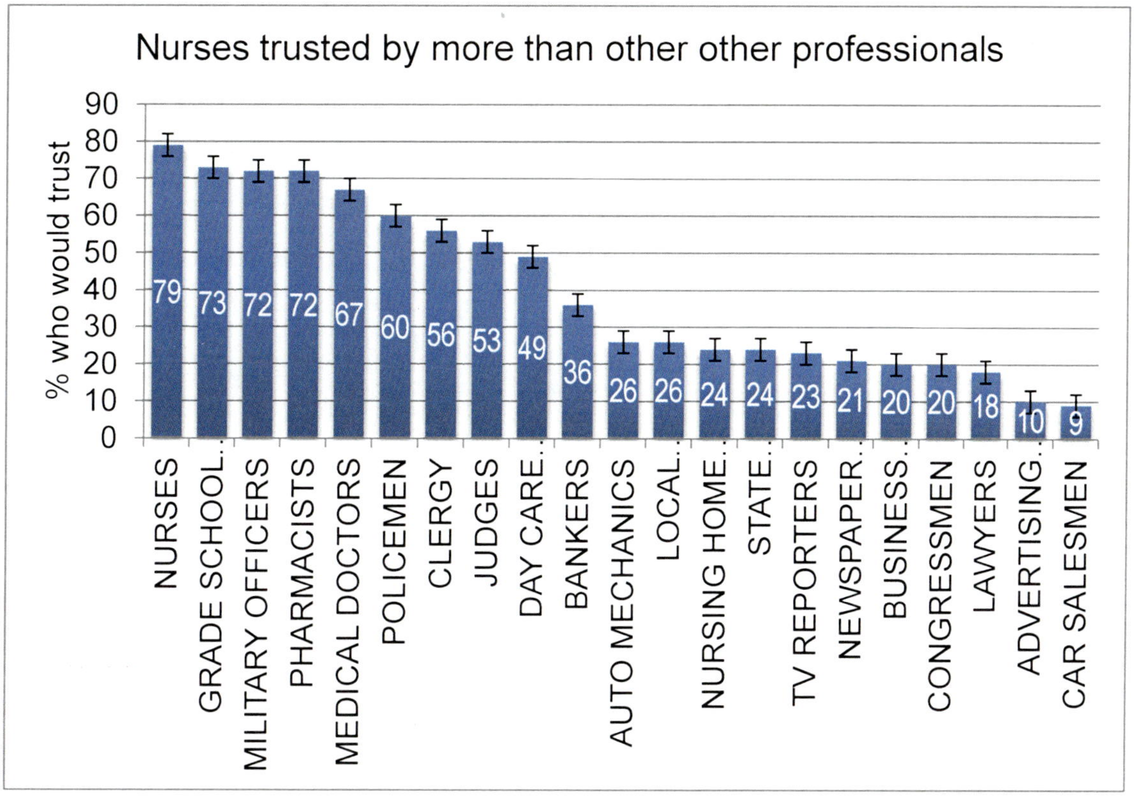

Figure 2.8 Pareto charts of the percents who judge professions honest

More Americans trust and respect nurses (79%, the *modal* response) than people in other professions, including doctors, clergy and teachers. Though a small minority judge business executives (20%) and advertising professionals (10%) as honest and ethical, most do not judge people in those fields to be honest (which highlights the importance of ethical business behavior in the future).

2.10 Descriptive Statistics Depend On The Data and Rely on Your Packaging

Descriptive statistics, graphics, central tendency and dispersion, depend upon the type of scale used to measure data characteristics (i.e., quantitative or categorical).

Table 2.3 summarizes the descriptive statistics (graph, central tendency, dispersion, shape) used for both types of data:

Table 2.3 Descriptive statistics (central tendency, disperson, graphics) for two types of data

	Quantitative	**Categorical**
Central tendency	*mean*	*mode*
	median	*proportion*
Dispersion	*range*	
	standard deviation	
Symmetry	*skewness*	
Graphics	*histogram*	*Pareto chart*
	cumulative distribution	*pie chart*
		column chart

If continuous data are Normally distributed, a dataset can be completely described with just the mean and standard deviation. We know from the *Empirical Rule* that 2/3 of the data will lie within one standard deviation of the mean and that 95% of the data will lie within two standard deviations of the mean.

Effective results are those which are remembered and used to improve decision making. Your presentation of results will influence whether or not decision makers remember and use your results. Round statistics to two or three significant digits to make them honest, digestible, and memorable. Title your graphics with the "bottom line," to guide and facilitate decision makers' conclusions.

Excel 2.1 Produce Descriptive Statistics

Executive Compensation. We will describe executive compensation packages by producing descriptive statistics, a histogram and cumulative distribution.

First, freeze the top row of **Excel 2.1 Executive Compensation** so that column labels are visible when you are at the bottom of the dataset.

From the first cell, **A1,**
Alt WFR
(The shortcuts, activated with **Alt** select the vie**W** menu , the **F**reeze panes menu, and then freeze **R**ows.)

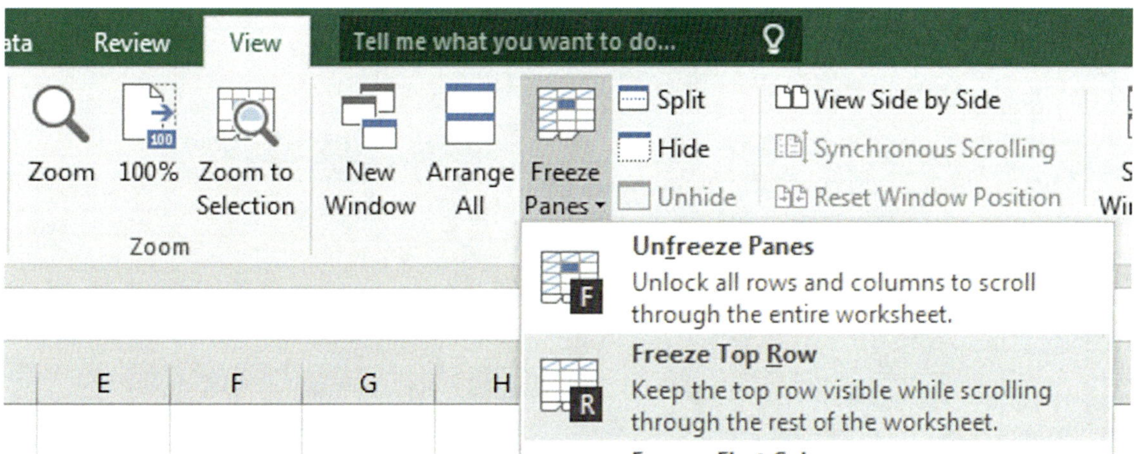

Descriptive Statistics.

Turn on the Excel statistics add-in, Analysis ToolPak.

Alt TI

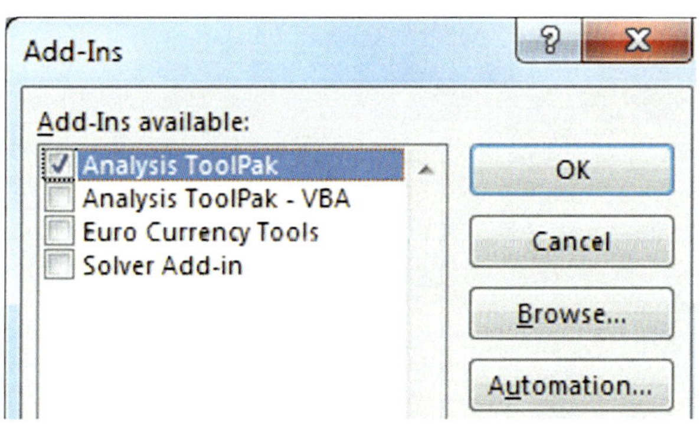

Request Descriptive Statistics.	**Alt AY*n* D** The number ***n*** varies. Enter what you see in the menu, which is 3 on this computer. **b1:b447 tab LS**

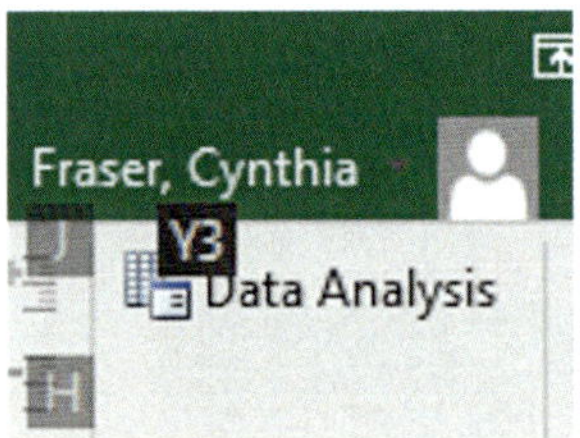

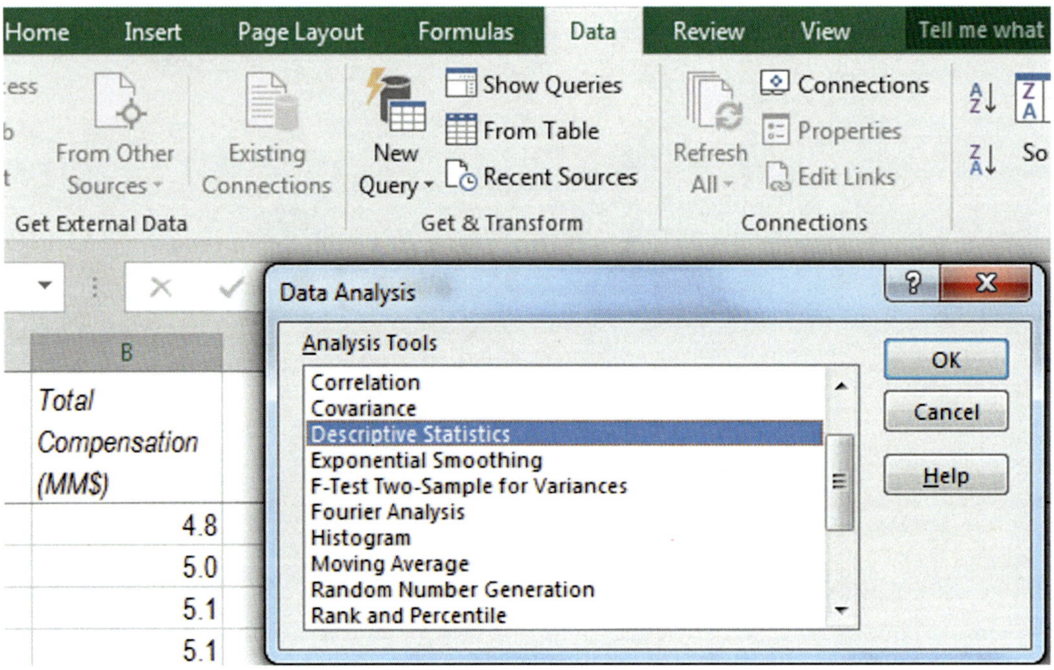

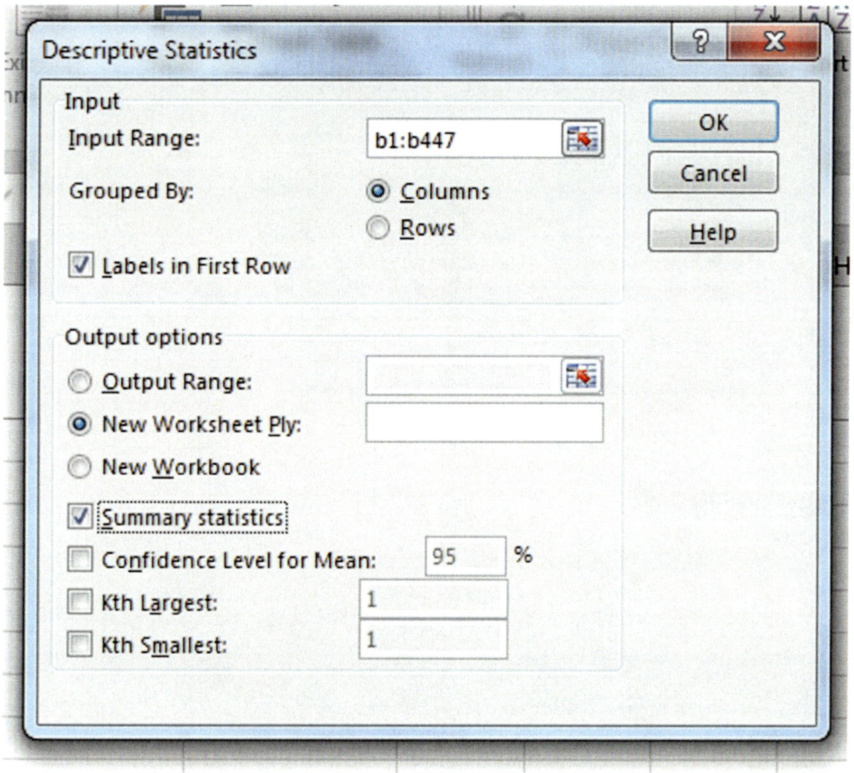

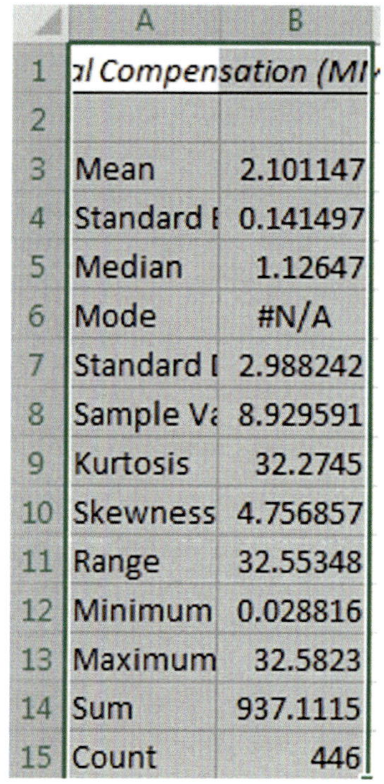

	A	B
1	al Compensation (MI	
2		
3	Mean	2.101147
4	Standard E	0.141497
5	Median	1.12647
6	Mode	#N/A
7	Standard I	2.988242
8	Sample Va	8.929591
9	Kurtosis	32.2745
10	Skewness	4.756857
11	Range	32.55348
12	Minimum	0.028816
13	Maximum	32.5823
14	Sum	937.1115
15	Count	446

Set up Histogram Bins. To make a histogram of compensation, Excel needs to know what ranges of values to combine. To take advantage of the *Empirical Rule,* create *bins*, or categories, using differences from the approximate sample mean that are in widths of approximate standard deviations. In this case, the mean is about 2 and the standard deviation is about 3.

Move back to the data page, and then move to the bottom of the data.	**Cntl+Page Down** In column B **Cntl+down arrow**

Excel uses bin values to set the upper limit for each category. Start with a bin with upper limit equal to 2, which will include compensation values that are at less than or equal to 2.

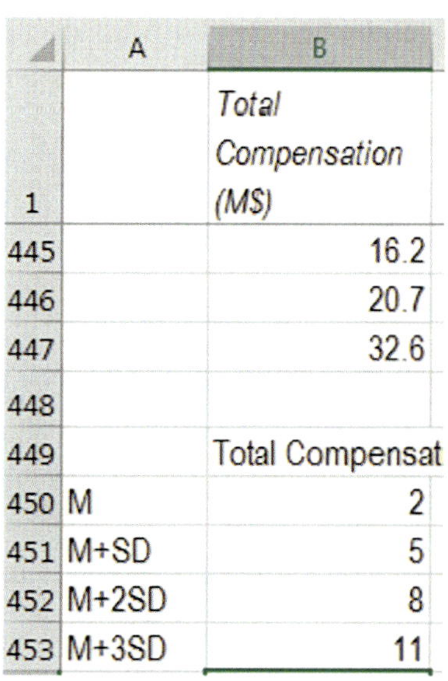

	A	B
1		*Total Compensation (M$)*
445		16.2
446		20.7
447		32.6
448		
449		Total Compensat
450	M	2
451	M+SD	5
452	M+2SD	8
453	M+3SD	11

This will be the first bin, since subtracting one standard deviation from the mean produces a negative number, and none of the executives earns negative salary dollars

In each of the three cells below this first bin, add one SD to the cell above, 3, creating bins with upper limits of M + 1SD, M + 2SD and M + 3SD.

Request a tabulation.

Alt AY*n* H
b1:b447 tab b449:b453 tab LM

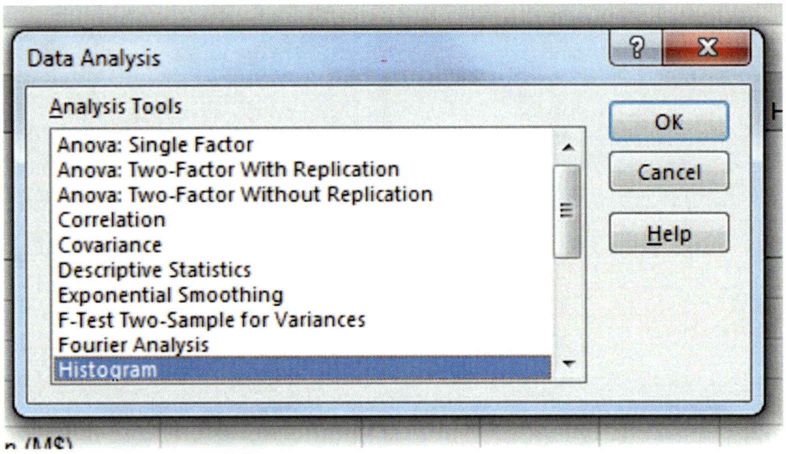

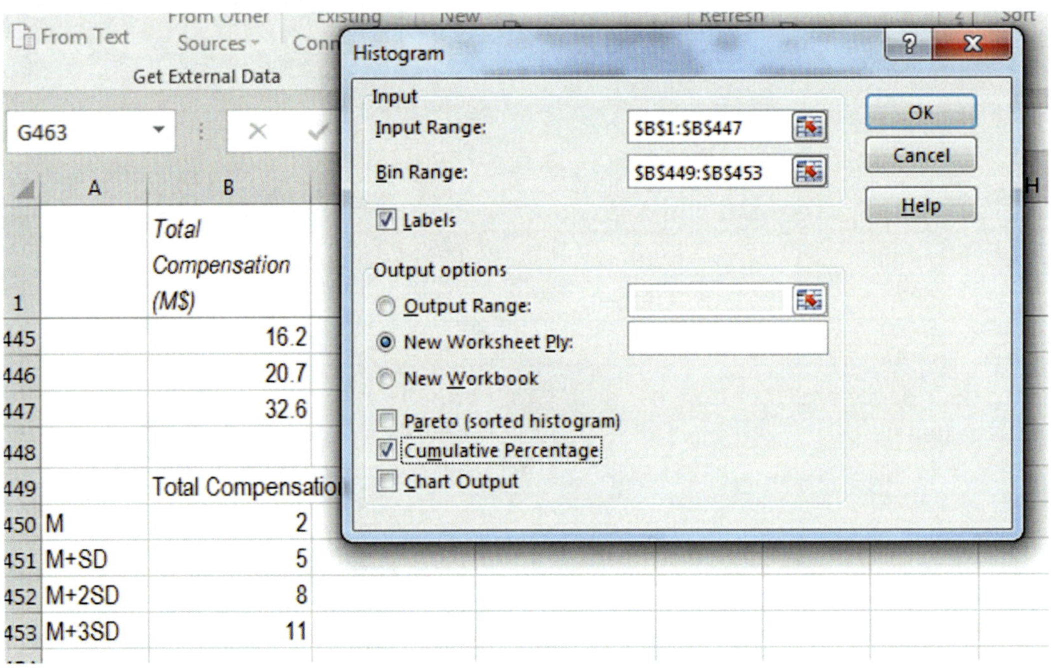

	A	B	C
1	›mpensati	Frequency	›mulative
2	2.0	312	69.96%
3	5.0	99	92.15%
4	8.0	17	95.96%
5	11.0	5	97.09%
6	More	13	100.00%

To produce a histogram showing percents of the sample in each compensation category, add a column D of the change in cumulative%.

d2=c2
d3=c3-c2
Select d3
Shift+down arrow to d6
Cntl+D

D3 | f_x | =C3-C2

	A	B	C	D
1	›mpensati	Frequency	›mulative	%
2	2.0	312	69.96%	69.96%
3	5.0	99	92.15%	22.20%
4	8.0	17	95.96%	3.81%
5	11.0	5	97.09%	1.12%
6	More	13	100.00%	2.91%

To increase readability of the histogram, change category labels in column A to indicate ranges and add M, M+SD, M+2SD, M+3SD:

	A	
1	Total Compensation (M$)	411 Fr
2	<2 M	
3	3 to 5 M+SD	
4	6 to 8 M+2SD	
5	9 to 11 M+3SD	
6	More	
7		

The histogram will plot percents in column D by categories in column A. Move column D to column B.

From column D
Cntl+spacebar,
Cntl+X
left arrow key to column B **Cntl+spacebar**
Alt HIE

B	C	D
%	*Frequency*	*Cumulative %*
69.96%	312	69.96%
22.20%	99	92.15%
3.81%	17	95.96%
1.12%	5	97.09%
2.91%	13	100.00%

Excel often shows more decimals than are desired. Select the percents in column B and reduce decimals.

From B2
Cntl+shift+down
Alt H9

B	
%	*Fre*
70%	
22%	
4%	
1%	
3%	

Produce the histogram by selecting data in columns A and B, then request a column chart.

From A1
Cntl+shift+down
Shift+right
Alt NC

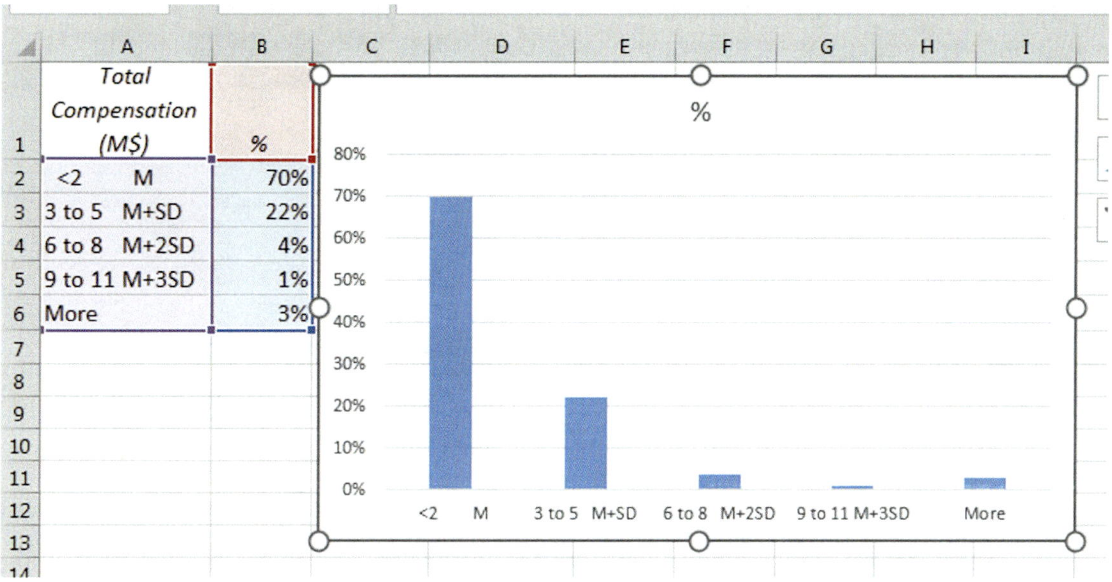

Add a horizontal axis title. | **Alt JCAAH**

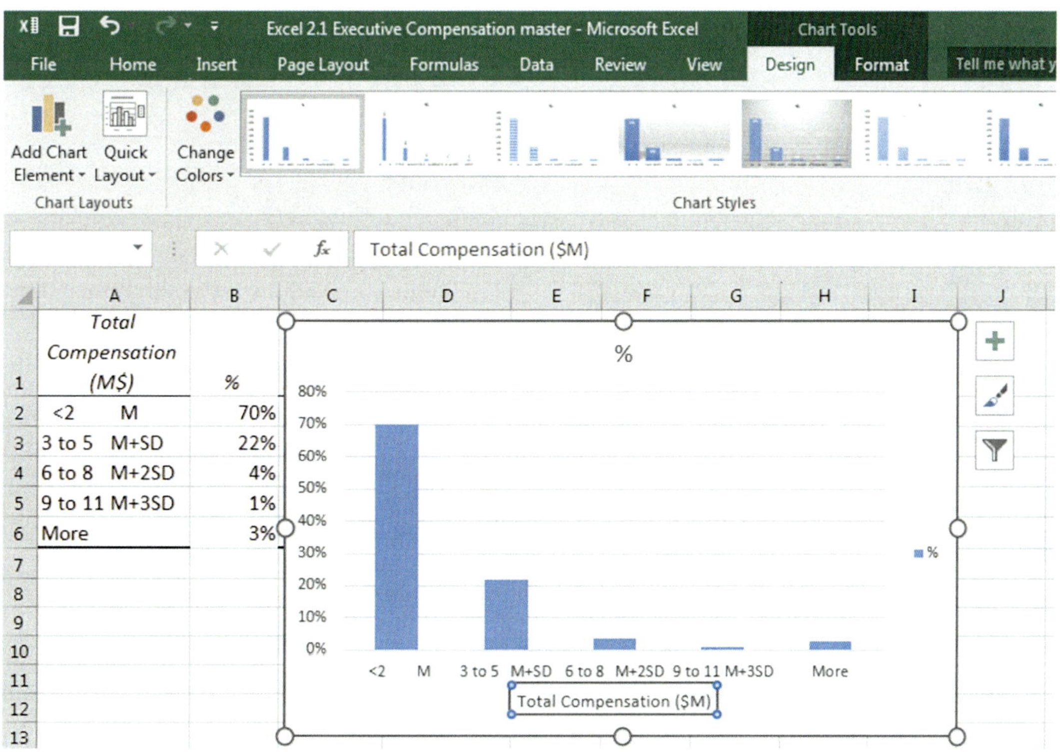

To make better use of the chart space, reformat the vertical axis, setting the maximum to 70%. | **Alt JAE down to Vertical (Value) Axis Alt JAM**

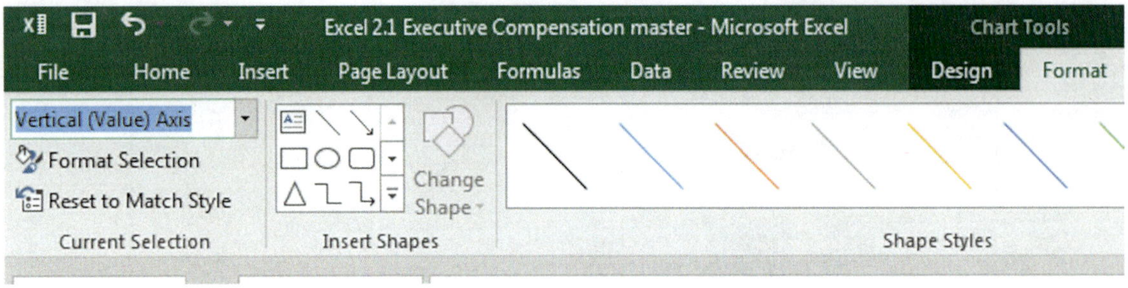

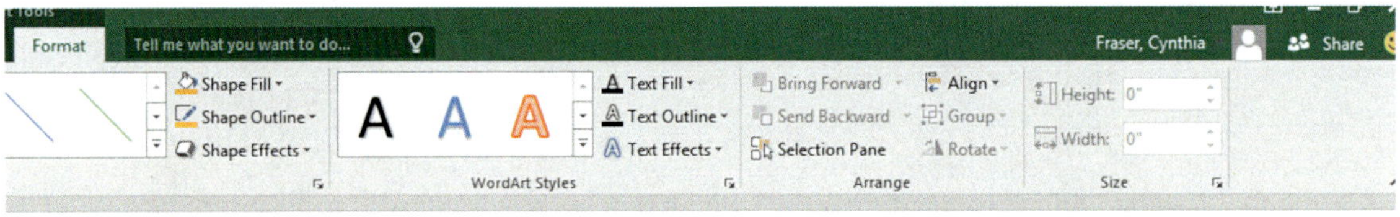

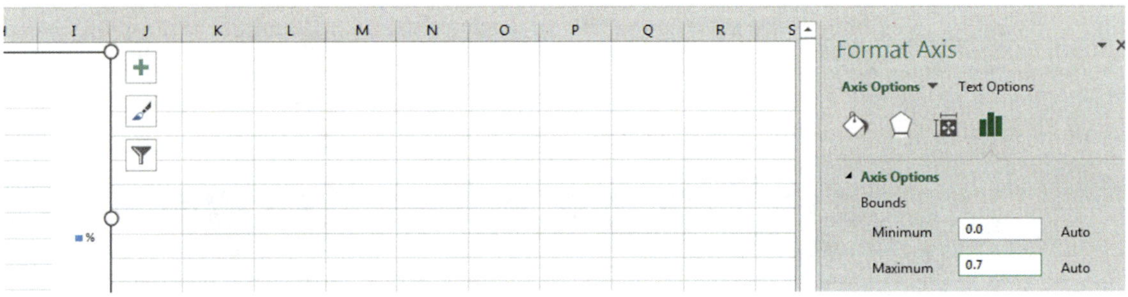

Increase the font size.

Click the outside edge of the chart, then **Alt HFS** 12

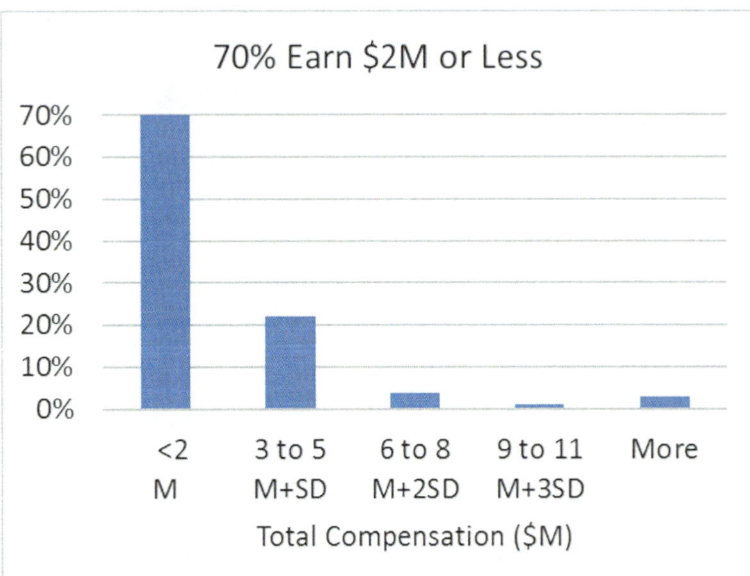

Delete the legend and replace the chart title with a stand alone title:

Excel 2.2 Sort to Produce Descriptives Without Outliers

To easily identify and remove outliers, sort the rows from lowest to highest *total compensation ($M)*. Move back to the data page, select *total compensation* data in column **B** (but not the histogram bins below the data), then use shortcuts to sort:

Cntl+Page Down
From B1,
Cntl+Shift+Down
Alt ASS

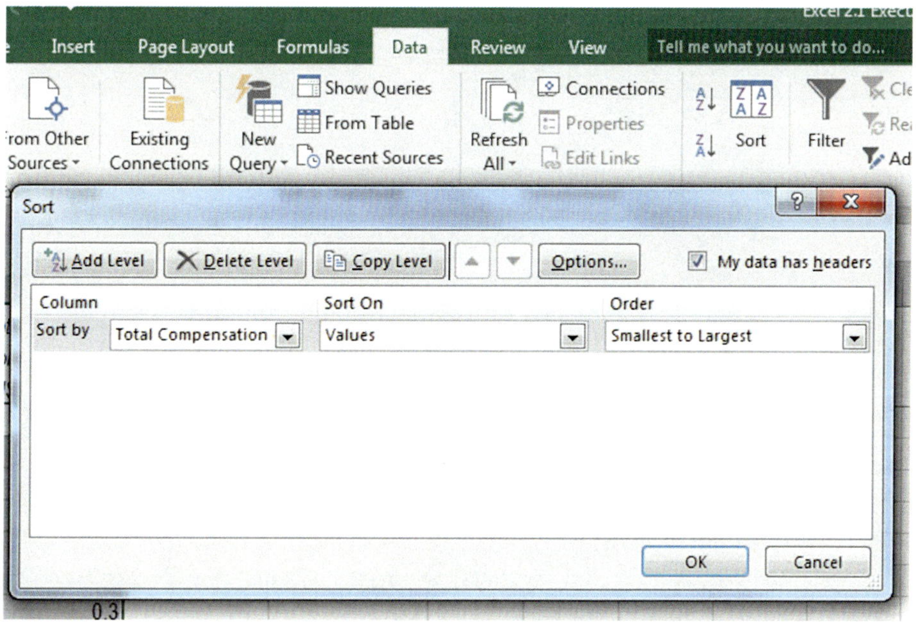

Move to the end of B, and then scroll up to identify the rows with compensation within 3SDs of the mean, 11.0 or less.

Cntl+Down
Up arrow

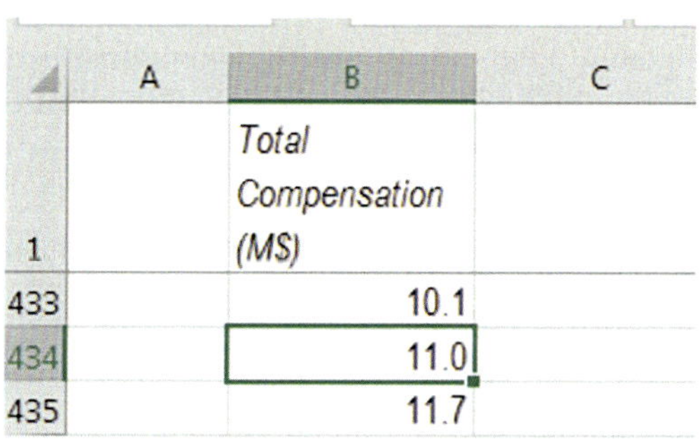

Run descriptives, again, changing the input range to b1:b434.

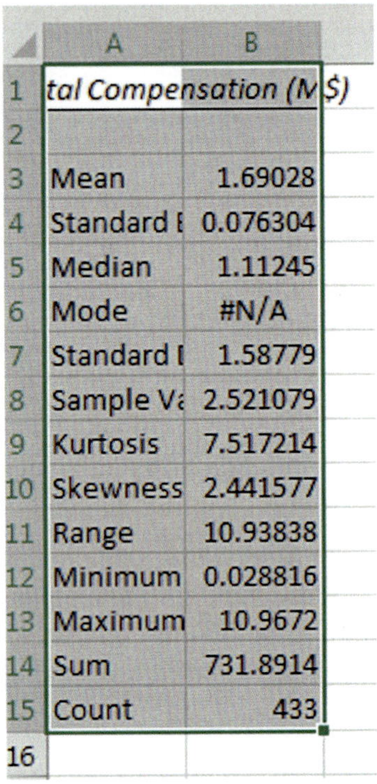

	A	B
1	tal Compensation (M$)	
2		
3	Mean	1.69028
4	Standard E	0.076304
5	Median	1.11245
6	Mode	#N/A
7	Standard I	1.58779
8	Sample Va	2.521079
9	Kurtosis	7.517214
10	Skewness	2.441577
11	Range	10.93838
12	Minimum	0.028816
13	Maximum	10.9672
14	Sum	731.8914
15	Count	433
16		

Excel 2.3 Plot a Cumulative Distribution

Return to the data page and request the cumulative distribution of total compensation.

Cntl+Page Dn
Alt AY*n*, R down
B1:b447 tab L

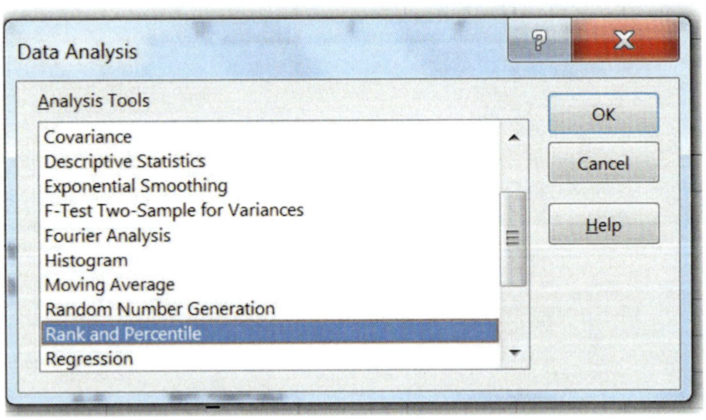

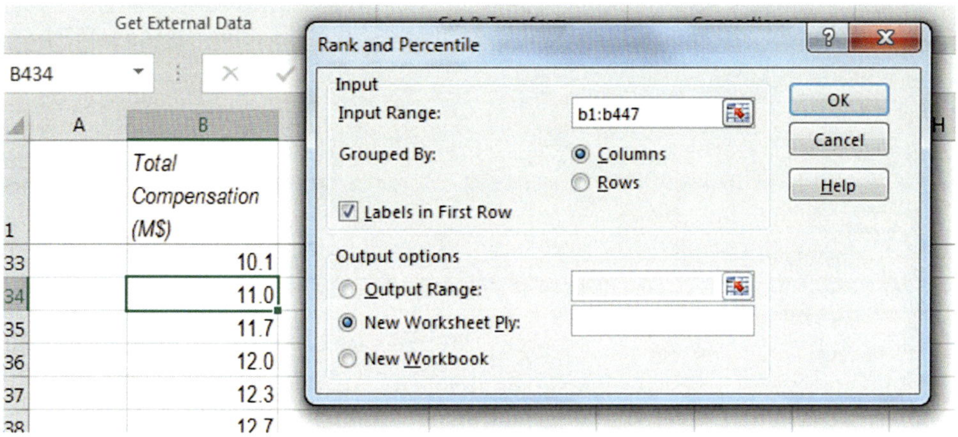

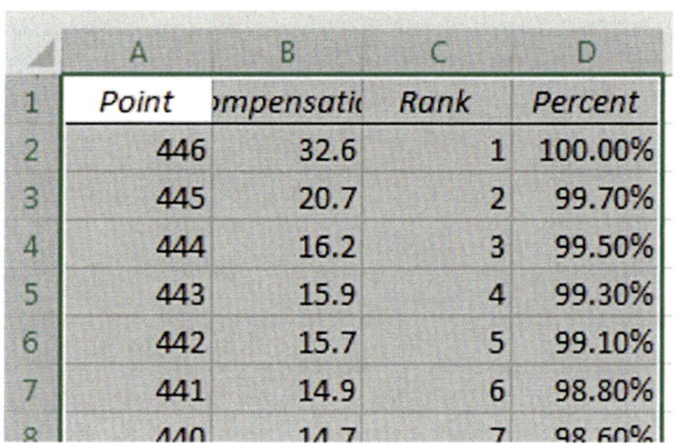

Excel will plot cumulative percents in column D by compensation in column B. For convenience, delete column C. Select the cumulative percent data now in column C and reduce decimals, and then select columns B and C and insert a scatterplot showing the cumulative distribution.

From column C,
Alt HDC.
(**H** selects the **H**ome menu, **D** selects the **D**elete menu, and **C** deletes the **C**olumn.)
From C2
Cntl+shift+down
Alt H9
From B1
Cntl+shift+down
Shift+right
Alt ND
Choose the last scatter option.

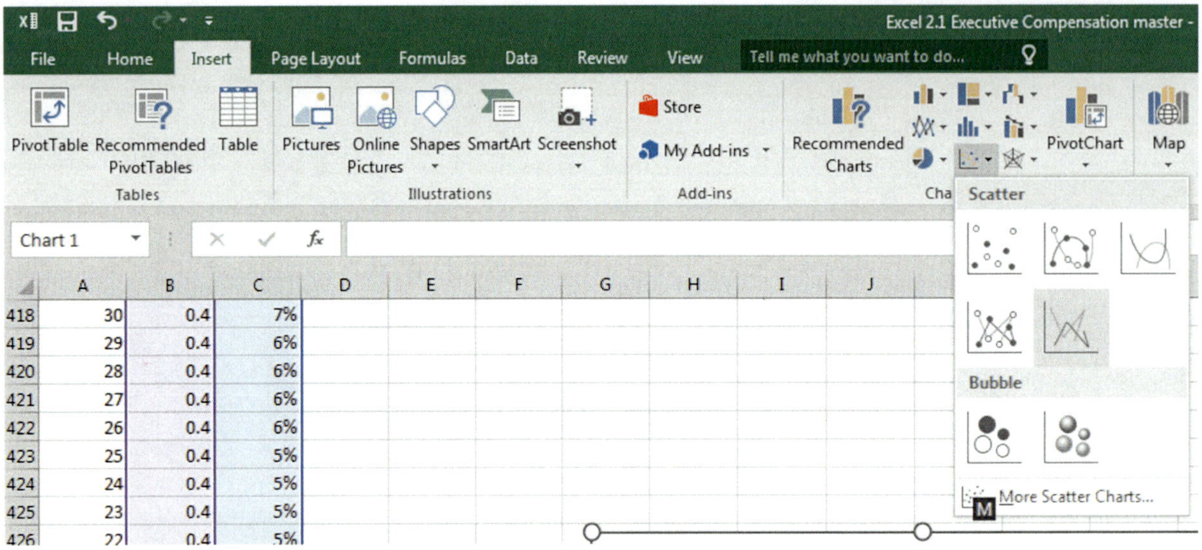

Use shortcuts to choose design layout 1 to add axes labels and title. Delete the legend and adjust font size to 12.	**Alt JCL** **Alt HFS** 12
Use shortcuts to select and format axes. Set the vertical axis maximum at 100%. Set the horizontal axis maximum at 11, with 0 decimals.	**Alt JAE** **Alt JAM**
Use shortcuts to add vertical gridlines:	**Alt JCAGV**

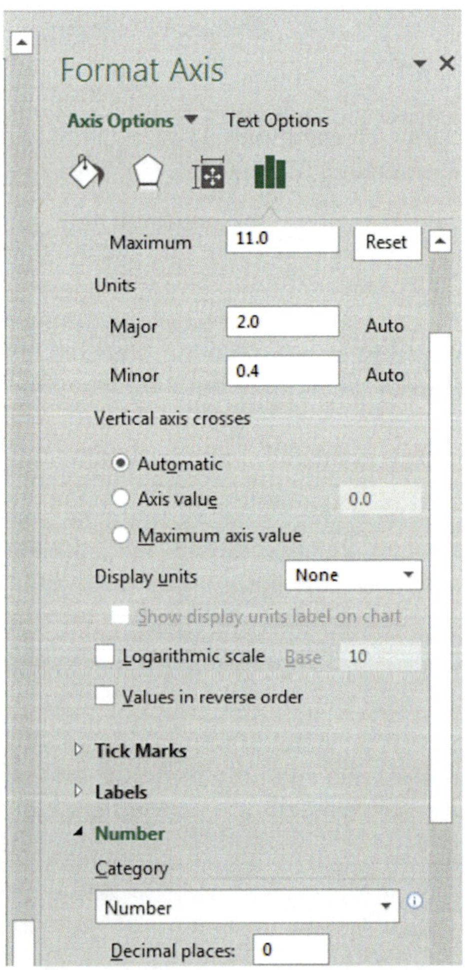

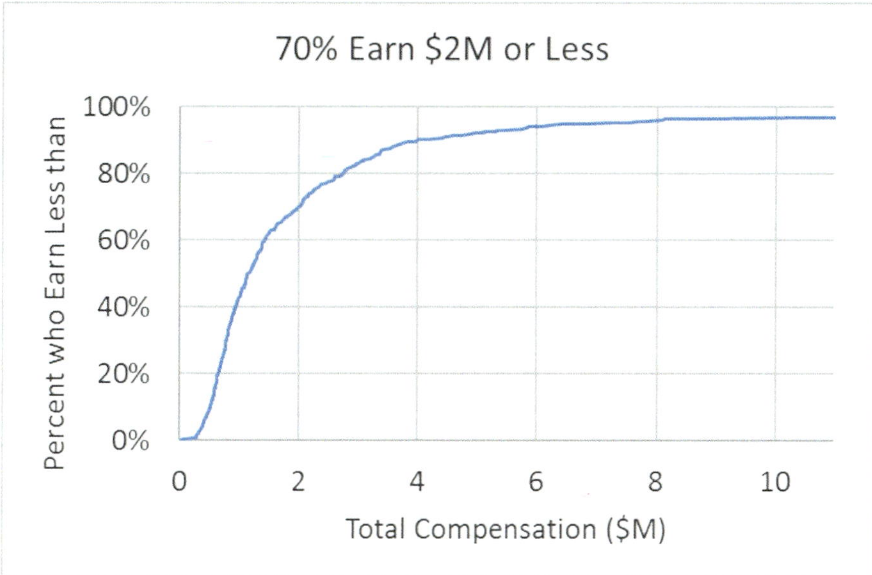

Excel 2.4 Use a PivotTable to Sort by Industry

Use a PivotTable to sort the 200 best paid CEOs by industry. Open **Lab 2 Highest Paid CEOs 2014**.

Select compensation and gender data in columns C and D. Insert a PivotTable to compare compensation by gender.

From C1,
Cntl+shift+down
Shift+right
Alt NV
Drag Gender to the Rows.
Drag Compensation to the Sum Values

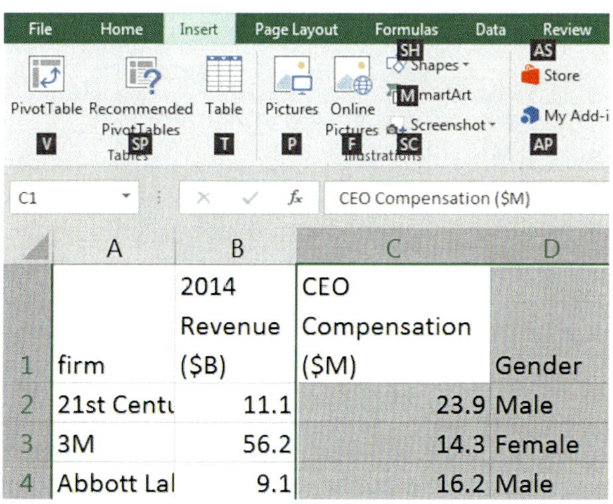

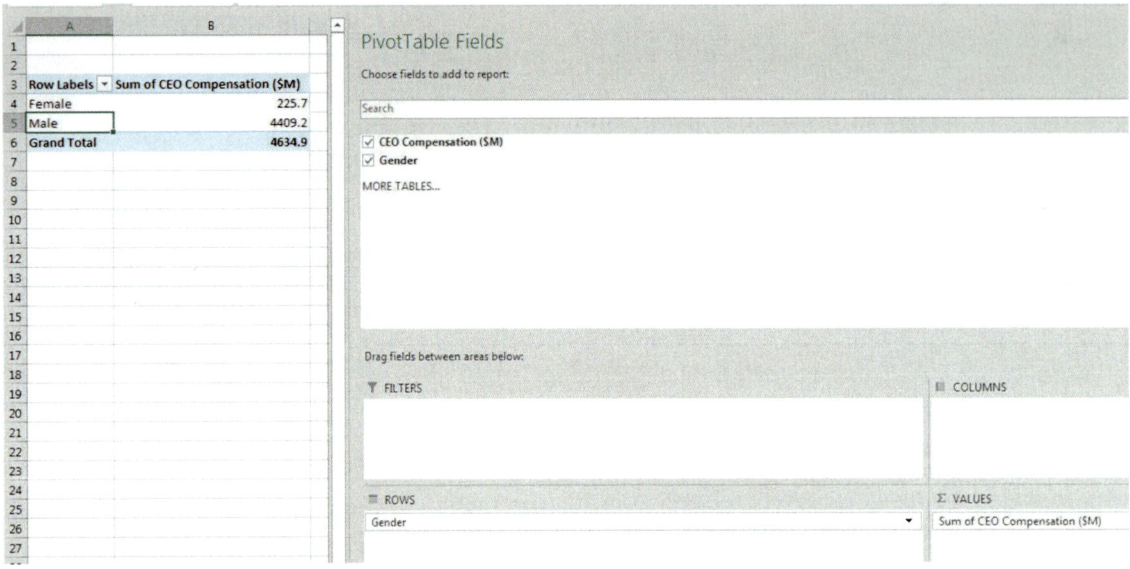

Excel shows the sums in PivotTables.

Convert the sums to averages.
Alt JTG tab tab down to Average

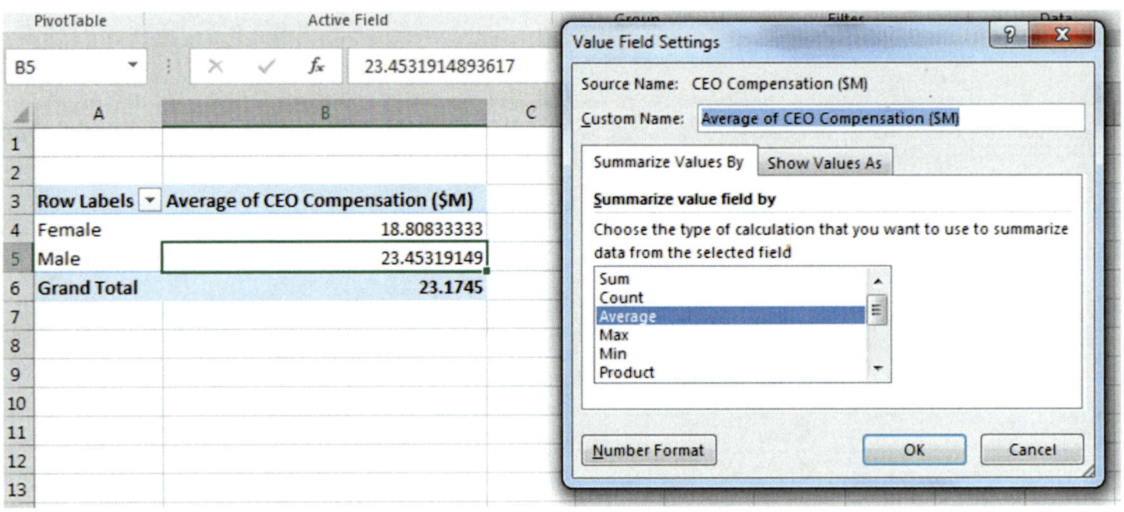

Reduce decimals.

From B4,
Shift+down
Alt H9

Show compensation by gender in a PivotChart.

Alt JTC

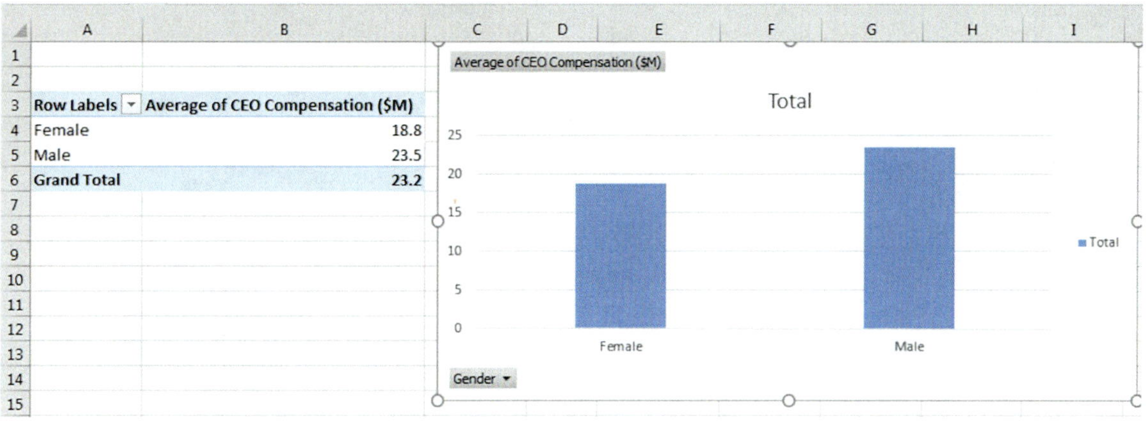

Choose the ninth design layout, add axis titles and a stand alone chart title, delete the legend, and adjust fontsize to 12.

Alt JCL
Alt HFS 12

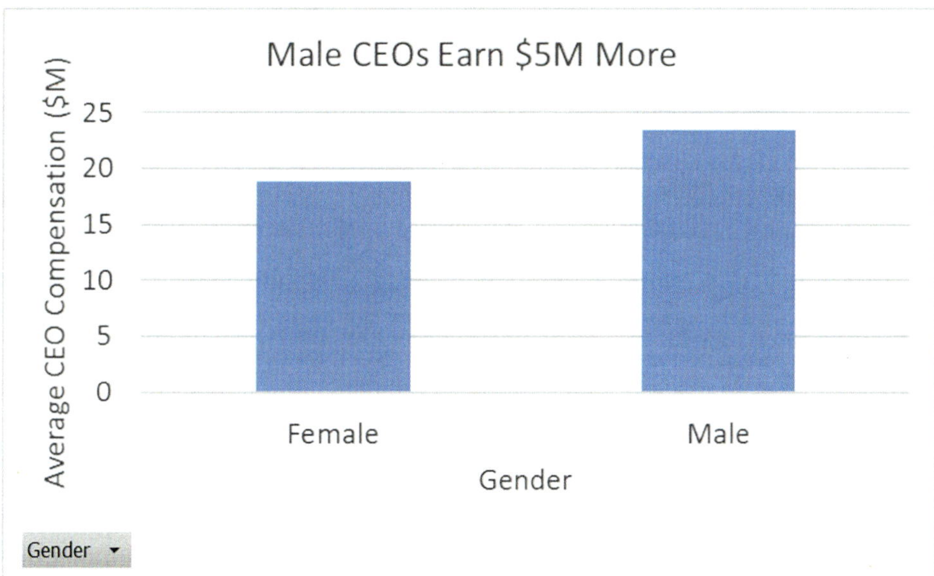

Excel 2.5 Produce a Column Chart of a Nominal Variable

To show percents who choose alternate media for daily news, produce a column chart from Gallup Poll of 992 Americans.

Open **Excel 2.2 News Sources,** select the *News Source* and *% Who Get Daily News Here* data**,** and insert a column chart.

Alt NC

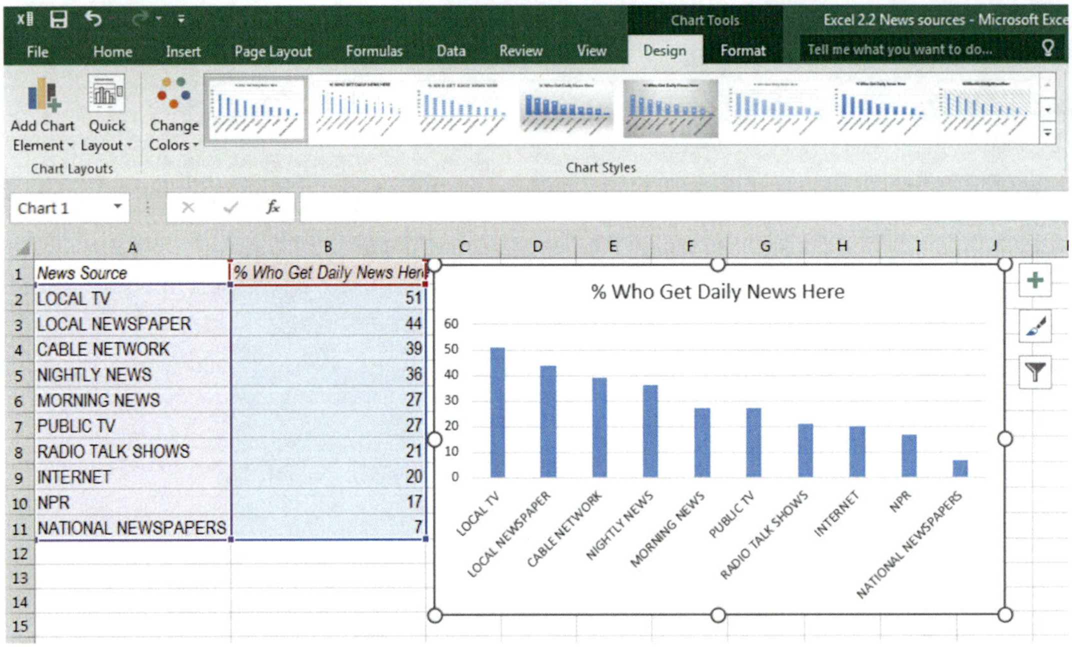

Choose **Design Chart Layout 9** and type in a stand alone title and axes titles. | **Alt JCL**

Add vertical margin of error bars fixed at the approximate margin of error, 3. | **Alt JCAEM**

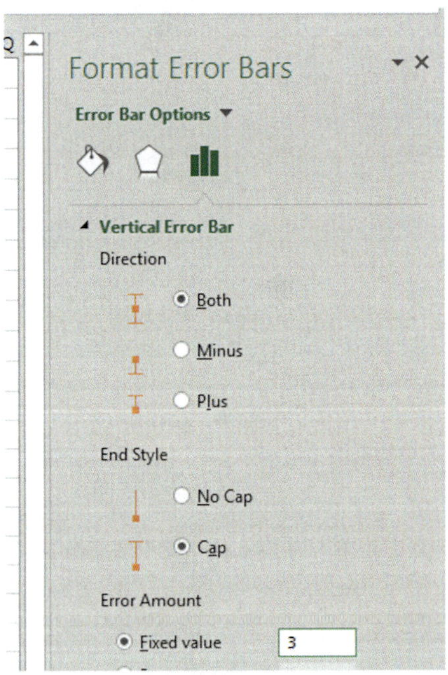

Set fontsize to 12. | **Alt HFS** 12

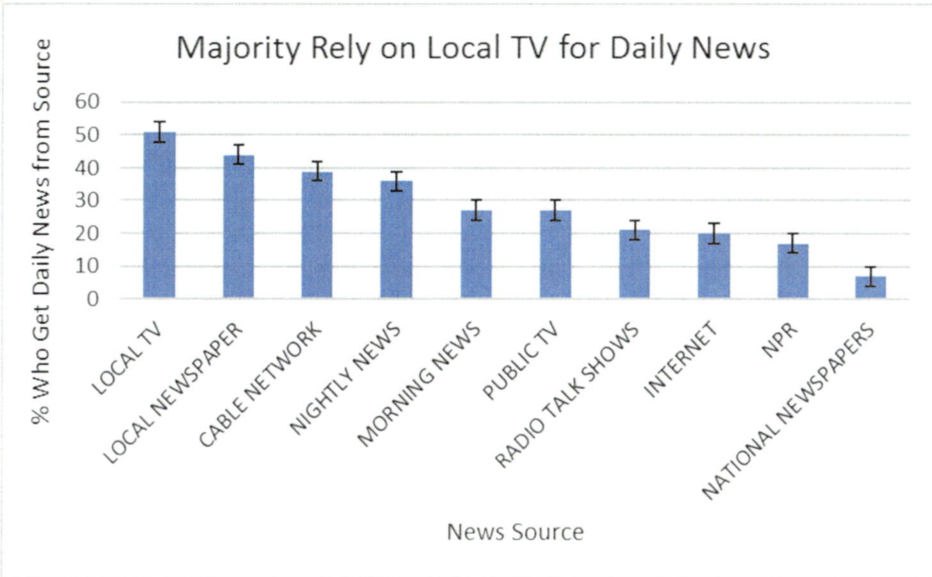
Majority Rely on Local TV for Daily News
% Who Get Daily News from Source
60
50
40
30
20
10
0
LOCAL TV
LOCAL NEWSPAPER
CABLE NETWORK
NIGHTLY NEWS
MORNING NEWS
PUBLIC TV
RADIO TALK SHOWS
INTERNET
NPR
NATIONAL NEWSPAPERS
News Source

Excel Shortcuts Used in Chapter 2

Home menu shortcuts	
Insert c**E**lls that were copied or cut	**Alt HIE**
set **F**ont**S**ize	**Alt HFS**
reduce decimals	**Alt H9**
Delete a **C**olumn	**Alt HDC**
Insert menu shortcuts	
i**N**sert a **C**olumn chart	**Alt NC**
i**N**sert a scatterplot	**Alt ND**
i**N**sert a Pi**V**otTable	**Alt NV**
Data menu shortcuts	
analyze d**A**ta	**Alt AY*n***
Sort selected d**A**ta	**Alt ASS**
View menu shortcuts	
Freeze top **R**ow	**Alt WFR**
Cntl+ to move, select an array, extend formula down an array, or cut selected array	
move to the page right	**Cntl+Page Down**
move to bottom of data array	**Cntl+down arrow**
select data below	**cntl+shift+down**
select a column	**cntl+spacebar**
fill down	**cntl+D**
cut selected cells	**cntl+X**
Shift+ to select adjacent cells	
select adjacent cells	**shift+down arrow**
	shift+right

Chart or scatterplot design	
add a horizontal axis title	**Alt JCAAH**
select a design layout	**Alt JCL**
add vertical gridlines	**Alt JCAGV**
add vertical margin of error bars	**Alt JCAEM**
Chart or scatterplot element selection, formatting	
select an axis	**Alt JAE**
format selected chart element	**Alt JAM**
Reformat or graph a PivotTable	
show averages instead of sums	**Alt JTG**
produce a PivotChart	**Alt JTC**
Other	
turn on Add-In	**Alt TI**

Alt activates shortcuts menus, linking keyboard letters to Excel menus. Press and press letters linked to the menus you want.

Alt Home:

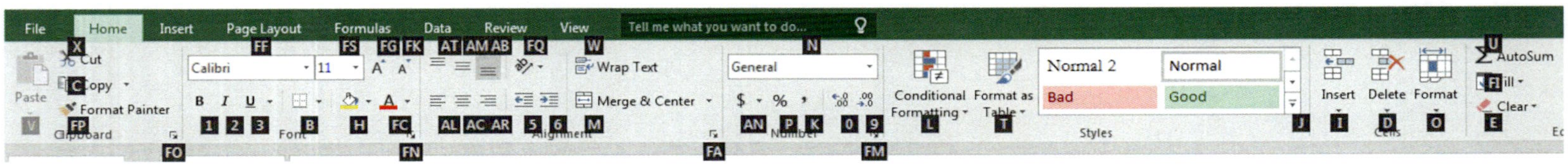

Home menu leys, from left to right, include:

V	Paste	FF	Choose a font	FS	Choose a fontsize	W	Wrap text	9	Reduce decimals	I	Insert
X	Cut	1	Bold	FC	Choose font color					D	Delete
C	Copy	2	Italicize								
		3	Underline								

Other useful menus activated with **Alt** include:

A	Data	N	Insert	W	View

From a chart or plot, **Alt** provides access to chart menus:

JC	Chart design	JA	Chart format	JT	Reformat PivotTable data or chart

Significant Digits Guidelines

The number of significant digits in a number are those which convey information. Significant digits include:

1. All nonzero numbers
2. Zeros between nonzero numbers, and
3. Trailing zeros.

Zeros acting as placeholders aren't counted.

The number 2061 has four significant digits, while the number 2610 has three, since the zero is merely a placeholder. The number 0.0920 has three significant digits, "9," "2," and the final, trailing "0." The first two zeros are placeholders that aren't counted.

In rare cases, it is not clear whether zero is a placeholder or a significant digit. The number 40,000 could represent the range 39,500 to 40,499. In that case, the number of significant digits is one, and the zeros are placeholders. Alternatively, 40,000 could represent the range 39,995 to 40,004. In this latter case, the number of significant digits is four, since the zeros convey meaning. When in doubt, a number could be written in scientific notation, which is unambiguous. For one significant digit, 40,000 becomes $4 \times E^4$. For four significant digits, 40,000 becomes $4.000 \times E^4$.

Lab 2 Description

Compensation of 200 Best Paid CEOs

The New York Times recently published the compensation packages of the 200 best compensated CEOs of publicly traded firms in the U.S. These data are in **Lab 2 Compensation of Best Paid CEOs**.

I. Describe the compensation of the best paid CEOs.

1. Find the average compensation (M) among the best compensated CEOs:__________

2. Find the standard deviation (SD) of compensation: _______

3. Is the distribution of compensation among the best paid CEOs approximately Normal?

   Y or N Evidence: ______________

II. Identify outlier(s) who earn(s) more than 3 SDs above the M and describe compensation of best paid CEOs excluding outliers.

1. Find average compensation, M, excluding outlier(s): ______

2. Find the standard deviation of compensation, SD, excluding outlier(s): _____

III. Make a histogram and cdf to illustrate distribution of CEO compensation

1. Make the histogram of compensation for top paid CEOs.

2. Plot the cumulative distribution of compensation.

3. What is median compensation among the best paid CEOs? _____

4. What is the Interquartile Range of compensation among the 25 best paid CEOs?

   _____

IV. Compare the Distribution of CEO Compensation to Normal

1. By the Empirical Rule, Normal is distributed as in second and third columns, below. Fill in the third through fifth columns to compare the distribution of CEO compensation to Normal:

	Normal		CEO compensation		
Range	%	Cum %	Range ($M)	%	Cumulative %
Below M-3SD	.1	.1			
M-3SD to M-2SD	2.1	2.2			
M-2SD to M-1SD	13.6	15.8			
M-1SD to M	34.1	50			
M to M+1SD	34.1	84.1			
M+1SD to M+2SD	13.6	97.7			
M+2SD to M+3SD	2.1	99.9			
Above M+3SD	.1	100			

2. By the Empirical Rule, Normal is distributed as in the second column, below. Fill in the third and fourth columns to compare the distribution of CEO compensation to Normal:

	Normal	CEO compensation	
Range	%	Range ($M)	%
Within 1SD of M			
Within 2SD of M			
Within 3 SD of M			

V. Identify Industries where CEOs are Best Compensated

Use a PivotTable to determine the best paid industry.

1. What is the best paid industry? _______

2. Average CEO compensation within industries range: ____ to ______

Assignment 2.1 Procter & Gamble's Global Advertising

Procter & Gamble spent $5,960,000 on advertising in 51 global markets. This data, from *Advertising Age*, Global Marketing is in **Assignment 2.1 P&G Global Advertising.**

P&G Corporate is reviewing the firm's global advertising strategy, which is the result of decisions made by many brand management teams. Corporate wants to be sure that these many brand level decisions produce an effective allocation when viewed together.

Describe *Procter & Gamble's advertising* spending across the 51 *countries* that make up the global markets.

Note: Be specific: label responses with appropriate units! Important! Also round responses to two or three significant digits. Points deducted for missing/incorrect units or too few/too many significant digits.

I. Describe P&G's global advertising.

1. Find the average advertising (M) in countries Worldwide:__________
2. Find the standard deviation (SD) of advertising in countries Worldwide: _______
3. Is the distribution of advertising across countries Worldwide approximately Normal?

Y or N Evidence: ______________

II. Identify *countries* which are outliers and list them here:

III. Illustrate the distribution of advertising levels in countries with a histogram and a cdf. Reduce decimals, add axis labels, and add a "stand alone" chart title. *Points deducted for titles that aren't "stand alone."*

1. What is median advertising across countries Worldwide? _____
2. What is the Interquartile Range of across countries Worldwide? _____

IV. Compare the Distribution of P&G's Advertising to Normal

1. By the Empirical Rule, Normal is distributed as in second and third columns, below. Fill in the third through fifth columns to compare the distribution of CEO compensation to Normal:

	Normal		P&G's Advertising		
Range	%	Cum %	Range ($M)	%	Cumulative %
Below M-3SD	.1	.1			
M-3SD to M-2SD	2.1	2.2			
M-2SD to M-1SD	13.6	15.8			
M-1SD to M	34.1	50			
M to M+1SD	34.1	84.1			
M+1SD to M+2SD	13.6	97.7			
M+2SD to M+3SD	2.1	99.9			
Above M+3SD	.1	100			

2. By the Empirical Rule, Normal is distributed as in the second column, below. Fill in the third and fourth columns to compare the distribution of P&G advertising to Normal:

	Normal	CEO compensation	
Range	%	Range ($M)	%
Within 1SD of M	68		
Within 2SD of M	95		
Within 3 SD of M	99.7		

V. Conclusions

1. Make a Pivot table of advertising by level of development, and then make a Pivot chart and paste here, after reducing decimals, adding axis labels, and a SAS title:
2. Which advertising strategy describes the P&G strategy best: (i) advertise at a moderate level in many countries, (ii) advertise heavily in a small number of key countries and spend much less in many other markets.

Assignment 2.2 Best Practices Survey

Firm managers use statistics to advantage. Sometimes when results are lackluster, more significant digits are used, since readers will spend less time digesting results, and results with more significant digits are less likely to be remembered. Sometimes when results are impressive, fewer significant digits are used to motivate readers to digest and remember.

Choose an Annual Report and cite the firm and the year.

1. In the body of the report, what range of significant digits are used to report numerical results? Cite two examples, one with the smallest number of significant digits, one with the largest number of significant digits.
2. In the Financial Exhibits at the end, what range of significant digits are used? Cite two examples, one with the smallest number of significant digits, and one with the largest number of significant digits.
3. Survey the graphics. Cite an example where stand alone title is used to help readers interpret. Cite an example where the title could be more effective, and provide a suggestion for a better title.

Assignment 2.3 Shortcut Challenge

Complete the steps in the first Excel page of Lab 2 (find descriptive statistics, sort to identify and remove outliers, find descriptive statistics without outliers, make a histogram, plot the cdf, make a PivotTable, make a PivotChart), and record your time. If your time is more than 5 minutes, repeat twice, and then record your best time.

Case 2.1 VW Backgrounds

Volkswagon management comissioned background music for New Beetle commercials. The advertising message is that the New Beetle is unique… "round in a world of squares." To be effective, the background music must support this message.

Thirty customers were asked to write down the first word that came to mind when they listened to the music. The clip is in **Case 2.1 VW background.MP3** and words evoked are contained in **Case 2-1 VW background.** Listen to the clip, then describe market response.

Create a PivotTable of the percent who associate each image with the music and sort rows so that the modal image is first.

1. Create a PivotChart to illustrate the images associated with the background music. (Add a stand alone title and round percentages to two significant digits.)
2. What is the modal image created by the VW commercial's background music?
3. Is this music is a good choice for the VW commercial? Explain.

Case 2.2 Global Smelter Costs at Alcoa

Faced with recent expansion in Chinese aluminum production, Alcoa seeks to identify smelters that are less profitable. Data in **Alcoa Smelter Costs** contain costs for nine smelters… four which have been closed or curtailed, four which are candidates for closure or curtailment, and one benchmark smelter in which management plans to continue operations. Profit drivers are largely cost based and include total unit costs, labor and power costs per unit. Describe smelter costs at the nine Alcoa smelters. Data are in **AlcoaSmelterCosts.**

Note: label responses with appropriate units! Important! Also round responses to two or three significant digits.

I. Describe smelter unit costs.

1. Find the average (M) total, labor and power costs per unit:__________
2. Find the standard deviation (SD) of total, labor and power costs per unit: _______
3. Are the distributions of total, labor and power costs per unit approximately Normal?

   Total: Y or N Evidence: ________ Labor: Y or N Evidence: ______

   Power: Y or N Evidence: _____

II. Identify *smelters* which are outliers in terms of total, labor or power costs per unit and list them here:

III. Illustrate the distributions of total, labor and power costs per unit with histograms.

1. *Round the unit cost bins to two significant digits.* Reduce decimals, add axis labels, and add a "stand alone" chart titles. Paste your three graphs here:
2. By the Empirical Rule, Normal is distributed as in second and third columns, below. Fill in the third through fifth columns to compare the distribution of units costs to Normal:

	Normal		Total Unit Cost		
Range	%	Cum %	Range ($/t)	%	Cumulative %
Below M-3SD	.1	.1			
M-3SD to M-2SD	2.1	2.2			
M-2SD to M-1SD	13.6	15.8			
M-1SD to M	34.1	50			
M to M+1SD	34.1	84.1			
M+1SD to M+2SD	13.6	97.7			
M+2SD to M+3SD	2.1	99.9			
Above M+3SD	.1	100			

	Normal		Unit Labor Cost		
Range	%	Cum %	Range ($/t)	%	Cumulative %
Below M-3SD	.1	.1			
M-3SD to M-2SD	2.1	2.2			
M-2SD to M-1SD	13.6	15.8			
M-1SD to M	34.1	50			
M to M+1SD	34.1	84.1			
M+1SD to M+2SD	13.6	97.7			
M+2SD to M+3SD	2.1	99.9			
Above M+3SD	.1	100			

	Normal		Unit Power Cost		
Range	%	Cum %	Range ($/t)	%	Cumulative %
Below M-3SD	.1	.1			
M-3SD to M-2SD	2.1	2.2			
M-2SD to M-1SD	13.6	15.8			
M-1SD to M	34.1	50			
M to M+1SD	34.1	84.1			
M+1SD to M+2SD	13.6	97.7			
M+2SD to M+3SD	2.1	99.9			
Above M+3SD	.1	100			

3. By the Empirical Rule, Normal is distributed as in the second column, below. Fill in the third and fourth columns to compare the distribution of units costs to Normal:

	Normal	Total Unit Cost	
Range	%	Range ($/t)	%
Within 1SD of M	68		
Within 2SD of M	95		
H1	99.7		

	Normal	Unit Labor Cost	
Range	%	Range ($/t)	%
Within 1SD of M	68		
Within 2SD of M	95		
Within 3 SD of M	99.7		

	Normal	Unit Power Cost	
Range	%	Range ($/t)	%
Within 1SD of M	68		
Within 2SD of M	95		
Within 3 SD of M	99.7		

IV. Illustrate the distributions of total, labor and power costs per unit with cdfs.

1. Paste your cdfs plots here:

2. What are median unit costs across smelters? Total: _____ Labor: _____

   Power: _____

3. What is are the Interquartile Ranges of unit costs across smelters?

   Total: _____ to _____ Labor: _____ to _____ Power: _____ to _____

V. Conclusions

1. Make Pivot tables of total, labor and power costs per unit by operating status, location, and process ($3 \times 3 = 9$ total), and then make Pivot charts (9 total) and paste here, after reducing decimals, adding axis labels, and stand alone titles:
2. Describe an ideal, smelter with differential advantages which may lead to lower unit costs:

Chapter 3
Hypothesis Tests, Confidence Intervals to Infer Population Characteristics and Differences

Samples are collected and analyzed to estimate population characteristics. Chapter 3 explores the practice of *inference*: how *hypotheses* about what may be true in the population are tested and how population parameters are estimated with *confidence intervals*. Included in this chapter are tests of hypotheses and confidence intervals for

(i) a population mean from a single sample,
(ii) the difference between means of two populations, or segments from two independent samples, and
(ii) the mean difference within one population between two time periods or two scenarios from two matched or paired samples.

3.1 Sample Means Are Random Variables

The descriptive statistics from each sample of a population are unique. In the example that follows, teams in a New Product Development class each collected a sample from a population to estimate population demand for their concept. Each of the team's statistics is unique, but predictable, since the sample statistics are random variables with a predictable sampling distribution. If many random samples of a given size are drawn from a population, the means from those samples will be similar and their distribution will be Normal and centered at the population mean.

Example 3.1 Thirsty on Campus: Is There Sufficient Demand? An enterprising New Product Development class has an idea to sell on campus custom-flavored, enriched bottles of water from dispensers which would add customers' desired vitamins and natural flavors to each bottle. To assess profit potential, they need an estimate of demand for bottled water on campus. If demand exceeds the breakeven level of seven bottles per week per customer, the business would generate profit.

The class translated breakeven demand into hypotheses which could be tested using a sample of potential customers. The entrepreneurial class needs to know whether or not demand exceeds seven bottles per consumer per week, because below this level of demand, revenues wouldn't cover expenses. Hypotheses are formulated as *null* and *alternative*. In this case, the null hypothesis states a limiting conclusion about the population mean. This default conclusion is cannot be rejected unless the data indicate that it is highly unlikely.
The null hypothesis is that of insufficient demand, which would lead the class to stop development:

H_0: Campus consumers drink no more than seven bottles of water per week on average:

C. Fraser, *Business Statistics for Competitive Advantage with Excel 2016*,
DOI 10.1007/978-3-319-32185-1_3

$$\mu \leq 7$$

Unless sample data indicates sufficient demand, the class will stop development.

In this case, the alternative hypothesis states a conclusion that the population mean exceeds the qualifying condition. The null hypothesis is rejected only with sufficient evidence from a sample that it is unlikely to be true.

In **Thirsty,** the alternate hypothesis supports a conclusion that population demand is sufficient and would lead to a decision to proceed with the new product's development:

H_1: Campus consumers drink more than seven bottles of water per week on average:

$$\mu > 7$$

Given sufficient demand in a sample, the class would reject the null hypothesis and proceed with the project.

Sample statistics are used to determine whether or not the population mean is likely to be less than seven, using the sample mean as the estimate. To test the hypotheses regarding mean demand in the population of customers on campus, each of the fifteen student teams in the class independently surveyed a random sample of thirty consumers from the campus. The distribution of means of many "large" ($N \geq 30$) random samples is Normal and centered on the unknown population mean as illustrated in Figure 3.1.

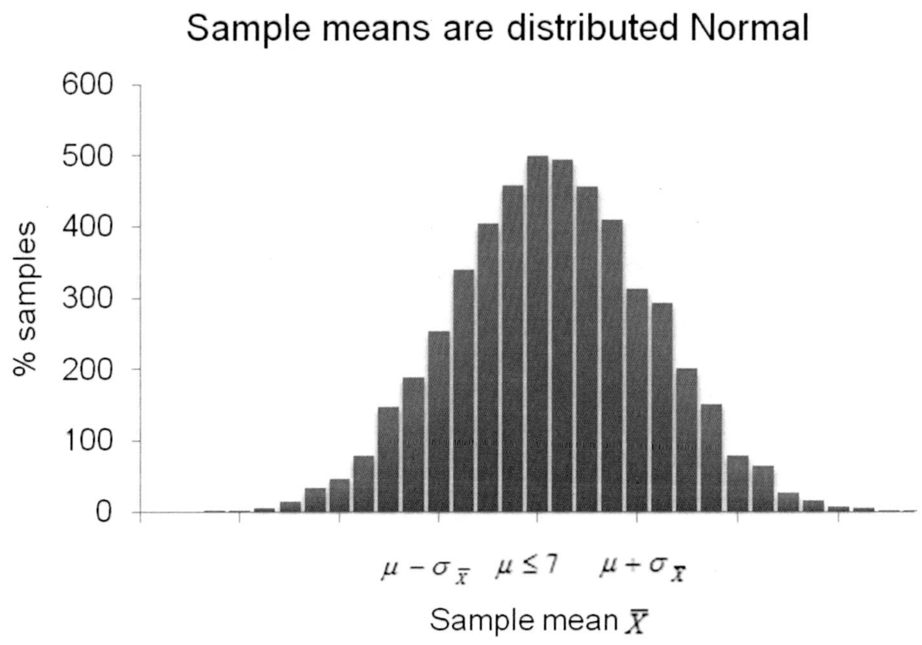

Figure 3.1 Distribution of sample means under the null hypothesis

On average, across all random samples of the same size N, the average difference between sample means and the population mean is the standard error of sample means:

$$\sigma_{\overline{X}} = \sigma / \sqrt{N}$$

where σ is the standard deviation in the population, and N is the sample size. The standard error is larger when there is more variation in the population and when the sample size is smaller.

With random samples of thirty, population mean μ=10.2 and standard deviation σ=4.0, the sampling standard error would be $s_{\overline{X}} = \sigma / \sqrt{30} = 4 / 5.5 = .7$. From the Empirical Rule introduced in Chapter 2, we would expect 2/3 of the teams' sample means to fall within one standard error of the population mean:

$$\mu - s_{\overline{X}} \le \overline{X} \le \mu + s_{\overline{X}}$$
$$10.2 - .7 \le \overline{X} \le 10.2 + .7$$
$$9.5 \le \overline{X} \le 10.9,$$

and we expect 95% of the teams' *sample means* to fall within two standard errors of the population mean:

$$\mu - 2s_{\overline{X}} \le \overline{X} \le \mu + 2s_{\overline{X}}$$
$$10.2 - 2(.7) \le \overline{X} \le 10.2 + 2(.7)$$
$$8.8 \le \overline{X} \le 11.6$$

Nearly all of sample means can be expected to fall within three standard errors of the mean, 8.1 to 12.3.

Each team calculated the sample mean and standard deviation from their sample. Team 1, for example, found that average demand in their sample is 11.2 bottles per week, with standard deviation of 4.5 bottles. Each of team's descriptive statistics from the fifteen samples is shown in Figure 3.2.

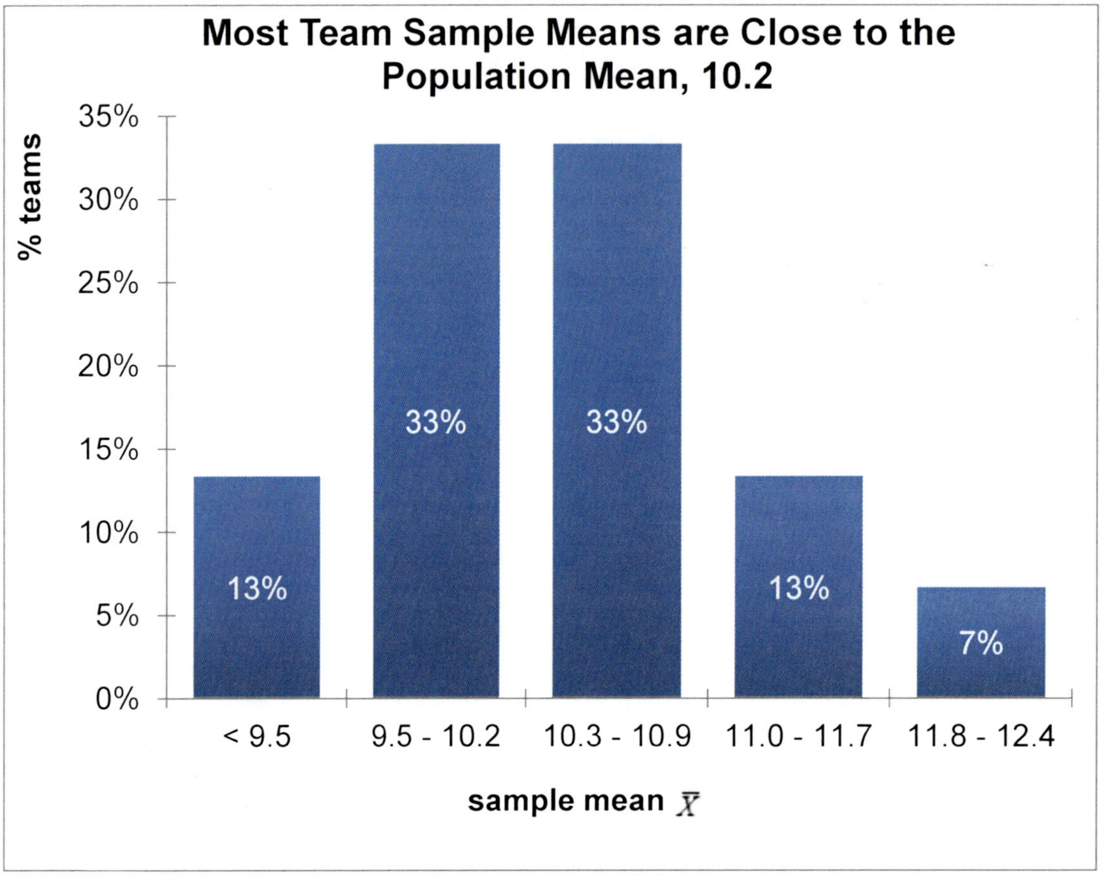

Figure 3.2 Fifteen teams' samples

Table 3.1 Fifteen teams' samples

	Sample Statistics	
Team	*Average Demand per consumer per week* $\bar{X}$	*Standard deviation* s_i
1	11.2	4.5
2	10.9	4.0
3	10.6	4.3
4	9.5	3.4
5	9.0	3.9
6	10.8	4.6
7	9.6	3.8
8	9.9	4.1
9	9.7	3.7
10	10.7	4.2
11	9.0	3.8
12	9.8	3.6
13	10.5	3.1
14	12.2	4.9
15	11.6	4.2

Sample means across the fifteen teams ranged from 9.0 to 12.2 bottles per week per consumer shown in Table 3.1. Each team's sample mean, $\bar{X}$, is close to the true, unknown, population mean, μ=10.2, and not as close to the hypothetical population mean of seven. Each of the sample standard deviations is close to the true, unknown population standard deviation σ=4. In addition, each team's sample statistics are unique.

Since the population standard deviation is almost never known, but estimated from a sample, the standard error is also estimated from a sample, using the estimate of the population standard deviation s:

$$s_{\bar{X}} = s / \sqrt{N}$$

When the standard deviation is estimated from a sample (which is nearly always), the distribution of standardized sample means $\bar{X} / s_{\bar{X}}$ is distributed as *Student t*, which is approximately Normal, shown in Figure 3.3.

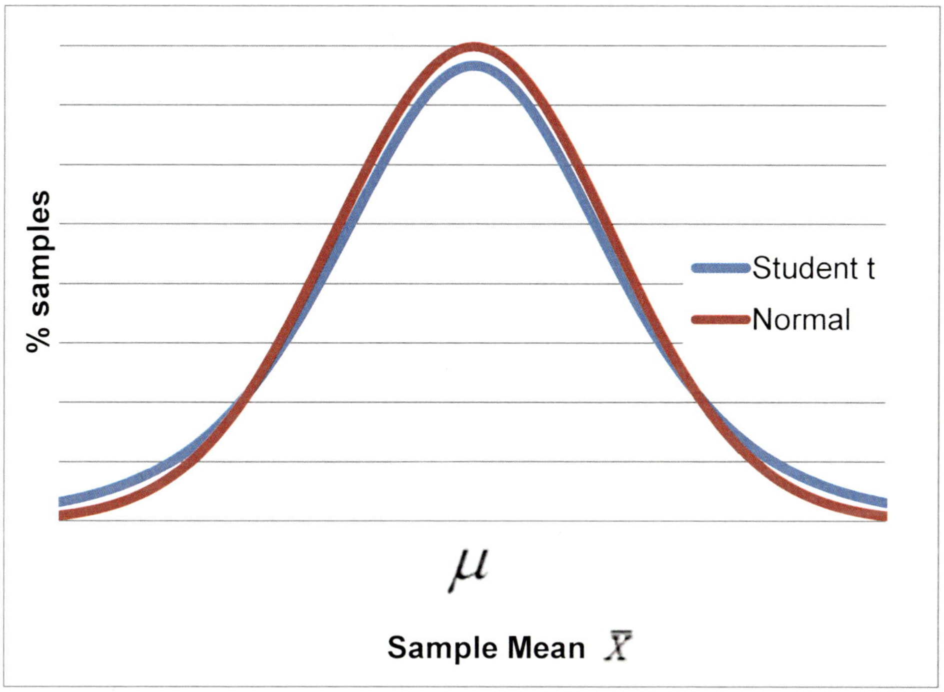

Figure 3.3 Distribution of sample means

Student t has slightly fatter tails than Normal, since we are estimating the standard deviation. How much fatter the tails are depend on the sample size. *Student t* is a family of distributions indexed by sample size. There is more difference from Normal if a sample size is small. For sample sizes of about thirty or more, there is little difference between Student t and Normal An estimate of the standard deviation from the sample is close to the true population value if the sample size meets or exceeds thirty.

3.2 Infer Whether a Population Mean Exceeds a Target

Each team asks, "How likely is it that we would observe this sample mean, were the population mean seven or less?" From the Empirical Rule, sample means are expected to fall within approximately two standard errors of the population mean 95% of the time.

Rearranging the Empirical Rule formula, we see that *Student t* counts the standard errors between a sample mean and the population mean:

$$\left|\overline{X} - \mu\right| / s_{\overline{X}} = t_{N-1}$$

A difference between a sample mean and the break-even level of seven that is more than approximately two standard errors ($t > 2$) is a signal that population demand is unlikely to be seven or less. In this case, the sample mean would lie to the extreme right in the hypothetical distribution of sample means with center at the hypothetical population mean of seven, where fewer than 5% of sample means are expected.

In the **Thirsty** example, each team calculated the number of standard errors by which their sample mean exceeded seven. Next, each referred to a table of Student *t* values or used statistical software to find the area under the right distribution tail, called the *p value*. Were true demand less than seven, it would be unusual to observe a sample mean more than $t_{2\alpha=.1;\, 29} = 1.7$ standard errors greater than seven. The larger a *t* value, the smaller the corresponding *p value* will be, and the less likely the sample statistics would be observed were the null hypothesis true:

p value > .05 … if the null hypothesis were true, it would not be unusual to observe the data.

The conclusion of insufficient demand H_0 cannot be rejected.

The Team recommends halting development.

p value $\leq$ .05 … if the null hypothesis were true, it would be unusual to observe the data.

Reject the null hypothesis.

The Team recommends proceeding with development.

Each team used software to test the hypothesis that demand exceeds seven. Team 8's analyses are illustrated in Figure 3.4, as an example:

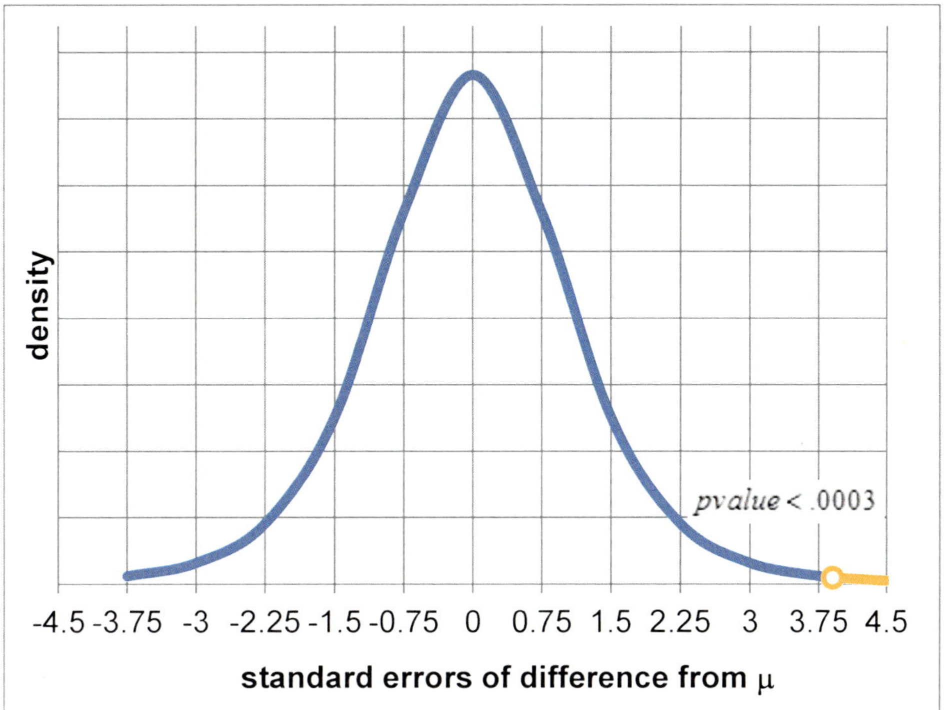

$$t_{29} = \frac{\bar{X} - \mu}{s_{\bar{X}}} = \frac{(9.9 - 7)}{.75} = 3.9$$

mean	9.9
sd	4.1
se	.75
t	3.9
p value	.0003

Figure 3.4 *t test* of the hypothesis that population demand is seven or less

Reviewing these results Team Eight would conclude:

Demand in our sample of thirty ranged from zero to nineteen bottles per person per week, averaging 9.9 bottles per person per week. With this sample of thirty, the standard error is .75 bottles per week. Our sample mean is 3.9 standard errors greater than breakeven of seven. (The t statistic is 3.9.) Were population demand seven or less, it would be unusual to observe demand of 9.9 in a sample of thirty. The p-value is .0003. We conclude that demand is not seven or less.

In a test of the level of demand for bottles of water, each team used a "one-tail" test. Regardless of how much demand exceeds seven bottles per consumer per week, a team would vote to proceed with development as long as they can be reasonably sure demand exceeds breakeven. They require only that the chance of observing the data be less than 5%, the *critical p value,* were true demand less than seven. Thus, it is only the area under the right tail that concerns them.

3.3 Critical t Provides a Benchmark

Before statistical software became popular and statistical calculations were done by hand, it was standard practice to conduct a *one tail t test* by finding the *critical t* value for a given sample size which cut off 5% of the t distribution right tail. *Critical t* values were published in the appendices of texts, indexed by sample size. Comparing a sample *t* statistic with the *critical t* enabled a yes-no test of the null hypothesis. If the sample *t* exceeded the *critical t*, the null hypothesis was rejected.

From the **Thirsty** example, for a sample of 30, the *critical t* value for 29 *df* (= N−1 = 30−1 = 29) is 1.70. Figure 3.5 illustrates *p values* returned by Excel for *t* values with a sample size of 30, or *df* of 29.

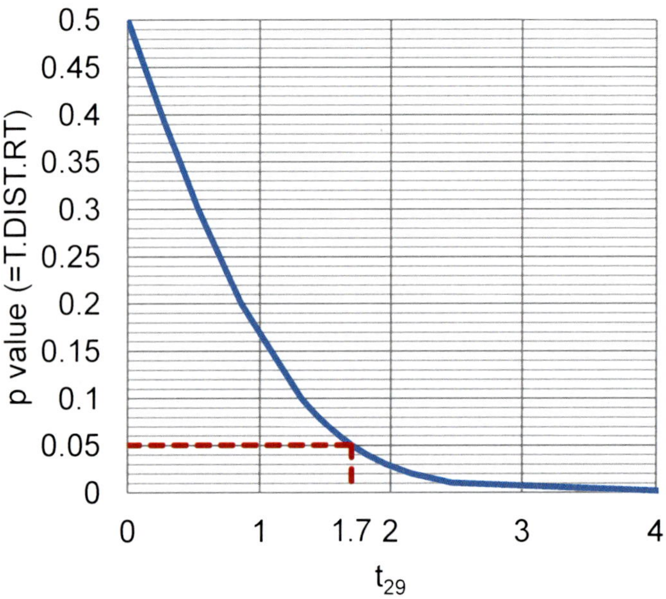

Student's t Distribution *critical t*	
df	Level of Significance for One Tail Test, α = .05
29	1.70

Figure 3.5 *p values* returned by Excel for *t* statistics from a sample of 30

Team Eight's sample *t* of 3.9 exceeds the *critical t*, so the null hypothesis is rejected with 95% (= 1 − α) confidence.

The wide availability of statistical software allows easy determination of the *p value*, providing a more informative estimate of the chance that the sample mean would be observed if the null hypothesis were true. Consequently, it has become standard practice to compare the sample *p value* to the *critical p value* of .05 to test the null hypothesis.

Whether you choose to compare the sample *p value* with the *critical p value* or, alternatively, the sample *t* to the *critical t value* for a given sample size will lead to the same conclusion. Both comparisons are correct choices.

3.4 Confidence Intervals Estimate the Population Mean

Since the class of entrepreneurs in the **Thirsty** example doesn't know that the population mean is 10.2 bottles per customer per week, each team will estimate this mean using their sample data. Rearranging the formula for a *t test*, we see that each team can use their sample standard error, the Student t value for their sample size and the desired level of confidence to estimate the range that is likely to contain the true population mean:

$$\bar{X} - t_{\alpha/2,N-1} \times s_{\bar{X}} < \mu < \bar{X} + t_{\alpha/2,N-1} \times s_{\bar{X}}$$

where α is the chance that a sample is drawn from one of the sample distribution tails, and $t_{\alpha/2, (N-1)}$ is the *critical Student t* value for a chosen level of certainty (1−α) and sample size *N*.

The *confidence level* (1−α) allows us to specify the level of certainty that an interval will contain the population mean. Generally, decision makers desire a 95% level of confidence (α=.05), insuring that in 95 out of 100 samples, the interval would contain the population mean. The *critical Student t* value for 95% confidence with a sample of thirty (*N*=30) is $t_{\alpha/2, (N-1)=29} =$ 2.05. In 95% of random samples of thirty drawn, we expect the sample means to be no further than 2.05 standard errors from the population mean:

$$\bar{X} - 2.05 s_{\bar{X}} \le \mu \le \bar{X} + 2.05 s_{\bar{X}}$$

Each team's sample standard error, margin of error, and 95% confidence interval from the **Thirsty** example are shown in Table 3.2:

Table 3.2 Confidence intervals from each team's sample

team i	*average demand/ consumer/week,* $\bar{X}_i$	*standard deviation* s_i	*standard error* $s_{\bar{X}}$	*margin of error* $2.05\, s_{\bar{X}}$	*95% confidence interval* $\bar{X} \pm 2.05 s_{\bar{X}}$	
1	11.2	4.5	.84	1.7	9.5	12.9
2	10.9	4.0	.74	1.5	9.4	12.4
3	10.6	4.3	.80	1.6	9.0	12.2
4	9.5	3.4	.63	1.3	8.2	10.8
5	9.0	3.9	.72	1.5	7.5	10.5
6	10.8	4.6	.85	1.7	9.1	12.5
7	9.6	3.8	.71	1.5	8.1	11.1
8	9.9	4.1	.75	1.5	8.4	11.4
9	9.7	3.7	.69	1.4	8.3	11.1
10	10.7	4.2	.78	1.6	9.1	12.3
11	9.0	3.8	.71	1.5	7.5	10.5
12	9.8	3.6	.67	1.4	8.4	11.2
13	10.5	3.1	.58	1.2	9.3	11.7
14	12.2	4.9	.91	1.9	10.3	14.1
15	11.6	4.2	.78	1.6	10.0	13.2

In practice, fifteen samples would not be collected. A single sample would be selected, just as each individual team did in their market research. Team 8's analysis is shown in Figure 3.6 as an example:

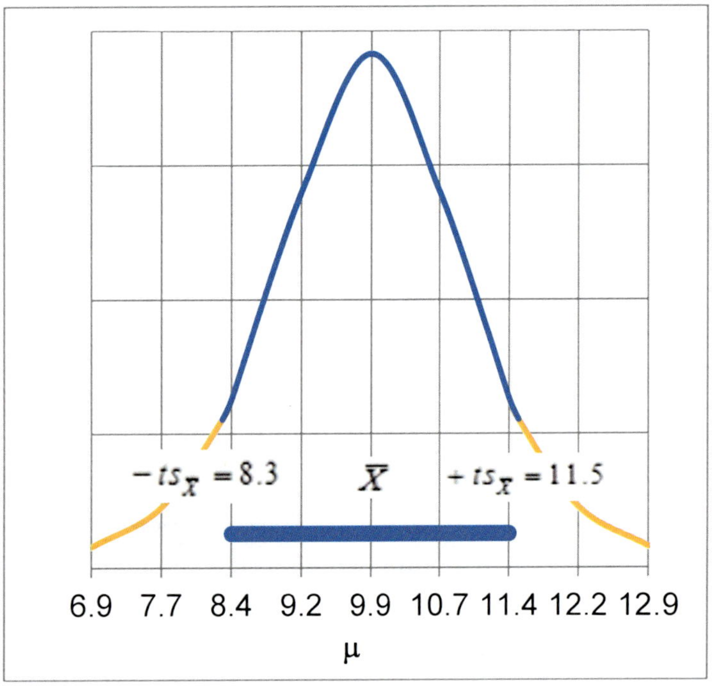

mean	9.9
standard error	.75
critical t	2.1
margin of error	1.5
95% lower	8.3
95% upper	11.5

Figure 3.6 Confidence interval for bottled water demand μ

Team 8 would conclude:

"Average demand in our sample of thirty is 9.9 bottles per person per week, with a margin of error of 1.5 bottles. It is likely that average campus demand is between 8.3 and 11.5 bottles per person per week."

3.5 Calculate Approximate Confidence Intervals with Mental Math

When the sample size is "large," $N \geq 30$, we can use an approximate $t \cong 2.0$ to produce approximate confidence intervals with mental math. Using $t \cong 2$ for an approximate 95% level of confidence, the fifteen student teams each calculated the likely ranges for bottled water demand in the population, shown in Table 3.3.

Table 3.3 Each team's approximate confidence interval

team$_i$	*Average customer demand/ week* $\overline{X}_i$	*standard error* $s_{\overline{X}}$	*margin of error* $2.05\, s_{\overline{X}}$	*95% confidence interval* $\overline{X} \pm 2.05 s_{\overline{X}}$		*approximate margin of error* $2s_{\overline{X}}$	*approximate 95% confidence interval* $\overline{X} \pm 2s_{\overline{X}}$	
1	11.2	.84	1.7	9.5	12.9	1.7	9.5	12.9
2	10.9	.74	1.5	9.4	12.4	1.5	9.4	12.4
3	10.6	.80	1.6	9.0	12.2	1.6	9.0	12.2
4	9.5	.63	1.3	8.2	10.8	1.3	8.2	10.8
5	9.0	.72	1.5	7.5	10.5	1.4	7.6	10.4
6	10.8	.85	1.7	9.1	12.5	1.7	9.1	12.5
7	9.6	.71	1.5	8.1	11.1	1.4	8.2	11.0
8	9.9	.75	1.5	8.4	11.4	1.5	8.4	11.4
9	9.7	.69	1.4	8.3	11.1	1.4	8.3	11.1
10	10.7	.78	1.6	9.1	12.3	1.6	9.1	12.3
11	9.0	.71	1.5	7.5	10.5	1.4	7.6	10.4
12	9.8	.67	1.4	8.4	11.2	1.3	8.5	11.1
13	10.5	.58	1.2	9.3	11.7	1.2	9.3	11.7
14	12.2	.91	1.9	10.3	14.1	1.8	10.4	14.0
15	11.6	.78	1.6	10.0	13.2	1.6	10.0	13.2

With the approximation, Team 8's conclusion remains: expected demand will range from 8.4 to 11.4 bottles per week per customer.

3.6 Margin of Error Is Inversely Proportional to Sample Size

The larger a sample N is, the smaller the *95% confidence interval* is,

$$\overline{X} - 2s_{\overline{X}} \leq \mu \leq \overline{X} + 2s_{\overline{X}}$$

since the standard error $s_{\overline{X}}$ and *margin of error,* roughly 2 $s_{\overline{X}}$, are inversely proportional to the square root of our size N, shown in Figure 3.7.

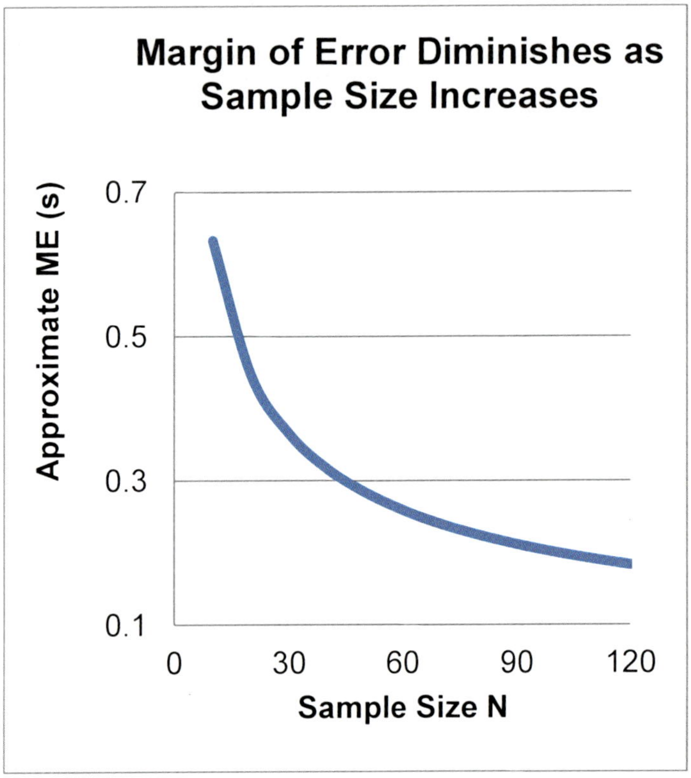

Sample Size N	*Approximate Margin of Error* $2s/\sqrt{N}$
25	$.4s$
100	$.2s$
400	$.1s$

Figure 3.7 Margin of error, given sample size

To double precision, the sample size must be quadrupled. Gains in precision become increasingly more expensive.

3.7 Determine Whether Two Segments Differ with Student t

Example 3.2 SmartScribe: Is Income a Useful Base for Segmentation? SmartScribe, manufacturers of a brand of smart pens, would like to identify the demographic segment with the highest demand for its new concept. Smart pens record presentation notes onto a file that can be downloaded. Since the new pens were being sold at a relatively high price, Adopters might have higher incomes. To test this hypothesis, customers at an office supply retail store where sorted into SmartScribe purchasers, which management refers to as The Adopters, and other Nonadopter customers. Random samples from these two segments were drawn and offered a store coupon in exchange for completion of a short survey, which included a measure of annual household income. Fifty six SmartScribe pen Adopters and forty one Nonadopters completed the survey.

The null hypothesis states the conclusion that the average annual household income of Adopters is not greater than that of Nonadopters.

H_0: Average annual household income of Adopters is equal to or less than that of Nonadopters of the new pen.

$$\mu_{Adopters} \leq \mu_{Nonadopters}$$

OR

$$\mu_{Adopters} - \mu_{Nonadopters} \leq 0.$$

Alternatively:

H_1: Average annual household incomes of Adopters exceeds that of Nonadopters of the new pen:

$$\mu_{Adopters} > \mu_{Nonadopters}$$

OR

$$\mu_{Adopters} - \mu_{Nonadopters} > 0.$$

If there is no difference in incomes between the two segment samples, or if Adopters earn lower incomes, the null hypothesis cannot be rejected based on the sample evidence.

Average income in the sample of Nonadopters was \$35K, and \$80K, in the sample of Adopters, shown in Figure 3.8:

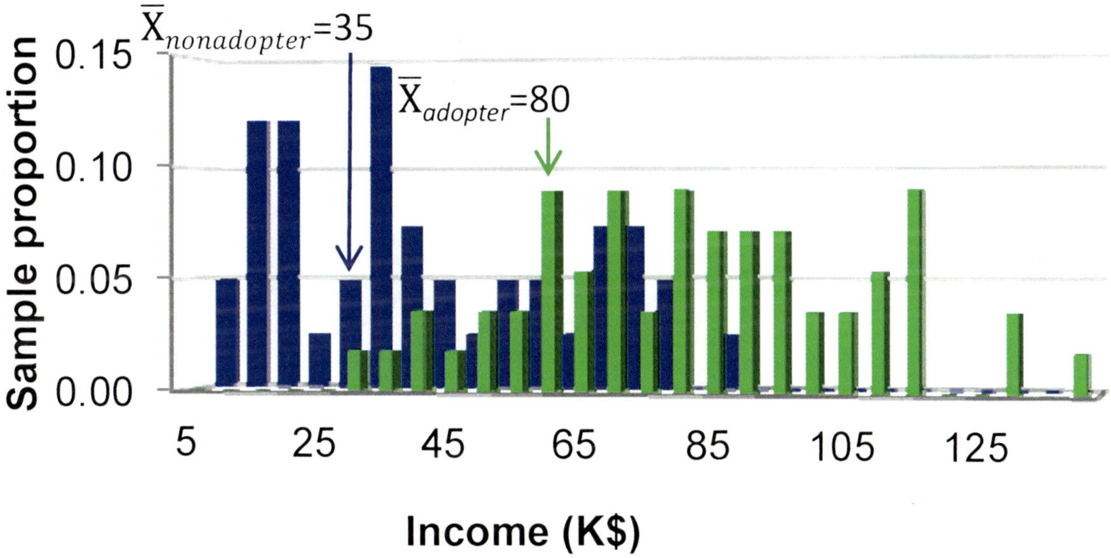

Figure 3.8 Incomes of samples from two segments

A test of the significance of the difference between the two segments' average annual household incomes is based on the difference between the two sample means, SmartScribe needs to determine whether or not this difference in average incomes,

$$\bar{X}_{Adopters} - \bar{X}_{Nonadopters} = \$80K - \$35K = \$45K$$

is large enough to be significant.

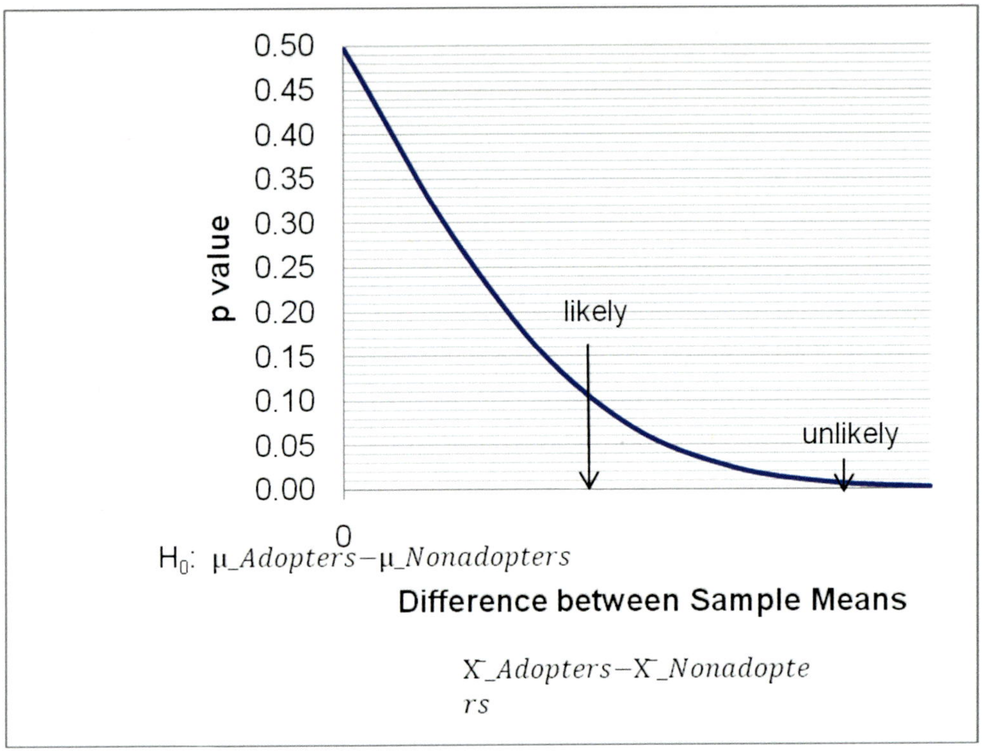

Figure 3.9 The null hypothesis adopters earn less or equivalent incomes to nonadopters

Whether the difference between two sample means is large enough to be significant depends on the amount of dispersion in the two populations, in this case $\sigma_{Adopters}$ and $\sigma_{Nonadopters}$, and the two sample sizes, $n_{Adopters}$ and $n_{Nonadopters}$ in this case shown in Figure 3.9. The standard error of the difference between two sample means,

$$s_{\bar{X}_{Adopters}-\bar{X}_{Nonadopters}} = \sqrt{\left(\frac{s_{Adopters}}{\sqrt{n_{Adopters}}}\right)^2 + \left(\frac{s_{Nonadopters}}{\sqrt{n_{Nonadopters}}}\right)^2}$$

$$= \sqrt{s^2_{\bar{X}_{Adopters}} + s^2_{\bar{X}_{Nonadopters}}}$$

captures both the dispersion in the two populations, as well as the two sample sizes.

The standard error of average difference in annual household income (in thousands) is the square root of the two sample standard errors squared, equal to \$4.9K in this case:

$$s_{\bar{X}_{Adopters}-\bar{X}_{Nonadopters}} = \sqrt{\left(\frac{25}{\sqrt{56}}\right)^2 + \left(\frac{23}{\sqrt{41}}\right)^2}$$

$$= \sqrt{(3.3)^2 + (3.6)^2}$$

$$= \sqrt{11.2 + 12.9}$$

$$= \sqrt{24.1}$$

$$= 4.9(\$K)$$

This estimate for the standard error of the difference between segment means assumes that the two segment standard deviations may differ. Since it is not usually known whether or not the segment standard deviations are equivalent, this is a conservative assumption.

The number of standard errors of difference between sample means is measured with Student t.

$$t_{90} = (\bar{X}_{Adopters} - \bar{X}_{Nonadopters})/s_{\bar{X}_{Adopters}-\bar{X}_{Nonadopters}}$$

$$= \$45K / \$4.9K$$

$$= 9.2$$

When the two samples have unique standard deviations, the degrees of freedom for a two segment t test depend on both standard deviations and both sample sizes:

$$df = \frac{(s_{Adopters}{}^2/N_{Adopters} + s_{Nonadopters}{}^2/N_{Nonadopters})^2}{(s_{Adopters}{}^2/N_{Adopters})^2/(N_{Adopters}-1) + (s_{NOnadopters}{}^2/N_{Nonadopters})^2/(N_{Nonadopters}-1)}$$

$$= \frac{(25^2/56+31^2/41)^2}{(25^2/56)^2/55+(23^2/41)^2/40}$$

$$= 90$$

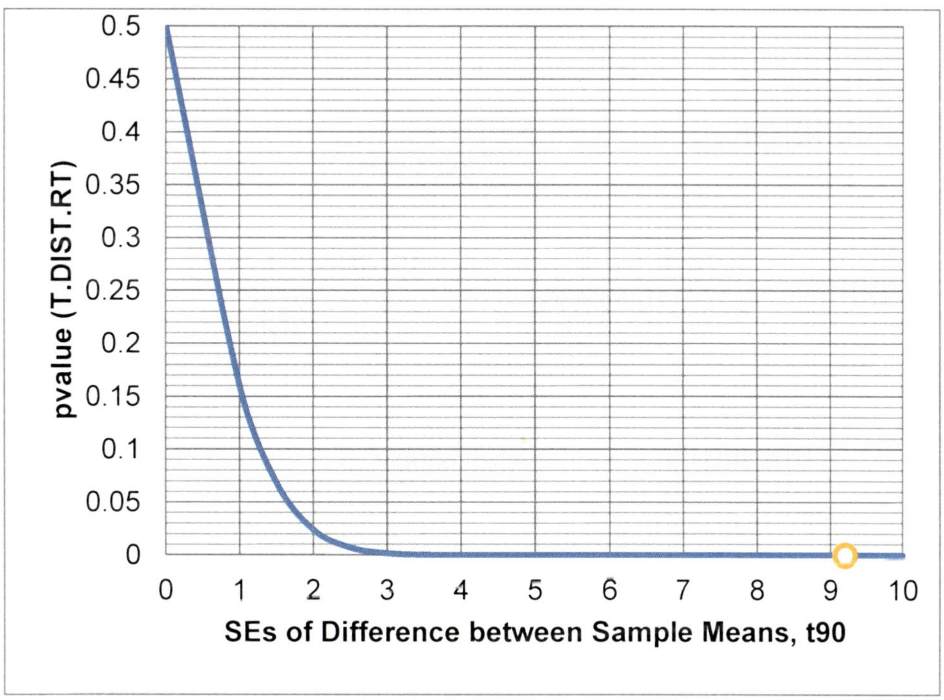

Figure 3.10 *t* test of difference between segment means

The *p value* for this *t* with 90 degrees of freedom is less than .0001.

From the *t test* of difference between segment incomes, shown in Figure 3.10, SmartScribe management could conclude:

"In segment samples of 56 Adopters and 41 Nonadopters, the corresponding average segment sample incomes are \$80K and \$35K, a difference of \$45K, more than 9 standard errors. Were there no difference in segment mean incomes in the population, it would be unusual to observe this difference in segment average incomes in the segment samples. Based on sample evidence, we conclude that average incomes of Adopters cannot be less than or equal to the average incomes of Nonadopters. Income is a useful basis for segmentation."

3.8 Estimate the Extent of Difference Between Two Segments

From the sample data, SmartScribe managers estimated the average annual household income difference (in thousands) between Adopters and Nonadopters:

$$\overline{X}_{Adopters} - \overline{X}_{Nonadopters} = \$80K - \$35K = \$45K$$

The 95% confidence interval around the difference in annual household incomes between Adopters and Nonadopters is made by adding and subtracting the *margin of error.*

The *margin of error* is equal to the two tail *critical t*, with degrees of freedom corresponding to the two sample sizes and α equal to .05 (for 95% confidence), times the standard error for the difference between sample means:

$$t_{df,\propto/2=.025} \times s_{\overline{X}_{Adopters}-\overline{X}_{Nonadopters}} \cong 2 \times \$4.9K$$

$$\cong \$9.8K$$

The approximate *t*, 2, is used in the example, instead of the *t* which corresponds to a confidence interval for the difference between segments.

The difference between means of the two samples would be no further than \$9.8K from the difference between means in the two populations.

The 95% confidence interval for the difference between means is \$35K to \$55K:

$$\left(\overline{X}_{Adopters} - \overline{X}_{Nonadopters}\right) - t_{df,\propto/2=.025} \times s_{\overline{X}_{Adopters}-\overline{X}_{Nonadopters}}$$

$$\lesssim \mu_{Adopters} - \mu_{Nonadopters}$$

$$\lesssim \left(\overline{X}_{Adopters} - \overline{X}_{Nonadopters}\right) + t_{df,\propto/2=.025} \times s_{\overline{X}_{Adopters}-\overline{X}_{Nonadopters}}$$

$$\$45K - 2 \times \$4.9K \lesssim \mu_{Adopters} - \mu_{Nonadopters} \lesssim \$45K + 2 \times \$4.9K$$

$$\$45K - \$9.8K \lesssim \mu_{Adopters} - \mu_{Nonadopters} \lesssim \$45K + \$9.8K$$

$$\$35K \lesssim \mu_{Adopters} - \mu_{Nonadopters} \lesssim \$55K$$

Management will conclude that annual household income can be used to differentiate the two market segments, and that Adopters are wealthier than Nonadopters.

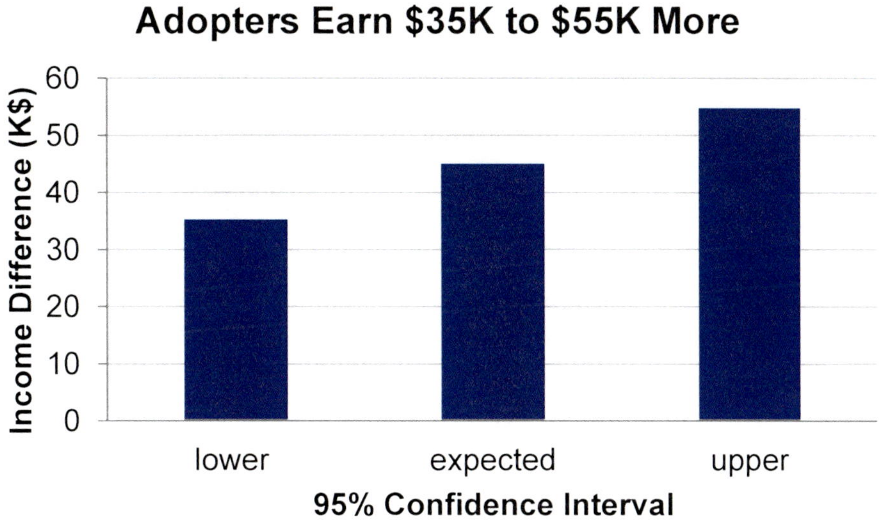

Figure 3.11 95% Confidence interval of the difference between segments

In our samples of 56 Adopters and 41 Nonadopters, the corresponding average difference in income between segment samples is $45K, *and the margin of error of the difference is approximately* $9.8K. *Relative to Nonadopters, we estimate that Adopters earn* $35K *to* $55K *more on average, annually.*

To construct confidence intervals for the difference in means of two samples, we assume that either (i) both segments' characteristics are bell-shaped (distributed approximately Normal) and we've randomly sampled both segments, or (ii) "large" random samples from both segments have been collected.

3.9 Estimate a Population Proportion from a Sample Proportion

Example 3.3 Guinea Pigs. A pharmaceutical company gauges reactions to their products by applying them to animals. An animal rights activist has threatened to start a campaign to boycott the company's products if the animal testing doesn't stop. Concerned managers have hired four public opinion polling organizations to learn whether medical testing on animals is accepted or not.

Four independent pollsters each surveyed thirty Americans and found the proportions shown in Table 3.4 agree that medical testing on animals is morally acceptable:

Table 3.4 Sample approval proportions by poll

Poll	*Sample Approval Proportion*
1	$P_1 = 16 / 30 = .53$
2	$P_2 = 19 / 30 = .63$
3	$P_3 = 17 / 30 = .57$
4	$P_4 = 21 / 30 = .70$

If numerous random samples are taken, sample proportions P will be approximately Normally distributed around the unknown population proportion $\pi = .6$, as long as this true proportion is not close to either zero or one.

The standard deviation of the sample proportions P, the *standard error of the sample proportion,* measures dispersion of samples of size N from the population proportion π:

$$\sigma_\pi = \sqrt{\pi \times (1-\pi)/N}$$

which is estimated with the sample proportion P:

$$s_P = \sqrt{P \times (1-P)/N}$$

The four poll organizations would each estimate the proportion of Americans who agree that medical testing on animals is morally acceptable, shown in Table 3.5.

Table 3.5 Confidence interval of approval proportion by poll, N=30

Poll ***i***	*Sample Proportion,* P_i	*Standard Error,* $s_{P_i}(N{=}30)$	*Margin of Error for 95% Confidence,* $Z \times s_{P_i} = 1.96 \times s_{P_i}$	*Interval containing the Population Proportion with 95% confidence* $P_i \pm Z \times s_{P_i}$
1	.57	.090	.18	.39 to .75
2	.61	.089	.17	.44 to 78
3	.58	.090	.18	.40 to .76
4	.63	.088	.16	.47 to .79

With samples of just thirty, margins of error are relatively large and we are uncertain whether a minority or a sizeable majority approves. In practice, polling organizations use much larger samples, which shrink margins of error and corresponding confidence intervals. Had samples of 1000 been collected instead, the poll results would be as shown in Table 3.6.

Table 3.6 Confidence interval of approval proportion by poll, N=1000

Poll i	*Sample Proportion,* P_i	*Standard Error,* s_{P_i} (N=1000)	*Margin of Error for 95% Confidence,* $Z \times s_{P_i} = 1.96 \times s_{P_i}$	*95% Confidence Interval* $P_i \pm Z \times s_{P_i}$
1	.57	.016	.031	.54 to .60
2	.61	.015	.029	.58 to .64
3	.58	.016	.031	.55 to .61
4	.63	.015	.029	.60 to .66

With much larger samples and correspondingly smaller margins of error, it becomes clear that the majority approves of medical testing on animals.

The second polling organization would report:

The majority of a random sample of 1,000 Americans approves of medical testing on animals. 61% believe medical testing on animals is morally acceptable, with a margin of error of 3%.

3.10 Conditions for Assuming Approximate Normality

It is appropriate to use the Normal distribution to approximate the distribution of possible sample proportions if sample size is "large" ($N \geq 30$), and both $N \times P \geq 5$ and $N \times (1-P) \geq 5$. When the true population proportion is very close to either zero or one, we cannot reasonably assume that the distribution of sample proportions is Normal. A rule of thumb suggests that $P \times N$ and $(1-P) \times N$ ought to be at least five in order to use Normal inferences about proportions. For a sample of thirty, the sample proportion P would need to be between .17 and .83 to use Normal inferences. For a sample of 1000, the sample proportion P would need to be between .01 and .99. Drawing larger samples allows more precise inference of population proportions from samples.

3.11 Conservative Confidence Intervals for a Proportion

Polling organizations report the sample proportion and margin of error, rather than a confidence interval. For example, "61% approve of medical testing on animals. (The margin of error from this poll is 3 percentage points.)" A 95% level of confidence is the industry standard. Because the true proportion and its standard deviation are unknown, and because pollsters stake their reputations on valid results, a *conservative* approach, which assumes a true proportion of .5, is used. This conservative approach

$$s_P = \sqrt{.5 \times (1 - .5)/N}$$

yields the largest possible standard error for a given sample size and makes the margin of error ($Z \times s_P$) a simple function of the square root of the sample size N.

With this conservative approach and samples of $N = 1000$, the pollsters' results are shown in Table 3.7.

Table 3.7 Conservative confidence intervals for approval proportions, N=1000

Poll i	*Sample Proportion,* P	*Conservative Margin of Error for 95% Confidence,* $Z \times s_P = 1.96 \times s_P$	*Conservative 95% Confidence Interval* $P - Z \times s_P \leq \pi \leq P + Z \times s_P$	
1	.57	.031	.54	.60
2	.61	.031	.58	.64
3	.58	.031	.55	.61
4	.63	.031	.60	.66

An effective display of proportions or shares is a *pie chart*. The second poll organization used Excel to create this illustration of their survey results, shown in Figure 3.12:

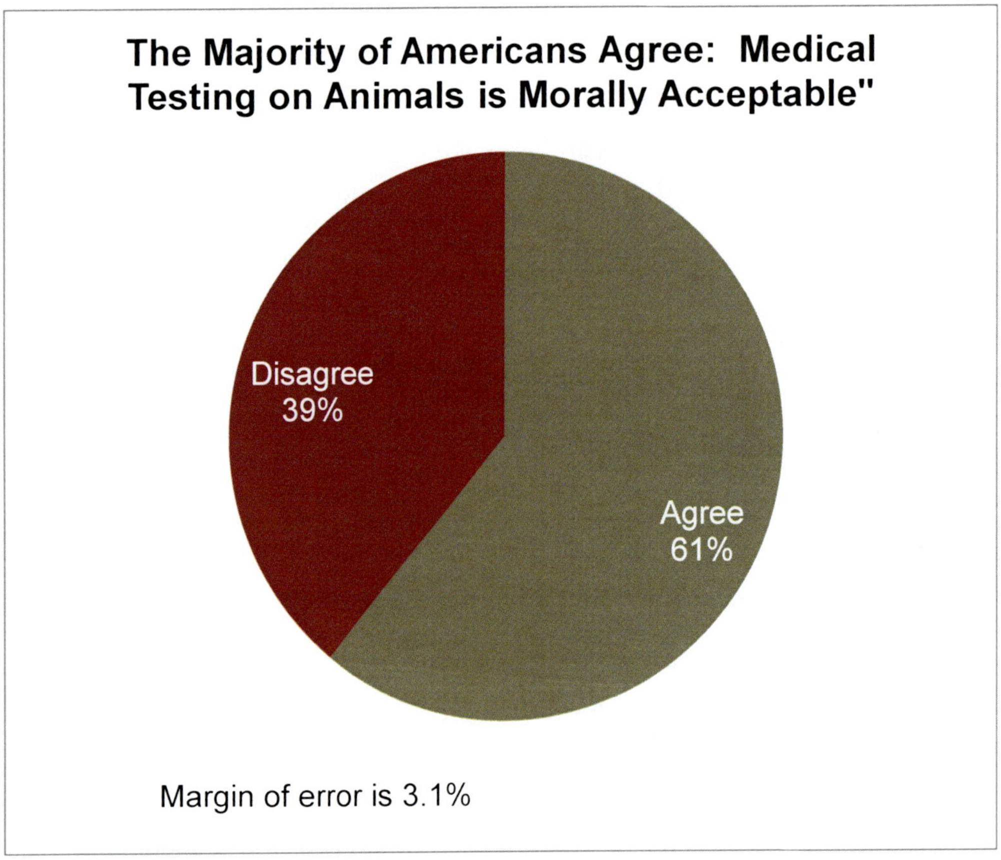

Figure 3.12 Pie chart of approval percentage

The second polling organization would report:

"Sixty-one percent of American adults agree that medical testing on animals is morally acceptable. Poll results have a margin of error of 3.1 percentage points. The majority of Americans supports medical testing on animals."

Other appropriate applications for confidence intervals to estimate population proportions or shares include:

- Proportion who prefer a new formulation to an old one in a taste test
- Share of retailers who offer a brand
- Market share of a product in a specified market
- Proportion of employees who call in sick when they're well
- Proportion of new hires who will perform exceptionally well on the job

3.12 Assess the Difference Between Alternate Scenarios or Pairs

Sometimes management is concerned with the comparison of means from a single sample taken under varying conditions—at different times or in different scenarios—or comparison of sample pairs, like the difference between an employee's opinion and the opinion of the employee's supervisor.

- Financial management might be interested in comparing the reactions of a sample of investors to "socially desirable" stock portfolios, excluding stocks of firms that manufacture or market weapons, tobacco, or alcohol, versus alternate portfolios which promise similar returns at similar risk levels, but which are not "socially desirable."
- Marketing management might be interested in comparing taste ratings of sodas which contain varying levels of red coloring—do redder sodas taste better to customers?
- Management might be interested in comparing satisfaction ratings following a change which allows employees to work at home.

These examples compare *repeated samples*, where participants have provided multiple responses that can be compared.

- Financial management might also be interested in comparing the risk preferences of husbands and wives.
- Marketing management might want to compare children and parents' preferences for red sodas.
- Management might also be interested in comparing the satisfaction ratings of those employees with their supervisors' satisfaction ratings.

In these examples, interest is in comparing means from *matched pairs*.

In either case of repeated or matched samples, a *t test* can be used to determine whether or not the difference is non-zero. Testing hypotheses that concern a difference between pairs is equivalent to a one sample *t* test. The difference is tested in the same way that a characteristic mean is tested, using a one sample test.

Example 3.4 Are "Socially Desirable" Portfolios Undesirable? An investment consulting firm's management believes that they have difficulty selling "socially desirable" portfolios because potential investors assume those funds are inferior investments. Socially Desirable funds exclude stocks of firms which manufacture or market weapons, tobacco or alcohol. There may be a perceived sacrifice associated with socially desirable investment which causes investors to avoid portfolios labeled "socially desirable." The null hypothesis is:

H_0: Investors rate "socially desirable" portfolios at least as attractive as equally risky, conventional portfolios promising equivalent returns:

$$\mu_{Socially\ Desirable} - \mu_{Conventional} \geq 0.$$

If investors do not penalize "socially desirable" funds, the null hypothesis cannot be rejected.

The alternative hypothesis is:

H_1: Investors rate "socially desirable" portfolios as less attractive than other equally risky portfolios promising equivalent returns:

$$\mu_{Socially\ Desirable} - \mu_{Conventional} < 0.$$

Thirty-three investors were asked to evaluate two stock portfolios on a scale of attractiveness (−3 = "Not At All Appealing" to 3="Very Appealing"). The two portfolios promised equivalent returns and were equally risky. One contained only "socially desirable" stocks, while the other included stocks from companies which sell tobacco, alcohol and arms. These are shown in Table 3.8.

Table 3.8 Paired ratings of other & socially desirable portfolios

appeal of conventional portfolio	*appeal of socially desirable portfolio*	*difference*	*appeal of conventional portfolio*	*appeal of socially desirable portfolio*	*difference*
−3	1	−4	2	−1	3
−3	2	−5	2	−1	3
−3	3	−6	2	−2	4
−3	3	−6	2	2	0
0	−1	1	2	1	1
0	1	−1	2	2	0
1	−3	4	2	2	0
1	−3	4	2	3	−1
1	−1	2	3	−3	6
1	−1	2	3	−3	6
1	−1	2	3	−3	6
1	1	0	3	−1	4
1	1	0	3	−1	4
1	2	−1	3	−3	6
2	−3	5	3	3	0
2	−3	5	3	3	0
2	−2	4			

From a random sample of 33 investors' ratings of conventional and Socially Desirable portfolios of equivalent risk and return, the average difference is 1.5 points on a 7-point scale of attractiveness.

$$\overline{X}_{dif} = \overline{X}_{SD} - \overline{X}_{C} = -.2 - 1.3 = -1.5$$

With this sample of 33, the standard error of the difference is .6.

$$s_{\overline{X}_{dif}} = \frac{s_{dif}}{\sqrt{N}} = \frac{3.4}{\sqrt{33}} = .6$$

The average difference in attractiveness between the Conventional and the Socially Desirable portfolio is 2.5 standard errors:

$$t_{32} = \frac{\overline{X}_{dif}}{s_{\overline{X}_{dif}}} = -\frac{1.5}{.6} = -2.5$$

The *p value* for $t_{32} = -2.5$, for a sample size of 33, is .009. Were the Socially Desirable portfolio at least as attractive as the Conventional portfolio with equivalent risk and return, it would be unusual to observe such a large sample mean difference in ratings. Based on sample evidence, shown in Figure 3.13, we conclude that a "socially desirable" label reduces portfolio attractiveness.

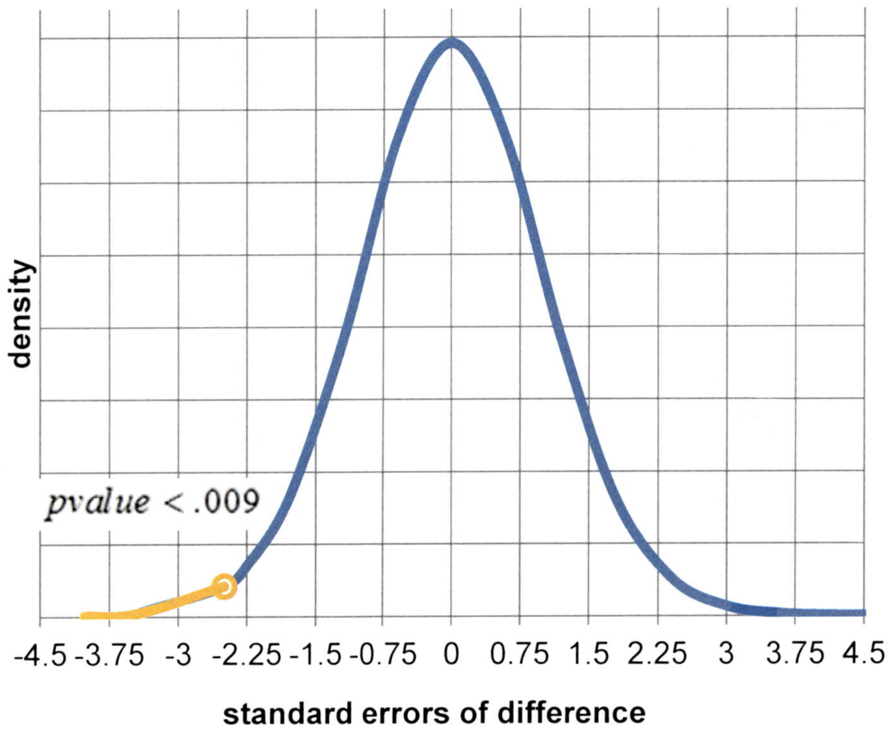

Figure 3.13 *t test* of differences between paired ratings of socially desirable and conventional portfolios

The 95% confidence interval for the difference is

$$\overline{X}_{dif} \pm t_{\alpha/2,N-1} s_{\overline{X}_{dif}}$$

$$-1.5 \pm 2.04\ (.6)$$

$$-1.5 \pm 1.2$$

OR −2.7 to −.3 on the 7 point scale, shown in Figure 3.14.

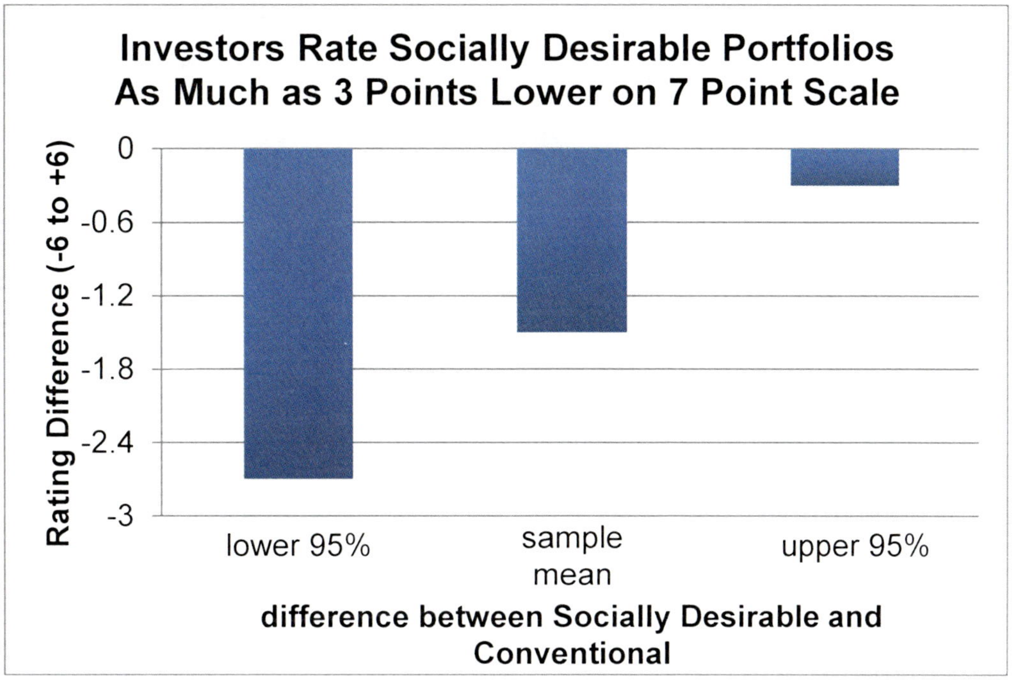

Figure 3.14 *Confidence interval* of mean difference between paired ratings of socially desirable & conventional portfolios

The investment consultants would conclude:

A "socially desirable" label reduces investors' judged attractiveness ratings. Investors downgrade the attractiveness of "socially desirable" portfolios by about 1 to 3 points on a 7 point scale, relative to equivalent, but conventional, portfolios.

3.13 Inference from Sample to Population

Managers use sample statistics to infer population characteristics, knowing that inference from a sample is efficient and reliable. Because sample standard errors are approximately *Normally* distributed, we can use the Empirical Rule to build confidence intervals to estimate population means and to test hypotheses about population means with *t tests*. We can determine whether a population mean is likely to equal, be less than, or exceed a target value, and we can estimate the range which is likely to include a population mean.

Our certainty that a population mean will fall within a sample based confidence interval depends on the amount of population variation and on the sample size. To double precision, sample size must be quadrupled, because the margin of error is inversely proportional to the square root of sample size.

Differences are important to managers, since differences drive decision making. If customers differ, segments are targeted in varying degrees. If employee satisfaction differs between alternate work environments, the workplace may be altered. Inference about differences between two populations is similar, and relies on differences between two independent samples. A *t test* can be used to determine whether there is a likely difference between two population means, and with a confidence interval, we can estimate the likely size of difference.

Confidence intervals and hypothesis tests are consistent and complementary, but are used to make different decisions. If a decision maker needs to make a qualitative Yes/No decision, a hypothesis test is used. If a decision maker instead requires a quantitative estimate, such as level of demand, confidence intervals are used. Hypothesis tests tell us whether demand exceeds a critical level or whether segments differ. Confidence intervals quantify demand or magnitude of differences between segments.

Sample statistics are used to estimate population statistics because it is often neither possible nor feasible to identify and measure the entire population. The time and expense involved in identifying and measuring all population elements is prohibitive. To survey the bottled water consumption of each faculty member, student, and staff member on campus would take many hours. An estimate of demand is inferred from a random, representative sample which includes faculty, students, and staff. Though sample estimates will not be exactly the same as population statistics because of sampling error, samples are amazingly efficient if properly drawn and representative of the population.

Excel 3.1 Test the Level of a Population Mean with a One Sample t test

Thirsty on Campus. Team 8 wants to know whether the demand for bottled water exceeds a breakeven level of 7 bottles per day. To compare the level of demand with to this critical level, use a one tail *t test* of *Bottles* purchased per day.

Open **Excel 3.1 Bottled Water Demand.**

Find the sample *mean* and *sample standard error.*

Alt AY*n* D
B1:B31tab LSN

	A	B
1	*bottles*	
2		
3	Mean	9.9
4	Standard E	0.749483
5	Median	10
6	Mode	10
7	Standard [	4.105085
8	Sample Vá	16.85172
9	Kurtosis	0.53141
10	Skewness	-0.2245
11	Range	19
12	Minimum	0
13	Maximum	19
14	Sum	297
15	Count	30
16	Confidenc	1.532864

Find he difference between the sample mean and the critical value, **7,** and then divide that difference by the standard error to find *t*.

In A17,
Sample difference from 7
In A18,
t
In A19,
pvalue
In B17,
=b3-7
In B18,
=b17/b4
In B19
=t.dist.rt(b18,b15-1)

	A	B
1	*bottles*	
2		
3	Mean	9.9
4	Standard Error	0.749483
5	Median	10
6	Mode	10
7	Standard Deviation	4.105085
8	Sample Variance	16.85172
9	Kurtosis	0.53141
10	Skewness	-0.2245
11	Range	19
12	Minimum	0
13	Maximum	19
14	Sum	297
15	Count	30
16	Confidence Level(95.0%)	1.532864
17	difference between sample mean and 7	2.9
18	t	3.869336
19	pvalue	0.000285

Excel 3.2 Make a Confidence Interval for a Population Mean

Determine the range which is likely to contain average demand in the population. Construct the 95% confidence interval for the population mean *Bottles* demanded.

In A20,
Upper 95% critical bound
In A21,
Lower 95% critical bound
In B20,
=b3+b16
In B21,
=b3-b16

	A	B
1	*bottles*	
2		
3	Mean	9.9
4	Standard Error	0.749483
5	Median	10
6	Mode	10
7	Standard Deviation	4.105085
8	Sample Variance	16.85172
9	Kurtosis	0.53141
10	Skewness	-0.2245
11	Range	19
12	Minimum	0
13	Maximum	19
14	Sum	297
15	Count	30
16	Confidence Level(95.0%)	1.532864
17	difference between sample mean and 7	2.9
18	t	3.869336
19	pvalue	0.000285
20	upper 95% critical bound	11.43286
21	lower 95% critical bound	8.367136

Excel 3.3 Illustrate Confidence Intervals with Column Charts

t-mobile's Service. t-mobile managers have conducted a survey of customers in 32 major metropolitan areas to assess the quality of service along three key areas: coverage, absence of dropped calls, and static. Customers rated t-mobile service along each of these three dimensions using a 5-point scale (1=poor to 5=excellent). Management's goal is to be able to offer service that is not perceived as inferior. This goal translates into mean metropolitan area ratings that exceed 3 on the 5-point scale in the national market across all three service dimensions. Make 95% confidence intervals to estimate the average perceived quality of service.

Open **Excel 3.2 t-mobile.**

95% Confidence Intervals.

Find the sample *mean, standard deviation, standard error, critical t* value, *margin of error (labeled as confidence level in Excel descriptives)*. Move labels above statistics and delete redundant columns C and E. Add *lower and upper 95%* confidence interval bounds for *coverage, dropped calls, and static ratings* below statistics.

Alt AY*n* D
C1:e33 tab LSN
From A1,
Alt HIII
From C1,
Alt HDC
From D1,
Alt HDC
In a17,
lower ci
In a18,
upper ci
In b17,
=b3-b4
In b18,
=b3+b4
In b17,
Shift+down
Shift+right right right right
Cntl+R

	A	B	C	D
1		*coverage rating (1=Poor to 5=Excellent)*	*dropped calls rating (1=Poor to 5=Excellent)*	*static rating (1=Poor to 5=Excellent)*
2				
3	Mean	2.25	3.375	2.9375
4	Standard E	0.17390209	0.1076696	0.099772928
5	Median	2.5	3	3
6	Mode	3	3	3
7	Standard I	0.98373875	0.60907121	0.56440091
8	Sample Va	0.96774194	0.37096774	0.318548387
9	Kurtosis	-1.3226667	1.12514178	0.441768284
10	Skewness	-0.1084299	1.42769055	-0.02691327
11	Range	3	2	2
12	Minimum	1	3	2
13	Maximum	4	5	4
14	Sum	72	108	94
15	Count	32	32	32
16	Confidenc	0.35467564	0.21959359	0.203488228
17	lower	1.89532436	3.15540641	2.734011772
18	upper	2.60467564	3.59459359	3.140988228
19				

Stacked column chart of confidence intervals.

Make a stacked column chart of the 95% upper and lower confidence interval bounds for the population means.

Copy the ratings labels and paste above the upper and lower 95% confidence interval bounds. Copy the margins of error (labelled confidence level) in row 16 and insert twice between the lower and upper ci bounds. Label these two new rows as M and upper ci. (When stacked on top of the lower ci, these will show the mean and the upper ci.)

From row 1
Shift+spacebar
Cntl+C
From A17,
Alt HIE
From A16,
Shift+spacebar
Cntl+C
From A19,
Alt HIE
Cntl+C
From A20,
Alt HIE
In A19,
M
In A20,
Upper ci

	A	B	C	D
1		coverage rating (1=Poor to 5=Excellent)	dropped calls rating (1=Poor to 5=Excellent)	static rating (1=Poor to 5=Excellent)
2				
3	Mean	2.25	3.375	2.9375
4	Standard E	0.17390209	0.1076696	0.099772928
5	Median	2.5	3	3
6	Mode	3	3	3
7	Standard D	0.98373875	0.60907121	0.56440091
8	Sample Va	0.96774194	0.37096774	0.318548387
9	Kurtosis	-1.3226667	1.12514178	0.441768284
10	Skewness	-0.1084299	1.42769055	-0.02691327
11	Range	3	2	2
12	Minimum	1	3	2
13	Maximum	4	5	4
14	Sum	72	108	94
15	Count	32	32	32
16	Confidenc	0.35467564	0.21959359	0.203488228
17		coverage rating (1=Poor to 5=Excellent)	dropped calls rating (1=Poor to 5=Excellent)	static rating (1=Poor to 5=Excellent)
18	lower ci	1.89532436	3.15540641	2.734011772
19	M	0.35467564	0.21959359	0.203488228
20	upper ci	0.35467564	0.21959359	0.203488228
21	upper	2.60467564	3.59459359	3.140988228

Then select the lower ci, M and upper ci labels and statistics and request a stacked column chart.

From A17,**Cntl+shift+right**
Cntl+shift+right
Shift+down down down
Alt NC

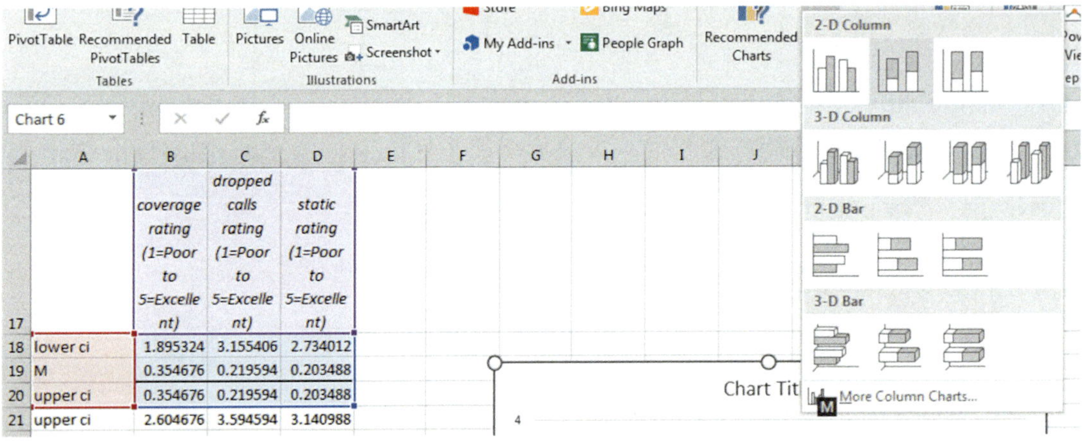

Remove the fill from zero to the lower ci so that only the areas from the lower ci to the mean and the area from the mean to the upper ci, the 95% confidence intervals, are filled.

Alt JAE down to Series lower ci
Alt JASFN

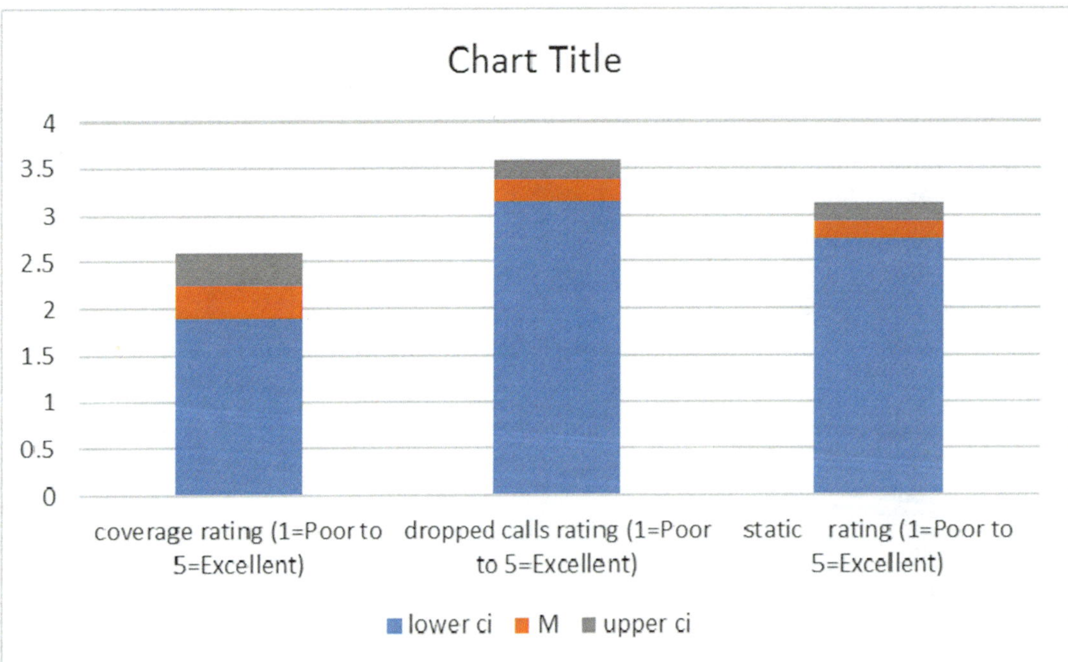

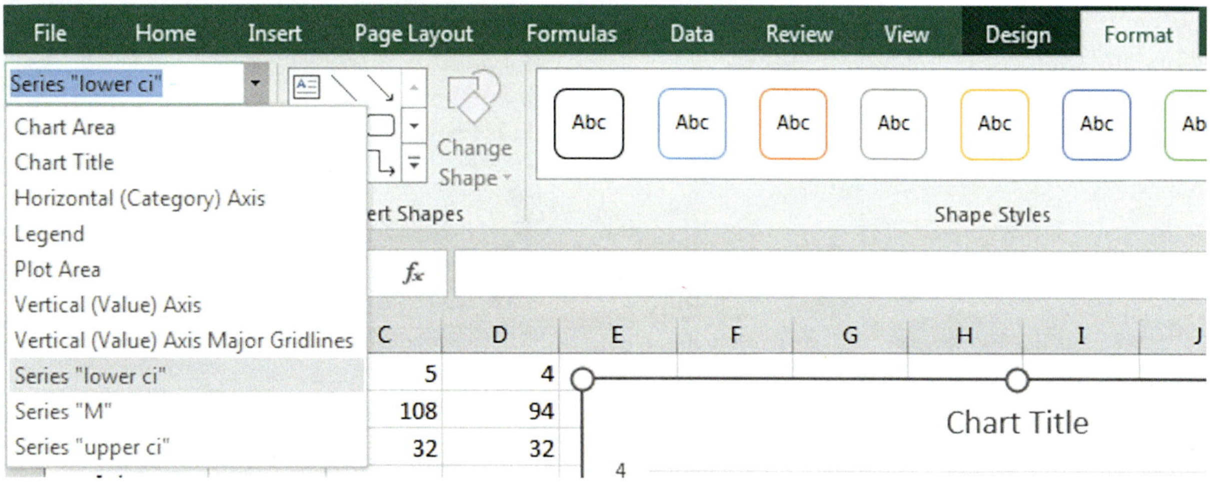
File
Home
Insert
Page Layout
Formulas
Data
Review
View
Design
Format
Series "lower ci"
Chart Area
Chart Title
Horizontal (Category) Axis
Legend
Plot Area
Vertical (Value) Axis
Vertical (Value) Axis Major Gridlines
Series "lower ci"
Series "M"
Series "upper ci"
Change Shape
Shape Styles
Chart Title

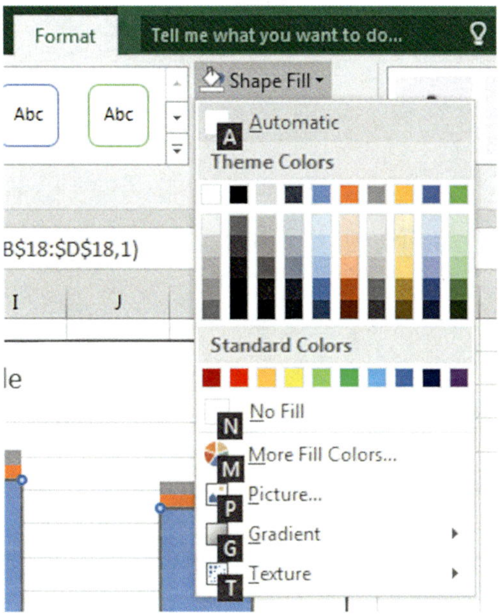
Format
Tell me what you want to do...
Shape Fill
Automatic
Theme Colors
Standard Colors
No Fill
More Fill Colors...
Picture...
Gradient
Texture

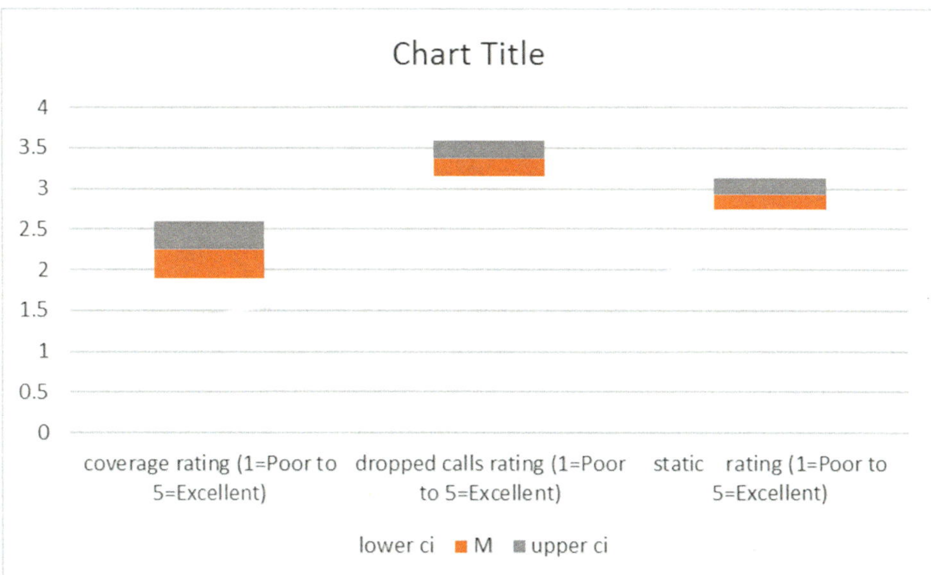
Chart Title
4
3.5
3
2.5
2
1.5
1
0.5
0
coverage rating (1=Poor to 5=Excellent)
dropped calls rating (1=Poor to 5=Excellent)
static rating (1=Poor to 5=Excellent)
lower ci
M
upper ci

Add a vertical axis label. Type in axis labels and chart title, adjust the vertical axis to make good use of space, and set fontsize to 12.	**Alt JCAAV** **Alt JAE** down to vertical axis **Alt JAM** **Alt HFS** 12

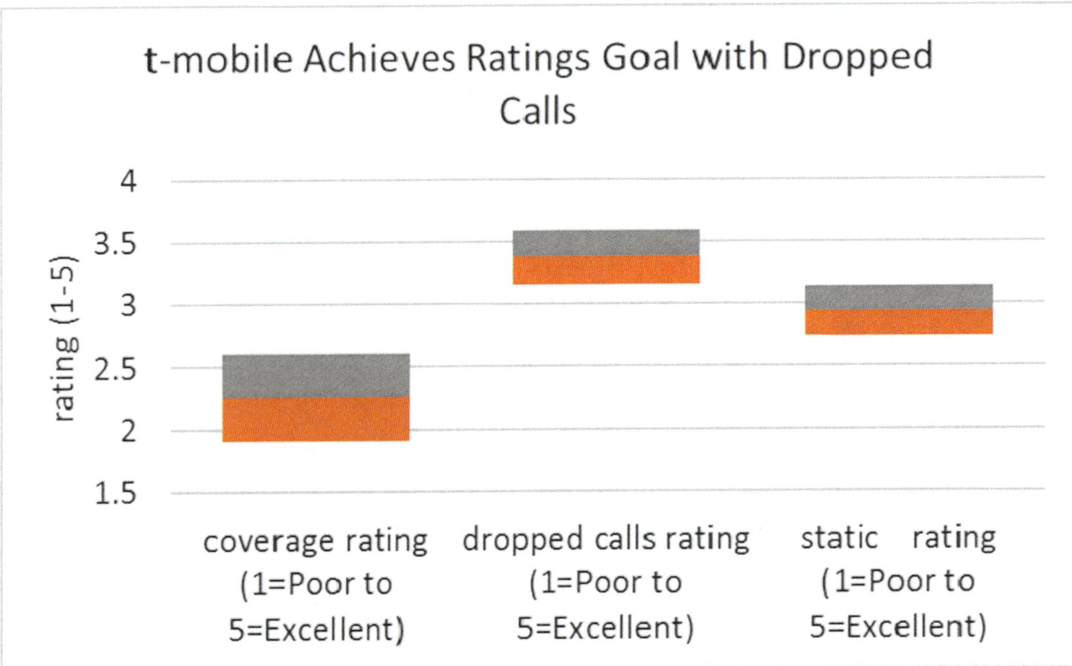

Excel 3.4 Test the Difference Between Two Segment Means with a Two Sample t Test

Pampers Preemies. Procter & Gamble management would like to know whether or not household income is a good base for segmentation in the market for their new preemie diaper. Test the hypothesis that average income is greater in the segment likely to try the new diapers than in the segment unlikely to try.

Open **Excel 3.3 Pampers Segment Income.** The first column **A** contains *likely trier income ($K)* and the second column **B** contains *unlikely trier income ($K).*

Use the Excel function **T.TEST(*array1,array2,tails,type*)** to find the *p value* from a *t test* of the difference between average incomes of the two segments. For *array1*, enter the sample *likely trier income* values. For *array2*, enter the sample *unlikely trier income* values. For *tails*, enter **1** for a *one tail* test, and for *type*, enter **3** to signal a two sample *t test* which allows the standard deviations to differ between segments.

In A58,
pvalue
In B58
=t.test(a2:a57,b2:b57,1,3)

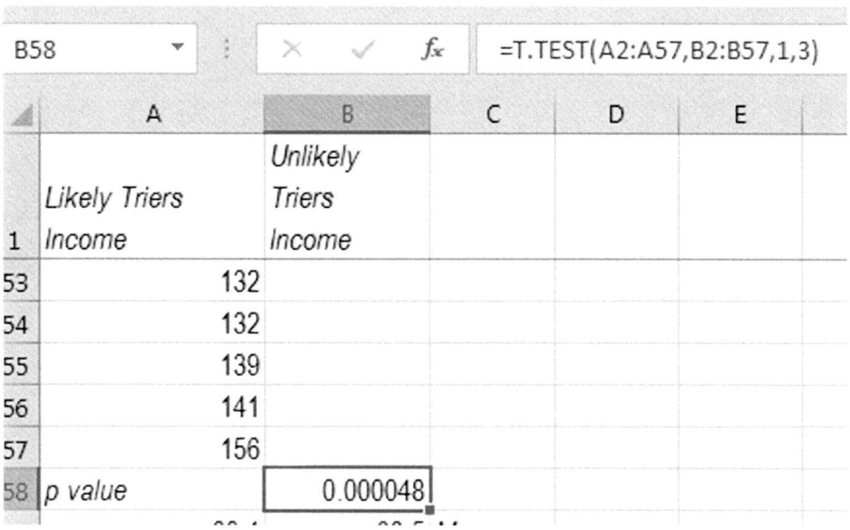

Excel 3.5 Construct a Confidence Interval for the Difference Between Two Segments

Estimate the difference in incomes between the Unlikely and Likely Trier segments.

Use descriptives to find the segment sample means, standard deviations, and standard errors.

Alt AY*n* D
A1:b57 tab LSN

	A	B	C	D
1	y Triers Inc	ome	*Unlikely Triers Income*	
2				
3	Mean	80.14286	Mean	38.53659
4	Standard E	6.90494	Standard E	7.489513
5	Median	67	Median	12
6	Mode	60	Mode	6
7	Standard I	51.67184	Standard I	47.95628
8	Sample Va	2669.979	Sample Va	2299.805
9	Kurtosis	-0.81613	Kurtosis	1.773522
10	Skewness	0.43472	Skewness	1.649431
11	Range	193	Range	176
12	Minimum	6	Minimum	6
13	Maximum	199	Maximum	182
14	Sum	4488	Sum	1580
15	Count	56	Count	41
16	Confidenc	13.83781	Confidenc	15.13687

Find the difference between segment means and the standard error of the difference from the segment sample means and standard errors.

In A17,
Difference between means
In A18,
Pooled s
In B17,
=b3-d3
In B18,
=sqrt(b4^2+d4^2)

B18 =SQRT(B4^2+D4^2)

	A	B	C	D	E
1	*Likely Triers Income*		*Unlikely Triers Income*		
2					
3	Mean	80.14286	Mean	38.53659	
4	Standard Error	6.90494	Standard E	7.489513	
5	Median	67	Median	12	
6	Mode	60	Mode	6	
7	Standard Deviatio	51.67184	Standard D	47.95628	
8	Sample Variance	2669.979	Sample Va	2299.805	
9	Kurtosis	-0.81613	Kurtosis	1.773522	
10	Skewness	0.43472	Skewness	1.649431	
11	Range	193	Range	176	
12	Minimum	6	Minimum	6	
13	Maximum	199	Maximum	182	
14	Sum	4488	Sum	1580	
15	Count	56	Count	41	
16	Confidence Level(	13.83781	Confidenc	15.13687	
17	difference between means	41.60627			
18	pooled s	10.18681			

Find the approximate margin of error of the difference between means, which is twice the standard error, and then make the 95% confidence interval for the difference by adding and subtracting the margin of error from the mean difference.

In A19,
Approximate me
In a20,
Lower 95% ci
In A21,
Upper 95% ci
In B19,
=2*b18
In B20,
=b17-b19
In B21,
=b17+b19

B21 =B17+B19

	A	B	C	D
1	*Likely Triers Income*	*Unlikely Triers Income*		
2				
3	Mean	80.14286	Mean	38.53659
4	Standard Error	6.90494	Standard E	7.489513
5	Median	67	Median	12
6	Mode	60	Mode	6
7	Standard Deviatio	51.67184	Standard I	47.95628
8	Sample Variance	2669.979	Sample Va	2299.805
9	Kurtosis	-0.81613	Kurtosis	1.773522
10	Skewness	0.43472	Skewness	1.649431
11	Range	193	Range	176
12	Minimum	6	Minimum	6
13	Maximum	199	Maximum	182
14	Sum	4488	Sum	1580
15	Count	56	Count	41
16	Confidence Level(	13.83781	Confidenc	15.13687
17	difference between means	41.60627		
18	pooled s	10.18681		
19	approximate me	20.37361		
20	lower 95% ci	21.23266		
21	upper 95% ci	61.97988		

Excel 3.6 Illustrate the Difference Between Two Segment Means with a Column Chart

Illustrate the difference between average incomes of Likely and Unlikely Triers. Select the *lower* and *upper 95%* confidence interval bounds and their labels, and request a column chart. Add the vertical axis title and chart title and set fontsize to 12.	From A20, **Shift+right** **Shift+down** **Alt NC** **Alt JCAAV** **Alt HFS** 12

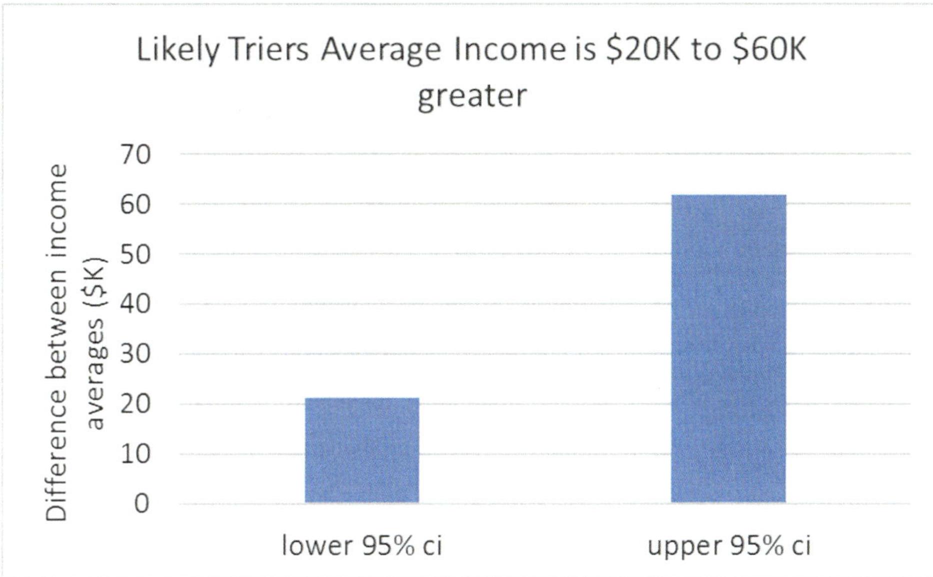

Excel 3.7 Construct a Pie Chart of Shares

Moral Acceptance of Medical Testing on Animals. Construct a pie chart to illustrate how sample ratings of the acceptability of medical testing on animals are split.

Open a new workbook and type in two new columns, *segment* and *%surveyed.* In the *segment* column, type in *acceptable* and *unacceptable.* In the *%surveyed* column, type in the sample proportions that found medical testing on animals acceptable, *61%* and unacceptable *39%.*

	A	B
1	*segment*	*%surveyed*
2	acceptable	61
3	unacceptable	39
4		

Find the conservative standard error of the proportion from P = .5 and sample size of 1000:

B4 f_x =0.5/SQRT(1000)

	A	B	C	D	E
1	*segment*	*%surveyed*			
2	acceptable	61			
3	unacceptable	39			
4	conservative s	0.0158114			

Find the margin of error from the *critical Z* for 95% confidence (1.96) and the *conservative standard error of the proportion*:

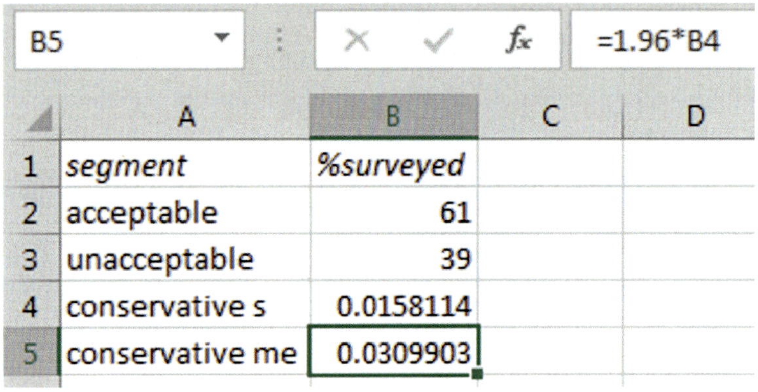

To make a pie chart, select the label and data cells, and then request a pie chart.	From A1, **Shift+right** **Shift+down** **Shift+down** **Alt NQ.**

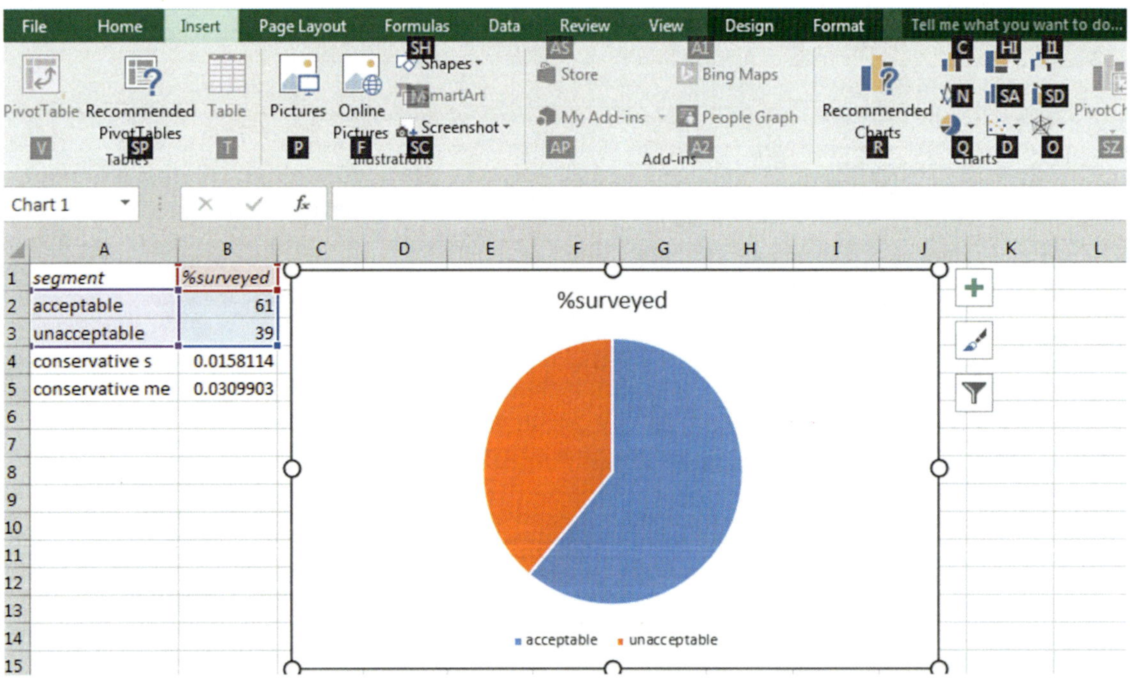

Choose Design Layout 1 and type in a stand alone title. Add the margin of error by inserting a text box, and set fontsize to 12.	**Alt JCL** **Alt NX** **Alt HFS** 12

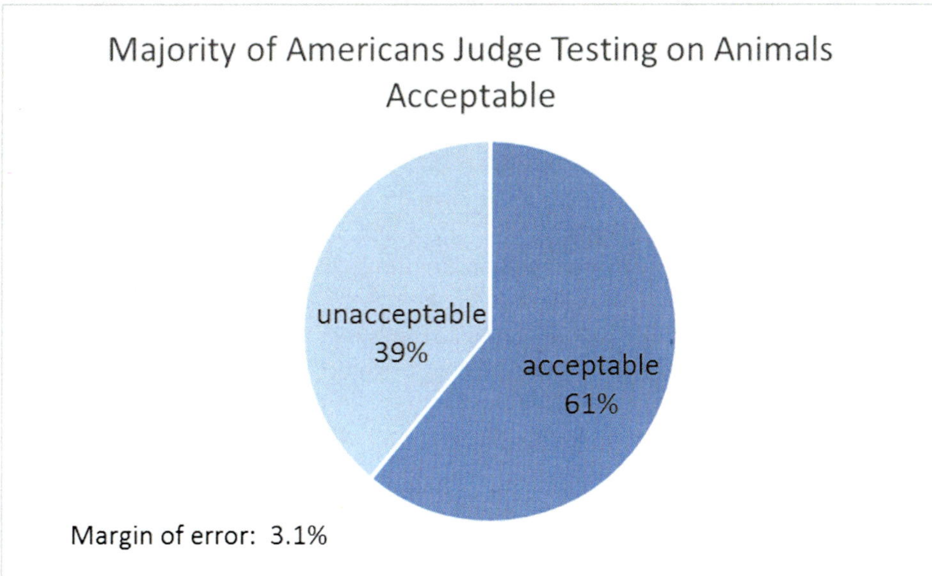

Excel 3.8 Test the Difference in Between Alternate Scenarios or Pairs with a Paired t Test

Difference between Conventional and Socially Desirable Portfolio Ratings. Test the hypothesis that the average difference between ratings of a Conventional portfolio and ratings of a Socially Desirable portfolio is greater than zero.

Open **Excel 3.4 SD Portfolio.**

Use the function **T.TEST(***array1, array2,tails,type***)** to calculate a paired *t test*. For *array1*, enter the *conventional portfolio ratings*. For *array2*, enter the *socially desirable portfolio ratings*. For *tails*, enter **1** for a *one tail* test, and for *type*, enter **1** to specify a paired *t test*.

In B35,
pvalue
In c35,
=t.test(a2:a34,b2:b34,1,1)

B35 =T.TEST(A2:A34,B2:B34,1,1)

	A	B	C	D	E
1	conventional rating	socially desirable rating	conventional minus socially desirable difference		
32	-3	2	-5		
33	-3	3	-6		
34	-3	3	-6		
35	p value	0.009664			

Excel 3.9 Construct a Confidence Interval for the Difference Between Alternate Scenarios or Pairs

To estimate the population difference in investors' ratings of Socially Desirable and Conventional portfolios from sample data, construct a confidence interval of the average rating difference.

Find the mean and standard deviation of the difference and the margin of error of the difference (labelled *Confidence*, in row 16 of descriptives). Subtract and add the margin of error from the mean difference to find the 95% confidence interval bounds for the difference.

Alt AY*n* D
C1:c34 tab LSN
In A17,
Lower 95% ci
In A18,
Upper 95% ci
In B17,
=B3-B16
In B18,
=B3+B16

B18 | f_x =B3+B16

	A	B	C
1	*conventional minus socially desirable difference*		
2			
3	Mean	1.454545	
4	Standard Error	0.590472	
5	Median	2	
6	Mode	0	
7	Standard Deviation	3.392003	
8	Sample Variance	11.50568	
9	Kurtosis	-0.10656	
10	Skewness	-0.64284	
11	Range	12	
12	Minimum	-6	
13	Maximum	6	
14	Sum	48	
15	Count	33	
16	Confidence Level(95.0%)	1.202752	
17	lower 95% ci	0.251794	
18	upper 95% ci	2.657297	

Lab 3.1 Inference

Cingular's Position in the Cell Phone Service Market

Cingular's managers have conducted a survey of customers in 21 major metropolitan areas to assess the quality of service along three key areas: *coverage, absence of dropped calls*, and *static*. Customers rated cingular service along each of these three dimensions using a 5-point scale (1=poor to 5=excellent). Data are in **Lab 3 cingular.**

Management's goal is to be able to offer service that is not perceived as inferior. This goal translates into mean ratings that are greater than 3 on the 5-point scale in the national market across all three service dimensions.

Management can conclude that they have achieved their goal along:

____ *coverage* _____*dropped calls* _____ *static*

Based on this sample, average ratings in all major metropolitan areas are

_________ to _________ for *coverage*,

_________ to _________ for absence of *dropped calls*,

_________ to _________ for *static*, with 95% confidence.

Value of a Nationals Uniform

The Nationals General Manager is concerned that his club may not be paying competitive salaries. He has asked you to compare Nationals' salaries with salaries of players for the closest team in the National League East, the Phillies. He suspects that the Phillies may win more games because they are attracting better players with higher salary offers. Data are in **Lab 3 Nationals.**

This is a _____tail *t test.*

p value from *t test* of difference in team *salary* means. ______________

The General Manager can conclude that, relative to the Phillies, the Nationals are paid ___ Less ___the same.

Extra Value of a Phillies Uniform. If you conclude that the Phillies do earn higher salaries, estimate the average difference at a 95% level of confidence.

On average, players for the Phillies earn _______ to ________ more than players for the Nationals.

The pooled standard error of the difference in mean salaries is: ____________

Illustrate the 95% confidence interval for the difference between the two teams' salaries with a column chart.

Confidence in Chinese Imports

Following the recall of a number of products imported from China, the Associated Press-Ipsos Poll asked 1005 randomly selected adults about the perceived safety of products imported from China. Poll results are below:

"When it comes to the products that you buy that are made in China, how confident are you that those products are safe … ?"

Confident	Not Confident	Unsure
%	%	%
42	57	1

Use this data to construct a *conservative 95% confidence interval* for the *proportion Not Confident* that Chinese imports are safe.

________ to ________ percent are not confident that products made in China are safe.

Illustrate your result with a pie chart which includes the margin of error in a text box. Add a "stand alone" title.

Lab 3.2 Inference: Dell Smartphone Plans

Managers at Dell are considering a joint venture with a Chinese firm to launch a new, competitively priced smartphone.

I. Estimate the percent of smartphone owners who will replace with Dell

In a concept test of 1000 smartphone owners, 20% indicated that they would probably or definitely replace their smartphone with the new Dell concept in the next quarter. Norms from past research suggest that 80% of those who indicate intent to replace actually will.

1. Expected Dell smartphone share = 80% × sample intent proportion: _________
2. Confidence interval for Dell smartphone share: ___ to ___

II. Distinguish Likely Dell Smartphone Adopters

Those who indicated that they were likely to switch to the Dell smartphone may be more price conscious than other smartphone owners. In the concept test, participants were asked to rate the importance of several smartphone attributes, including price. These data are in **Lab 3 Inference Dell smartphone.**

1. Do Likely Adopters rate price higher in importance than Unlikely Adopters?
   a. State the null and alternative hypotheses:
   b. State your conclusion in <u>one sentence</u> that a technically savvy manager would understand, including the statistic that you relied upon to form your conclusion and its p value.
   c. State your conclusion in <u>one sentence</u> with words that a manager, not necessarily statistically savvy, would understand.
2. Illustrate approximate 95% confidence interval for the difference between Likely and Unlikely Adopters' average price importances with a column chart. Add a "stand alone" title and label axes. Copy and paste below:

Assignment 3.1The Marriott Difference

There are 51 branded hotels in Washington, DC, owned or managed by Marriott or competitors. The hotel industry in Washington, DC is representative of the hotel industry in cities throughout the U.S. Differences in quality and price distinguish the hotels. Marriott would like to claim that its hotels offer higher average quality lodging than competing hotels and that Marriott's average *starting room price* is no greater than competitors' average *starting room price*. The dataset **Assignment 3.1 DC Hotels** contains *Guest rating*, a measure of quality, and *starting room price* for Marriott hotels and for competitors' hotels.

3. Can Marriott claim that Marriott hotels are rated higher in quality than competitors' hotels? (Assume a 95% level of confidence.)
   a. State the null and alternative hypotheses.
   b. State your conclusion in one sentence with words that a technically savvy manager would understand.
   c. State your conclusion in one sentence with words that a manager, not necessarily statistically savvy, would understand.

4. Can Marriott claim that Marriott hotels are priced no higher than competitors? (Assume a 95% level of confidence.)
   a. State the null and alternative hypotheses.
   b. State your conclusion in one sentence that a technically savvy manager would understand.
   c. State your conclusion in one sentence with words that a manager, not necessarily statistically savvy, would understand.

Assignment 3.2 Immigration in the U.S.

The FOX News/Opinion Dynamics Poll of (N=) 900 registered voters nationwide, reports public opinion concerning immigrants and proposed immigration legislation:

Join Society/ Give	Stay Separate/ Take	Depends (vol.)	Unsure
%	%	%	%
41	36	17	6
Increase	Decrease	No Change (vol.)	Unsure
%	%	%	%
24	51	17	8

Use this data to construct ***conservative*** *95% confidence intervals* for the *proportions* who (i) agree that immigrants joint society/give and (ii) agree that the U.S. should increase the number of legal immigrants.

Briefly summarize the opinions of **all registered voters** using language that American adults would understand.

Illustrate your summary with pie charts embedded in your report.

Be sure to include the margins of error in your pie charts.

Assignment 3.3 McLattes

McDonalds recently sponsored a blind taste test of lattes from Starbucks and their own McCafes. A sample of thirty Starbucks customers tasted both lattes from unmarked cups and provided ratings on a −3 (=worst latte I've ever tasted) to +3 (=best latte I've ever tasted) scale. These data are in **Assignment 3.3 Latte.**.

Can McDonalds claim that their lattes taste every bit as good as Starbucks' lattes? (Please use 95% confidence.)

What evidence allows you to reach this conclusion?

Assignment 3.4 A Barbie Duff in Stuff

Mattel recently sponsored a test of their new Barbie designed by Hillary Duff**.** The Duff Barbie is dressed in Stuff, Hillary Duff clothing designs, and resembles Hillary Duff. Mattel wanted to know whether or not the Duff Barbie could compete with rival MGA Entertainment's Bratz dolls.

A sample of thirty 7-year-old girls attended Barbie parties, played with both dolls, then rated both on a −3 (=Not At All Like Me) to +3 (=Just Like Me) scale. These data are in **Assignment 3.4 Barbie.**.

Do the 7-year-olds identify more strongly with the Duff Barbie in Stuff than the Bratz? (Please use 95% confidence.)

What evidence allows you to reach this conclusion?

Assignment 3.5 Alcoa Smelters

Alcoa executives believe that older "legacy" smelters, those with lower capacity, and those which use coal or gas may have higher costs. Determine whether or not their hypotheses are true, comparing total, labor, and power units costs ($/ton) between smelters first in operation in the 1950s or 1960s with those first in operation in the 1970s or 1980s, those with lower capacity with those with higher capacity, and those powered by coal or gas with those powered by hydroelectric. Data are i**n AlcoaSmelters by Year**.

1. Are costs lower in smelters with initial operation in 1970s or 1980s?

Total unit cost ($/ton): ___Y or ___N Labor Cost per Unit ($/ton): ___Y ___N

Power Cost per Unit ($/ton): ___Y ___N

2. Are costs lower in smelters with higher capacity?

Total unit cost ($/ton): ___Y or ___N Labor Cost per Unit ($/ton): ___Y ___N

Power Cost per Unit ($/ton): ___Y ___N

3. Are costs lower in smelters powered by hydroelectric?

Total unit cost ($/ton): ___Y or ___N Labor Cost per Unit ($/ton): ___Y ___N

Power Cost per Unit ($/ton): ___Y ___N

4. Illustrate any difference(s) that you find with (a) column chart(s). Copy and paste below.

Case 3.1 Yankees v Marlins: The Value of a Yankee Uniform[1]

The Marlins General Manager is disgruntled because two desirable rookies accepted offers from the Yankees instead of the Marlins. He believes that Yankee salaries must be noticeably higher—otherwise, the best players would join the Marlins organization. Is there a difference in salaries between the two teams? If the typical Yankee is better compensated, the General Manager is planning to chat with the Owners about sweetening the Marlins' offers. He suspects that the Owners will argue that the typical Yankee is older and more experienced, justifying some difference in salaries.

Data are in **Case 3.1 Yankees v Marlins Salaries.**
Determine:

- whether or not Yankees earn more on average than Marlins, and
- whether or not players for the Yankees are older on average than players for the Marlins.

If you find a difference in either case, construct a *95% confidence interval* of the difference between means in any season.

Briefly summarize your results using language that the General Manager and Owners would understand, and illustrate with a column chart.

Case 3.2 Gender Pay

The Human Resources manager of Slam's Club was shocked by the revelations of gender discrimination by WalMart and wants to demonstrate that there is no gender difference in average salaries in his firm. He also wants to know whether levels of responsibility (measured with the Position variable) and experience differ between men and women, since this could explain a difference in salaries.

Case 3.2 GenderPay contains *salaries*, *positions*, and *experience* of men and women from a random sample of the company records.
Determine

- whether or not the sample supports a conclusion that men and women are paid equally,
- whether average level of *responsibility* differs across genders,
- whether average *experience* differs across genders.

If you find that the data support the alternate hypothesis that men are paid more, on average, construct a 95% confidence interval of the difference between means.

If either average level of *responsibility* or average years of *experience* differs, construct *95% confidence intervals* of the difference between means.

[1] This example is a hypothetical scenario using actual data.

Briefly summarize your results using language that a businessperson (who may not remember quantitative analysis) could understand.

Illustrate your results with column charts. Choose bottom line titles that help your audience see the results.

Be sure to round your statistics to two or three significant digits.

Case 3.3 Polaski Vodka: Can a Polish Vodka Stand Up to the Russians?

Seagrams management decided to enter the premium vodka market with a Polish vodka, suspecting that it would be difficult to compete with Stolichnaya, a Russian vodka and the leading premium brand. The product formulation and the package/brand impact on perceived taste were explored with experiments to decide whether the new brand was ready to launch.

The taste. First, Seagrams managers asked, "Could consumers distinguish between Stolichnaya and Seagrams' Polish vodka in a *blind* taste test, where the impact of packaging and brand name were absent?"

Consultants designed an experiment to test the null and alternative hypotheses:

> H_0: The taste rating of Seagram's Polish vodka is at least as high as the taste rating of Stolichnaya. The average difference between taste ratings of Stolichnaya and Seagrams' Polish vodka does not exceed zero:
>
> $$\mu_{STOLICHNAYA} - \mu_{POLISH} \leq 0$$
>
> H_1: The taste rating of Seagram's Polish vodka is lower than the taste rating of Stolichnaya. The average difference between taste ratings of Stolichnaya and Seagram's Polish vodka is positive:
>
> $$\mu_{STOLICHNAYA} - \mu_{POLISH} > 0$$

In this first experiment, each participant tasted two unidentified vodka samples and rated the taste of each on a 10-point scale. Between tastes, participants cleansed palates with water. Experimenters flipped a coin to determine which product would be served first: if heads, Seagrams' polish vodka was poured first; if tails, Stolichnaya was poured first. Both samples were poured from plain, clear beakers. The only difference between the two samples was the actual vodka.

These experimental data in **Case 3.3 Polaski Taste** are repeated measures. From each participant, we have two measures whose difference is the difference in taste between the Russian and Polish vodkas.

Test the difference between taste ratings of the two vodkas.

Construct a *95% confidence interval* of the difference in taste ratings.

Illustrate your results with a PivotChart and interpret your results for management.

The brand and package. Seagrams management proceeded to test the packaging and name, Polaski. The null hypothesis was:

> H_0: The taste rating of Polaski vodka poured from a Polaski bottle is at least as high as the taste rating of Polaski vodka poured from a Stolichnaya bottle. The mean difference between taste ratings of Polaski vodka poured from a Stolichnaya bottle and Polaski vodka poured from the Seagrams bottle bearing the Polaski brand name is not exceed zero.

Alternatively, if the leading brand name and distinctive bottle of the Russian vodka affected taste perceptions, the following could be true:

> H_1: The mean difference between taste ratings of Polaski vodka poured from Stolichnaya bottle and Polaski vodka poured from the Seagrams bottle bearing the Polaski brand name is positive.

In this second experiment, Polaski samples were presented to participants twice, once poured from a Stolichnaya bottle, and once poured from the Seagrams bottle, bearing the Polaski name. Any minute differences in the actual products were controlled for by using Polaski vodka in both samples. Differences in taste ratings would be attributable to the difference in packaging and brand name.

Thirty new participants again tasted two vodka samples, cleansing their palates with water between tastes. As before, a coin toss decided which bottle the first sample would be poured from: Stolichnaya if heads, Polaski if tails. Each participant rated the taste of the two samples on a 10-point scale.

These data are in **Case 3.3 Polaski Package.**

Test the difference in ratings due to packaging.
Construct a *95% confidence interval* of the difference in ratings due to the packaging.
Illustrate your results with a PivotChart.
Interpret your results for management.

Chapter 4
Simulation to Infer Future Performance Levels Given Assumptions

Decision makers deal with uncertainty when considering future scenarios. Performance levels depend on multiple influences with uncertain future values. To estimate future performance, managers make assumptions about likely future scenarios and uncertain future values of performance components. To evaluate decision alternatives, the *"best"* and *"worst" case* outcomes are sometimes compared. Alternatively, *Monte Carlo simulation* can be used to simulate random samples using decision makers' assumptions about performance components, and those random samples can then be combined to produce a distribution of likely future scenarios and outcomes that are less extreme than the often unlikely "best" and "worst" cases. Inferences from a simulated distribution of outcomes can then be made to inform decision making. "Best" and "worst" case comparisons are contrasted with inferences from Monte Carlo simulation in this chapter.

4.1 Specify Assumptions Concerning Future Performance Drivers

Example 4.1 The **Thirsty** Team 8 partners were concerned that they might either pass up a profitable opportunity or invest in an unprofitable business. Their estimate of average bottles of water demanded per customer per week seemed promising, though they realized that success of the business depended on several factors, each with uncertain values in the future. An estimate of potential revenues from the first year of operation was desired.

Potential customers, bottles per customers, and revenues, shown in green, below, were outcomes that depended on uncertain factors: the market growth rate, access (restricted or not) to those potential customers, share of customers that the new business could capture, and bottles demanded per customer, shown with blue fill, below. Price would be determined from the access outcome and weeks in which units were stocked, shown in gold, below, were assumed known. Thus, if the Team were unusually fortunate, the market growth rate would be "best," access would be unrestricted, share of customers would be "best," and bottles demanded per customer would be "best." Alternatively, if the Team were unusually unfortunate, the market growth rate would be "worst," access would be restricted, share of customers would be "worst," and bottles demanded per customer would be "worst."

C. Fraser, *Business Statistics for Competitive Advantage with Excel 2016*,
DOI 10.1007/978-3-319-32185-1_4

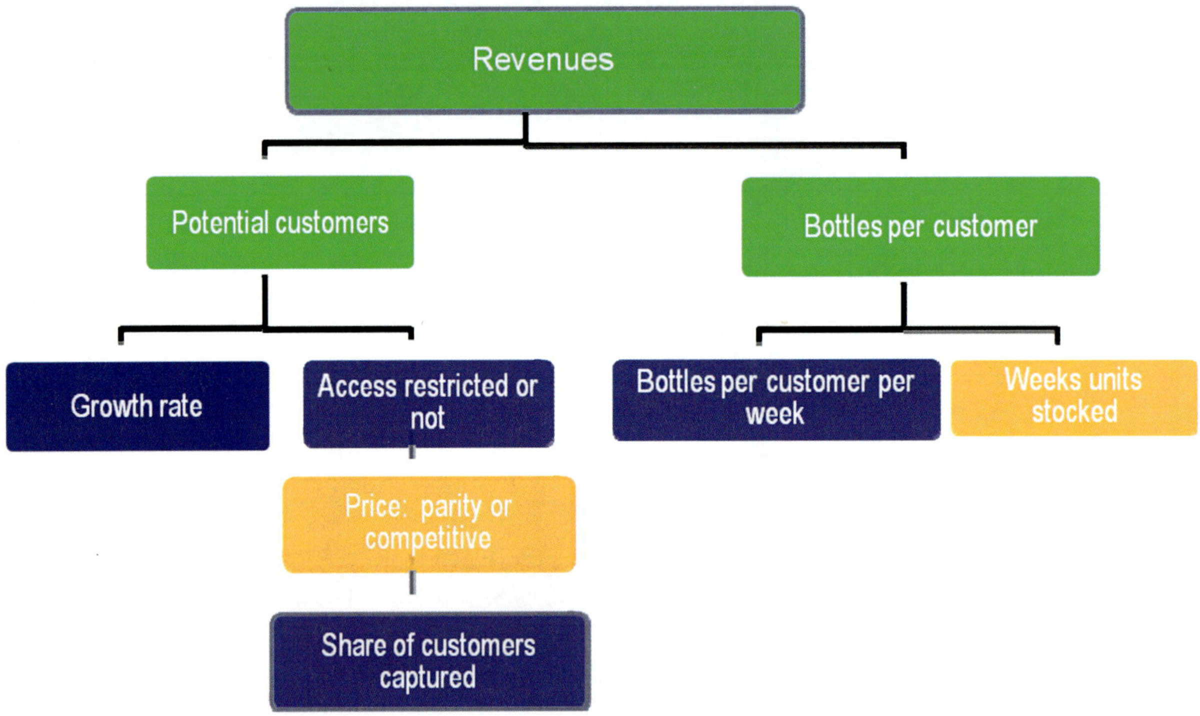

Spreadsheet. The Team created a spreadsheet linking each of the uncertain revenue influences to weekly revenues, given their assumptions about the potential market, shown in Table 4.1.

Assumptions concerning possible outcomes for uncertain influences were highlighted in blue, with three scenarios considered: the two extremes, "worst case" and "best case", and the expected, "best guess." The chance of an uncertain driver level worse than "the worst" or better than "the best" was assumed to be about 5%, making the interval from worst case to best case a 95% confidence interval. Outcomes resulting from uncertain drivers were highlighted in green.

Table 4.1 Spreadsheet for bottled water revenue

		Assumptions		
		95% confidence interval		
		Worst	*Expected*	*Best*
(1)	*Potential Customers Last Year (K)*	34.1	34.1	34.1
(2)	*Annual Growth %*	2.5%	3.5%	4.5%
(3)	*Potential Customers in Year 1 (K)* = (1) × (100% + (2))	35.0	35.3	35.6
(4)	*P(Unrestricted Access)*	0%	67%	100 %
(5)	*% Customers w Access* = (4) × 100% + (1 − (4)) × 80%	80%	93%	100 %
(6)	*Customers Accessed (K)* = (3) × (5)	28.0	33.0	35.6
(7)	*Price per Bottle ($) given Access* = $1.5 − (4) × $.25	$1.50	$1.33	$1.25
(8)	*Share captured at parity price*	10%	15%	20%
(9)	*Share captured at competitive price*	20%	35%	50%
(10)	*Share captured given price* = (100%−(4)) × (8) + (4) × (9)	10%	28%	50%
(11)	*Customers captured (K)* = (6) × (10)	2.8	9.4	17.8
(12)	*Bottles sold per customer per week*	8	10	12
(13)	*Bottles sold per week (K)* = (11) × (12)	22.4	93.6	213.8
(14)	*Weeks in business*	38	38	38
(15)	*Bottles sold in year 1 (M)* = (13) × (14)/1000	.85	3.6	8.1
(16)	*Revenues in Year 1 ($M)* = (7) × (13)	$1.28	$ 4.74	$ 10.2

Potential customers in year 1. Potential customers include faculty, staff, and students on campus, currently 34.1K.

University admissions had been growing between 3 and 4% in recent years, so future growth between 2.5 and 4.5% is anticipated with 95% confidence. Hiring of faculty and staff is expected to grow at similar rates to accommodate the student population.

The potential market in the first year of business is:

Potential customers (K) = *Potential customers last year (K)* × (100% + *annual growth%*)

= 34.1K × (100% + *annual growth%*)

= 34.1K × 102.5% = 35.0K in the worst case,
= 34.1K × 104.5% = 35.6K in the best case,
= 34.1K × 103.5% = 35.3K in the expected case.

Access. If the new business is successful in gaining approval to place units in dorms, 100% of the potential market would have access. Without this approval, restricted access for vending units would reach an about 80% of the potential market.

Customers accessed (K) = *%Accessed* × *Potential customers (K)*

= 80% × 35.0K = 28.0K in the worst case,
= 100% × 35.6K = 35.6K in the best case,

The Team assumed that chance of unrestricted access, *P*(*Unresticted Access*), was about 67%:

Customers accessed (K) = 67% × (100% × 35.6K)

+ (100% − 67%) × (80% × 35.0K) in the expected case.

Price. Bottled water on campus sells for \$1.50 from vending units and in campus eateries. If access is unrestricted, the Team assumes that the volume of business to be great enough to enable volume discounts on plastic bottles and natural flavorings. In this case, a lower price of \$1.25 could be charged, which would be assumed to stimulate trial and repeat sales.

Share. With restricted access and a parity price, the Team assumes that the business could capture at least 10% of the market, and possibly as much as 20%. With unrestricted access and the lower price, they assume that at least 20% of the market would be captured, and that 50% share would be possible.

Customers captured (K) = *Share captured* × *Customers accessed (K)*

= 10% × 28.0K = 2.8K in the worst case,
= 50% × 35.6K = 17.8K in the best case,

= (67% × 35% + (100%−67%) × 15%) × 33.0K = 9.4K expected.

From their market research, the Team estimates that the average number of bottles of water demanded per customer per week falls within the range of 8 to 12, with 95% confidence, and an average of 10 bottles per customer per week is expected.

Given this level of demand per customer, weekly sales would be

Bottles sold per week (K) = *Bottles per customer per week* × *Customers captured (K)*

$= 8 \times 2.8K = 22.4K$ in the worst case,
$= 12 \times 17.8K = 213.8K$ in the best case,
$= 10 \times 9.4K = 93.6K$ expected.

The Team assumes that the business will operate during the 38 weeks in which classes are in session. Therefore, volume in the first year, in millions (M), would be:

Bottles sold (M) = 38 × *Bottles sold per week (K)*/ 1000

$= 38 \times 22.4K/1000 = .85M$ in the worst case,
$= 38 \times 213.8K/1000 = 8.1M$ in the best case,
$= 38 \times 93.6K/1000 = 3.6M$ expected.

At those potential volumes, with the two alternative prices, revenue in the first year would be:

Potential revenue ($M) = Price × Bottles sold (M),

= \$1.50 × .85M = \$ 1.3M in the worst case,
= \$1.25 × 8.1M = \$10.2M in the best case,
= \$1.33 × 3.6M = \$ 4.7M expected.

4.2 Compare Best and Worst Case Performance Outcomes

Best versus Worst. If worst case outcomes occurred (slower growth, restricted access, parity price at \$1.50, 10% share, low demand per customer), revenue would be just \$1.3M in the first year, making the investment unattractive. However, if best case outcomes occurred (faster growth, unrestricted access, competitive price at \$1.25, 50% share, high demand per customer), revenue would be \$10.2M, making the investment extremely attractive.

These extreme outcomes differ widely. How likely are these two extremes?

Based on the Team's assumptions, the chance of the worst case outcome is equal to the joint probability assumed for the four uncertain influences:

P(the worst case outcome) = *P(annual growth* ≤ *2.5%)* × *P(access restricted)*

× *P(Share* ≤ *10%)*

× *P(Demand* ≤ *8 bottles per customer per week)*

The chance that annual market growth would be as low as 2.5%, *P(annual growth* ≤ *2.5%)*, the low end of the 95% confidence interval, is 2.5%.

The chance that share would be as low as 10%, *P(Share* ≤ *10%)*, the low end of the 95% confidence interval, is 2.5%.

The chance that demand would be as low as 8 bottles per customer per week, *P(demand* $\le$ *8)*, the low end of the 95% confidence interval, is 2.5%.

Therefore, considering the chance of each of these unfortunate outcomes, the chance the revenue could be as low as \$1.3M is:

P(the worst case outcome) = 2.5% × 33% × 2.5% × 2.5% = .00052% = .0000052, or one in 200,000 (= 1/.0000052), making the worst case extremely unlikely. The Team could be 95% certain that, given their assumptions, the worst case would not occur.

Based on the Team's assumptions, the chance that the best case outcome would occur is equal to the joint probability of four fortunate circumstances:

P(best case outcome) = *P(annual growth* $\ge$ *4.5%)* × *P(access unrestricted)*

× *P(Share* $\ge$ *50%)*

× *P(Demand* $\ge$ *12 bottles per customer per week)*

The chance that annual market growth would be as high as 4.5%, *P(annual growth* $\ge$ *4.5%)*, the high end of the 95% confidence interval, is 2.5%.

The chance that share would be as high as 50%, *P(Share* $\ge$ *50%)*, the high end of the 95% confidence interval, is 2.5%.

The chance that demand would be as high as 12 bottles per customer per week, *P(demand* $\ge$ 12*)*, the high end of the 95% confidence interval, is 2.5%.

Therefore, considering the chance of each of these fortunate outcomes, the chance the revenue could be as high as \$10.2M is:

P(best case outcome) = 2.5% × 67% × 2.5% × 2.5% = .0011% = .000011, or one in 100,000 (= 1/.000011), making the best case extremely unlikely, as well. The Team could be 95% certain that the best case outcome would also not occur.

Both the worst case and the best case outcomes were clearly not likely enough to warrant consideration. What range of revenues actually was likely?

To quantify the risks and produce a range of likely revenues that could actually occur, the Team decided to use Monte Carlo simulation. They could then incorporate the uncertainty, given their assumptions, into their forecast. Results would show the distribution of possible outcomes and their likelihoods under the Team's assumptions, and they would be able to determine a 95% confidence interval for possible outcomes.

4.3 Spread and Shape Assumptions Influence Possible Outcomes

Spread and Shape Assumptions. The Team updated their revenue spreadsheet, specifying the spread and shape for each of the uncertain influences in Table 4.2:

Table 4.2 Updated spreadsheet for bottled water revenue

		Assumptions		
			SD = 95% CI /4 or range	Distri-bution
		Expected		
(1)	*Potential Customers Last Year (K)*	34.1		
(2)	*Annual Growth %*	3.5%	2.5% to 4.5%	Normal
(3)	*Potential Customers in Year 1 (K)* = (1) × (100% + (2))	35.3		
(4)	*P(Unrestricted Access)*	67%		binomial
(5)	*% Customers w Access* = (4) × 100% + (1 − (4)) × 80%	93%		
(6)	*Customers Accessed (K)* = (3) × (5)	33.0		
(7)	*Price per Bottle ($) given Access* = $1.5 − (4) × $.25	$ 1.33		
(8)	*Share captured at parity price*	15%	2.5%	Normal
(9)	*Share captured at competitive price*	35%	7.5%	Normal
(10)	*Share captured given price* = *(100%−(4)) × (8) + (4) × (9)*	28%		
(11)	*Customers captured (K)* = (6) × (10)	9.4		
(12)	*Bottles sold per customer per week*	10	1	Normal
(13)	*Bottles sold per week (K)=* (11) × (12)	93.6		
(14)	*Weeks in business*	38		
(15)	*Bottles sold in year 1 (M)* = (13) × (14)/1000	3.6		
(16)	*Revenues in Year 1 ($M)* = (7) × (15)	$ 4.7		

4.4 Monte Carlo Simulation of the Distribution of Performance Outcomes

The distribution of performance outcomes, *revenues in year 1*, in the **Thirsty** case, depend on the distributions of performance influences. With assumptions for center, spread, and shape of each influence now specified in their spreadsheet, the Team drew simulated samples for each. Formulas in their spreadsheet then combined the simulated samples to produce the distribution of possible revenues in year 1.

Growth possibilities. The Team assumed a normal distribution for growth in the next year within the range of likely possibility, 2.5 to 4.5%. A random sample of 1000 simulated possible growth values was drawn and is shown in Figure 4.1.

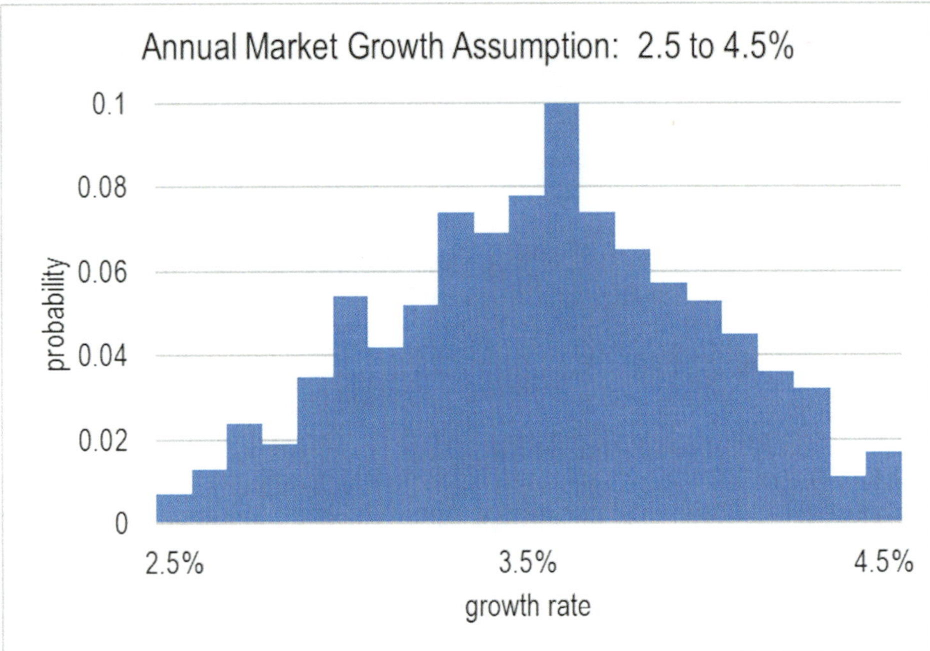

Figure 4.1 Simulated sample of possible annual growth values

With the assumption of a Normal distribution of potential growth rates, the distribution of possible values for potential customers, shown in Figure 4.2 would be Normal.

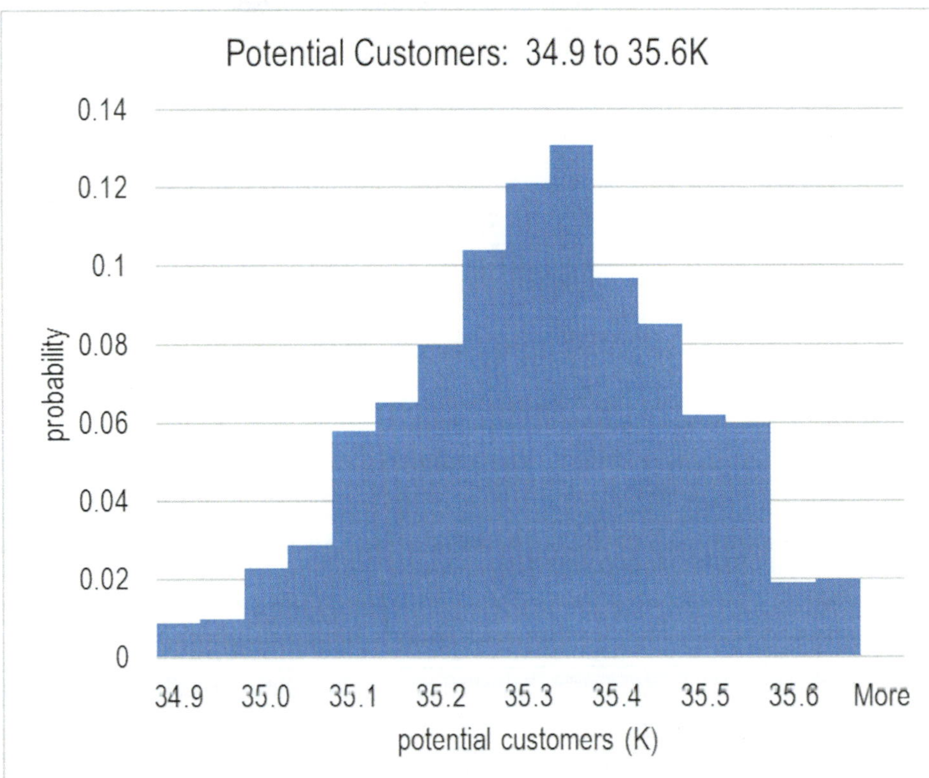

Figure 4.2 Simulated sample of possible potential customers

Access. A random sample of 1000 possible access outcomes was drawn, with the probability for a favorable unrestricted outcome set at 67%. With the random sample of access outcomes, possible outcomes for customers accessed is bimodal, as shown in Figure 4.3. The 95% confidence interval is 28 to 36K.

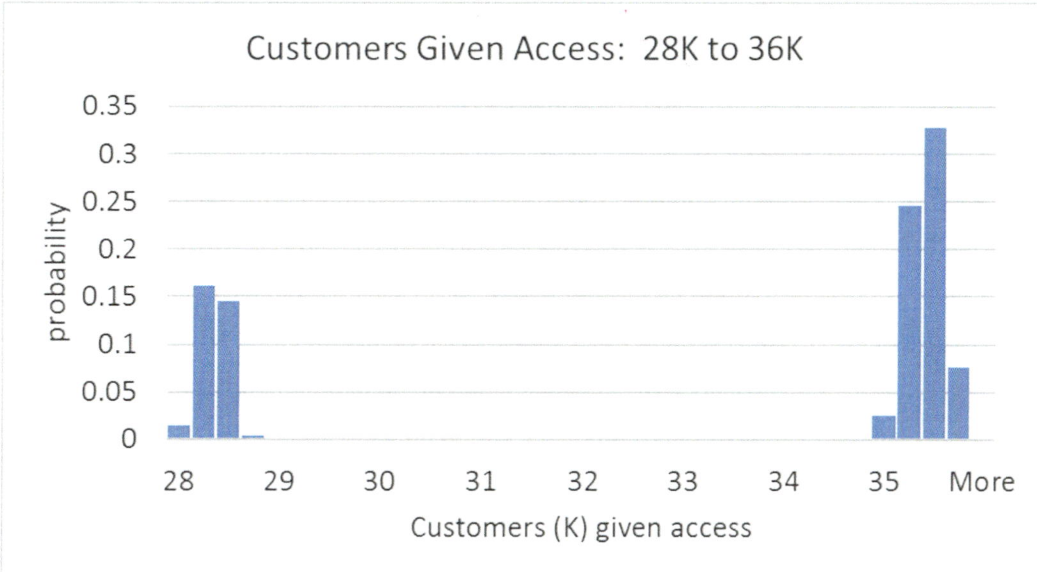

Figure 4.3 Simulated sample of customers accessed

Share. With restricted access, the parity price of \$1.50 would be charged. With unrestricted access, the competitive price of \$1.25 could be charged, which the Team assumes would yield a higher share of customers captured. A random sample of 1000 possible shares at each price was drawn, and the share corresponding to the each randomly selected *access* outcome and price was chosen, yielding the bimodal distribution shown in Figure 4.4, with 95% confidence interval 11% to 50%.

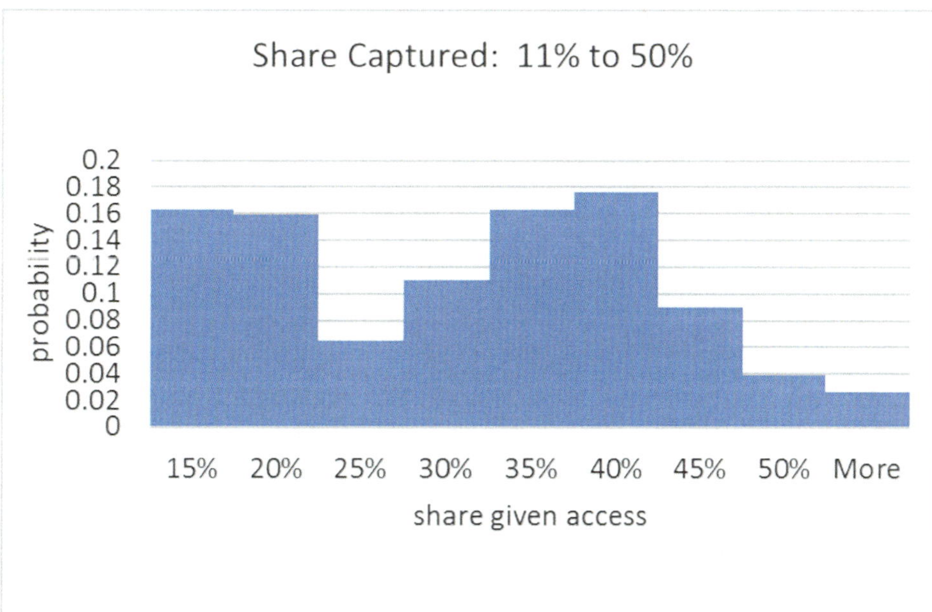

Figure 4.4 Simulated distribution of possible shares

Combining the random sample of *customers accessed* with the random sample of possible *shares of customers*, given access outcome and price, yields the random sample of *customers captured* shown in Figure 4.5, with 95% confidence interval 4K to 18K.

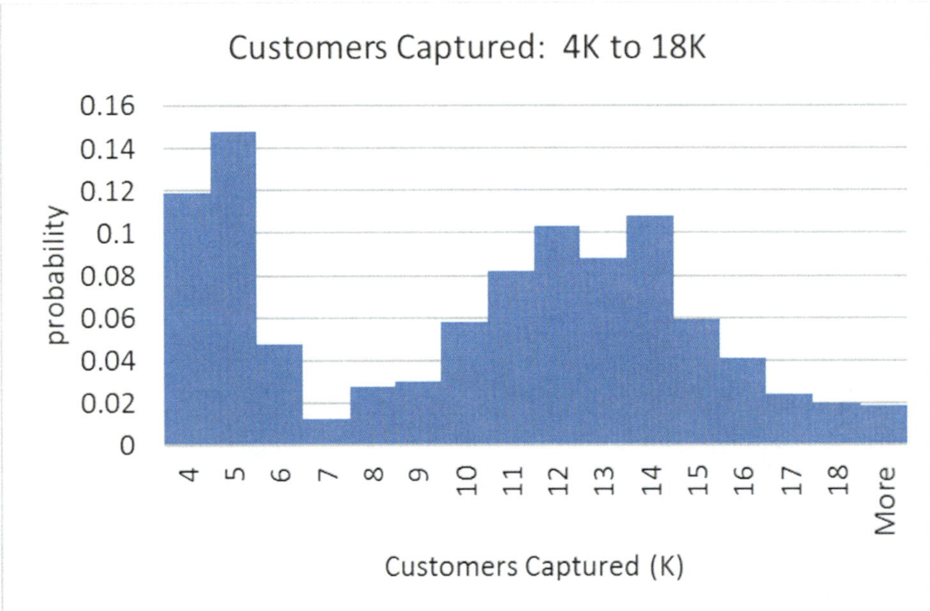

Figure 4.5 Simulated sample of possible customers captured

Demand per customer. A random sample of 1000 Normally distributed *bottles per customer per week* was drawn, producing a 95% confidence interval 8.1 to 11.9 bottles per customer per week, shown in Figure 4.6:

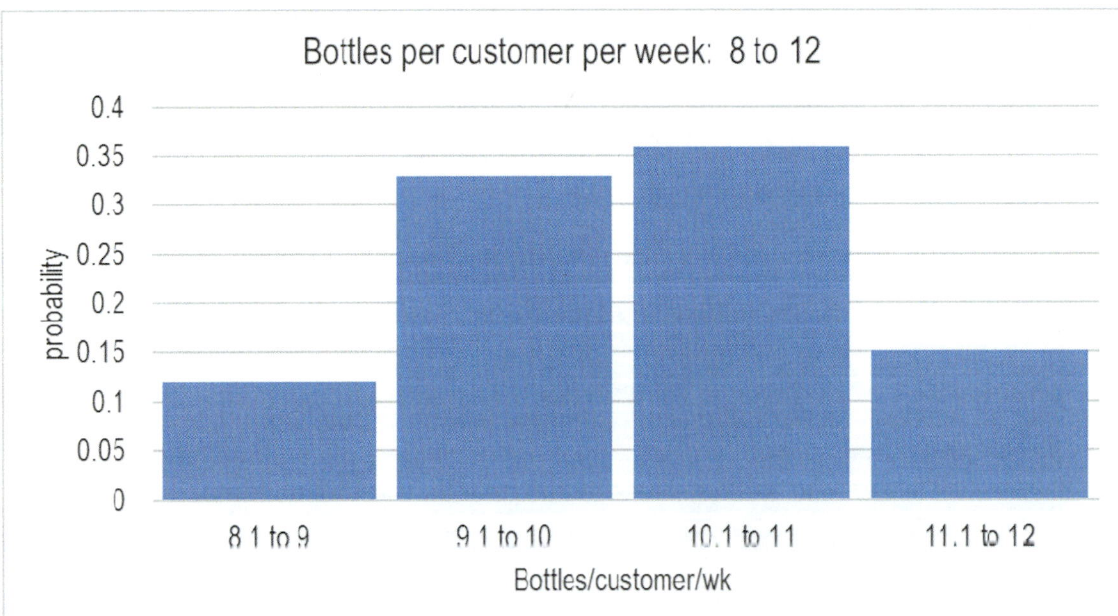

Figure 4.6 Simulated sample of bottles per customer per week

Bottles sold in year 1. Combining the random sample of *customers captured* with the random sample of *bottles per customer per week* and 38 weeks in the operating year provides the distribution of possible volumes in *bottles sold* in year 1, shown in Figure 4.7, with 95% confidence interval 1 to 7M.

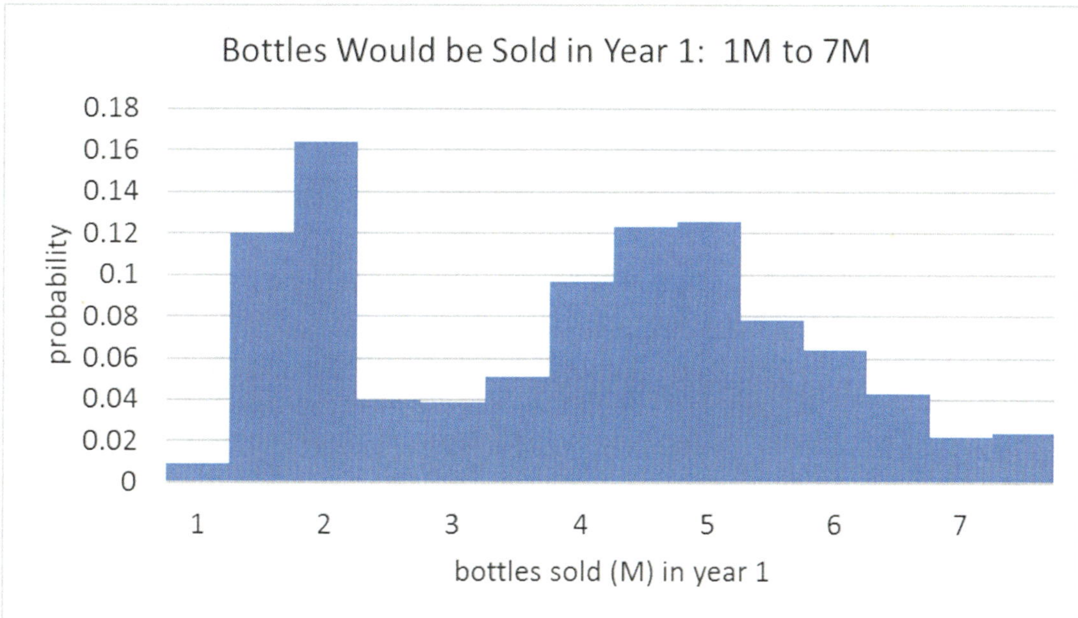

Figure 4.7 Simulated sample of bottles sold in year 1

Revenues in year 1. Combining the random sample of *bottles sold in year 1* with the sample of prices, given *access* outcomes, the Team could see the distribution of possible *revenues in year 1,* shown in Figure 4.8, with 95% confidence interval \$1.7M to \$8.6M.

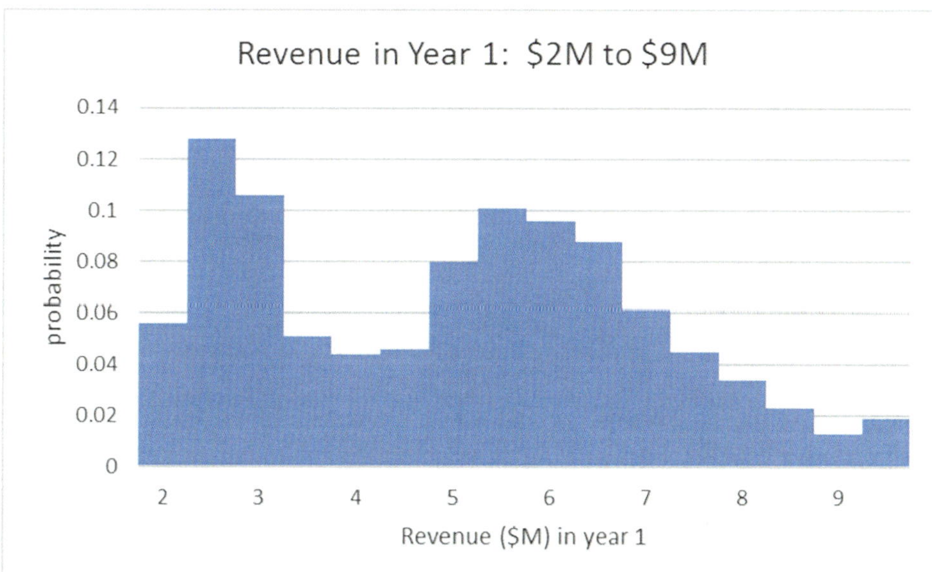

Figure 4.8 Simulated sample of revenues in year 1

This range in likely *revenues*, \$6.9M (= \$8.6M − \$1.7M) is narrower and, therefore, more informative, than the comparison of the extremely unlikely best and worst case outcomes, whose wider range was \$8.9M (= \$10.2M − \$1.3M).

If their assumptions were correct, the Team could expect business of at least \$1.7M. Expected revenues in year 1 are \$4.7M, given the Team's assumptions, and the median of possible revenues is \$4.9M, given assumptions: There was a 50% chance that the business would produce revenues of at least \$4.9M in year 1.

4.5 Monte Carlo Simulation Reveals Possible Outcomes Given Assumptions

Decisions concerning investments or allocation of resources depend on inference of likely future performance outcomes. Those performance outcomes hinge on the values that multiple uncertain influences will take on in the future. Monte carlo simulation offers a view of the possibilities, given the assumptions we make about each of those uncertain influences.

It is naïve and misleading to focus on the "best" and "worst" case scenarios. Multiple influences are always at play. The chance that all influences will take on the least favorable value, producing the "worst" case outcome is virtually zero. The chance that all influences will assume the most favorable value is also virtually zero. While these two extreme outcomes provide a range of possibility, it is an exaggerated range. Attractive decision alternatives may appear to be unattractive in a "worst" case. Unattractive decision alternatives may similarly appear to be attractive in a "best" case. It is much more productive and realistic to link performance drivers together in a spreadsheet, specify assumptions about the center, spread, and shape of each influence, then simulate a distribution of likely outcomes for each. Together, these reflections of our assumptions enable us to see the results of those influences on a distribution of future performance outcomes, with corresponding descriptive statistics. With a 95% confidence interval for outcomes, based on assumptions, decision makers are much more likely to choose favorable investments or resource allocations and to avoid unfavorable outcomes.

Monte carlo simulation is a powerful means to generate data when actual data is not available… either because it has not yet occurred, or because it is unaccessible. Simulation offers the additional advantage of allowing us to see how our multiple assumptions will together combine to produce possible outcomes. Decision making hinges on assumptions, and simulation provides a reflection of those assumptions.

Excel 4.1 Set Up a Spreadsheet to Link Simulated Performance Components

Use Team 8's assumptions to link revenue influences together in a spreadsheet.

Potential customers. Potential customers are the product of the existing market, 34.1K faculty, staff and students on campus, and the annual growth rate, which has been 3 to 4% in recent years.

Team 8 assumes that growth in the next year could be slightly lower or higher than in recent years, 2.5 to 4.5%.

Generate a random sample of 1000 possible market growth values using Excel's Random Number Generation, specifying a Normal distribution with mean .035 and standard deviation .005 (which, by the Empirical Rule is equal to one quarter of the assumed 95% range of possibilities, .025 to .045). Ask for output in B2.

Alt AY*n* R
1 tab 1000 tab N tab tab tab .035 tab .005 tab tab B2

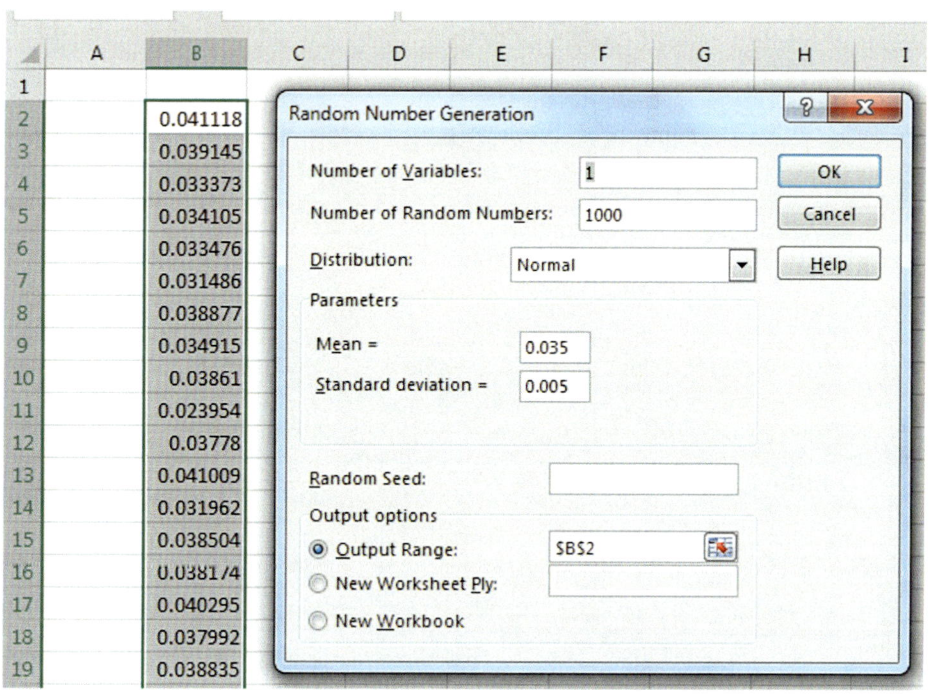

Add the label *market growth 2.5 to 4.5%* in B1.

Make a random sample of *customers in year 1* from existing customers, 34.1K, and each of the simulated market growth rates. (Your numbers will differ from those shown here since samples are random.)

In C1,
Customers (K) in year 1
In C2,
=34.1*(1+B2)
In C2,
Double click the lower right corner to fill in the new column.

C2 f_x =34.1*(1+B2)

	A	B	C	D	E
1		market growth 2.5% to 4.5%	customers (K) in year 1		
2		0.041	35.5		
3		0.039	35.4		
4		0.033	35.2		
5		0.034	35.3		

Find the 95% prediction interval for *customers (K) in year 1* using the PERCENTILE.INC(*array,k*).

Enter .975 for *k* to find the upper 95% prediction interval bound.
Enter .025 for *k* to find the lower 95% prediction interval bound.

C1003 f_x =PERCENTILE.INC(C2:C1001,0.025)

	A	B	C	D	E
1		*market growth 2.5 to 4.5%*	*customers (K)*		
1001		0.033	35.2		
1002		97.50%	35.62		
1003		2.50%	34.98		

Excel 4.2 View a Simulated Sample with a Histogram

To see the simulated sample of potential customers, make a histogram.

To identify the range of values that will be shown in the histogram, use the percentile function with .975 and .025 to see the 95% confidence interval of likely values. Find the mean and standard deviation.

In B1002,
Upper ci
In B1003,
Lower ci
In B1004,
M
In B1005,
SD
In C1002,
=percentile(c2:c1001,.975)
In C1003,
=percentile(c2:c1001,.025)
In C1004,
=average(c2:c1001)
In C1005,
=stdev(c2:c1001)

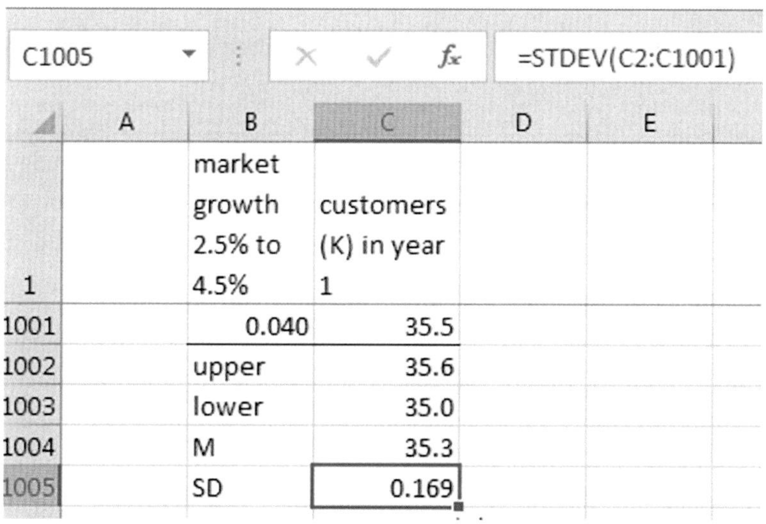

C1005 =STDEV(C2:C1001)

	A	B	C	D	E
1		market growth 2.5% to 4.5%	customers (K) in year 1		
1001		0.040	35.5		
1002		upper	35.6		
1003		lower	35.0		
1004		M	35.3		
1005		SD	0.169		

Make histogram bins that cover the 95% confidence interval, beginning with 35.0, in increments that are approximately equal to the standard deviation, which is about .15 (K).

In C1006,
Customers (K)
In C1007,
35.0
In C1008,
=c1007+.15
In C1008,
Shift+down to row 1012
Cntl+D

C1008 | f_x | =C1007+0.15

	A	B	C	D	E
1		market growth 2.5% to 4.5%	customers (K) in year 1		
1004		M	35.3		
1005		SD	0.169		
1006			customers (K)		
1007			35		
1008			35.15		
1009			35.3		
1010			35.45		
1011			35.6		
1012			35.75		

Create a histogram of *customers(K),* using the *customers (K)* bins and direct the output to C1014.

Alt AY*n* H
C1:c1001 tab c1006:c1012 tab L tab O C1014

	A	B	C	D
1		market growth 2.5% to 4.5%	customers (K) in year 1	
1014			*customers (K*	*Frequency*
1015			35	39
1016			35.15	163
1017			35.3	320
1018			35.45	294
1019			35.6	146
1020			35.75	38
1021			More	0

Convert frequencies to probabilities.

In E1014,
Probability
In E1015,
=d1015/1000
In E1015, click lower right to fill in probabilities

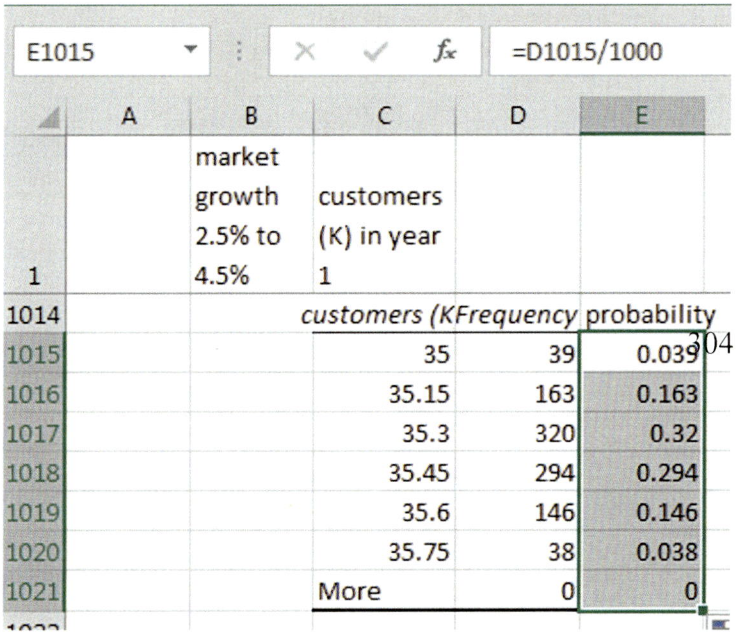

Plot probabilities by customer bins to see the distribution of customers in year 1, given market growth assumptions. Move probability cells from column E to D, so that they are adjacent to bins in column C, and then select bins and probabilities and request a column chart.

From E1014,
Cntl+shift+down
Cntl+X
From D1014,
Alt HIE
From C1014,
Cntl+shift+down
Shift+right
Alt NC

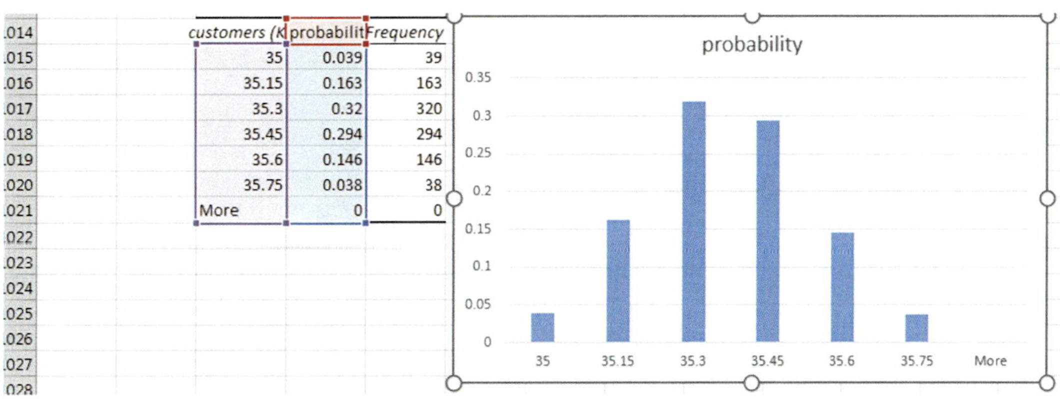

Add a vertical and horizontal axis titles and a stand alone chart title and set fontsize to 12. To present a more continuous distribution, reduce the bar gap width to 0.

From the chart,
Alt JAE down to Series probability
Alt JAM

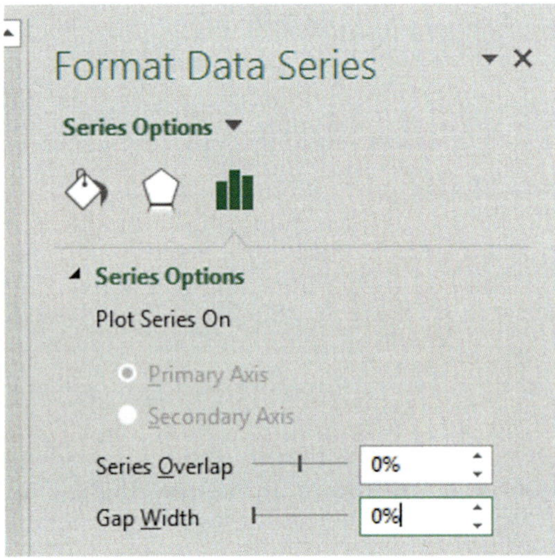

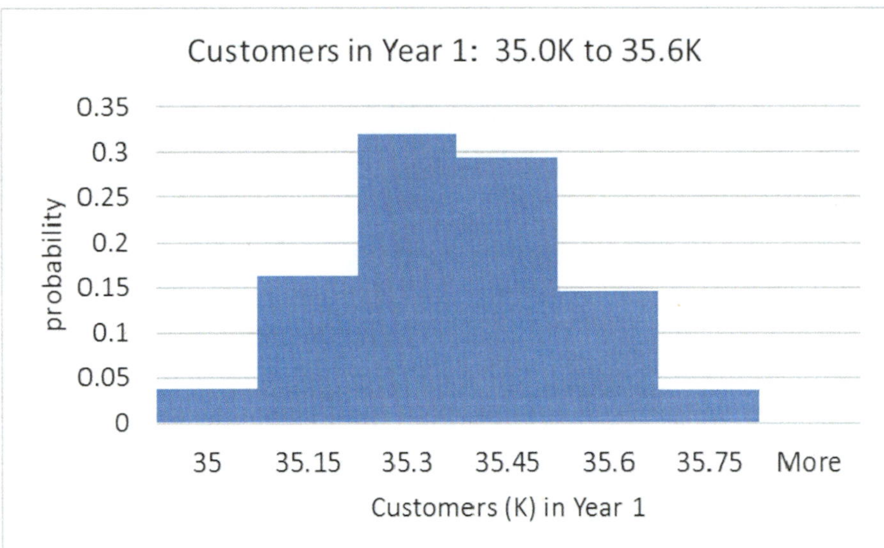

Access. The Team assumes that there is a 67% chance that they will be able to place vending units in any campus location, including dorms

Insert a new column D, with label *Access,* and generate a random sample of *access* outcomes using Excel Random Number Generation. Set the distribution to binomial and the p value to .67, with number of trails at 1. Direct output to D2.

Alt AY*n* R
1 tab 1000 tab bi tab tab tab 1 tab .67 tab tab O tab d2

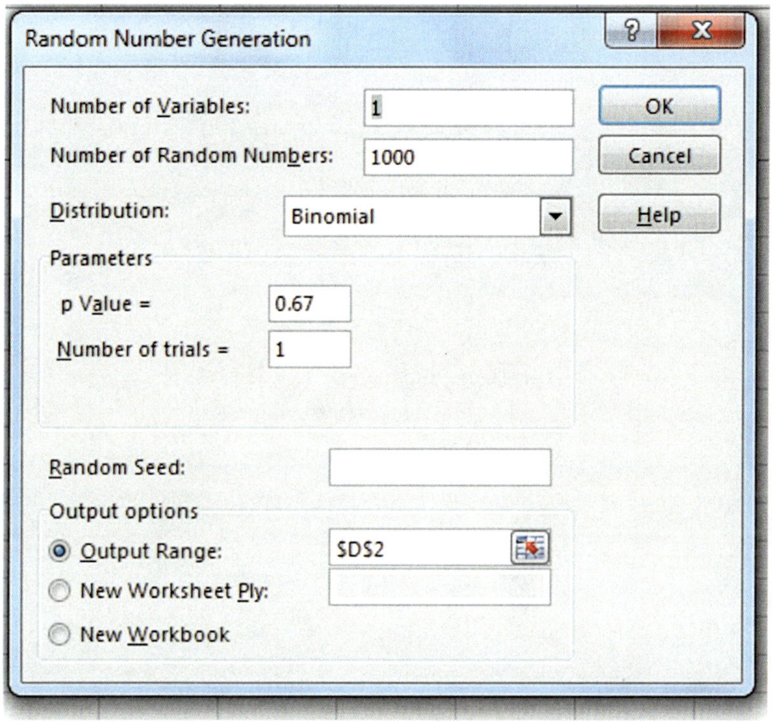

If access is restricted, dorms would be off limits, and the Team assumes that only 80% of potential customers could be reached.

Find the sample of *customers (K) given access* from the product of access and customers.

Customers (K) given access
In F2,
=(.8+.2*D2)*C2
In F2,
Click the lower right corner to fill in the column

In F1,

Find the 95% prediction interval, mean and standard deviation by reusing those formulas in column C by filling right.

In C1002,
Shift+down down down
Shift+right right right
Cntl+R

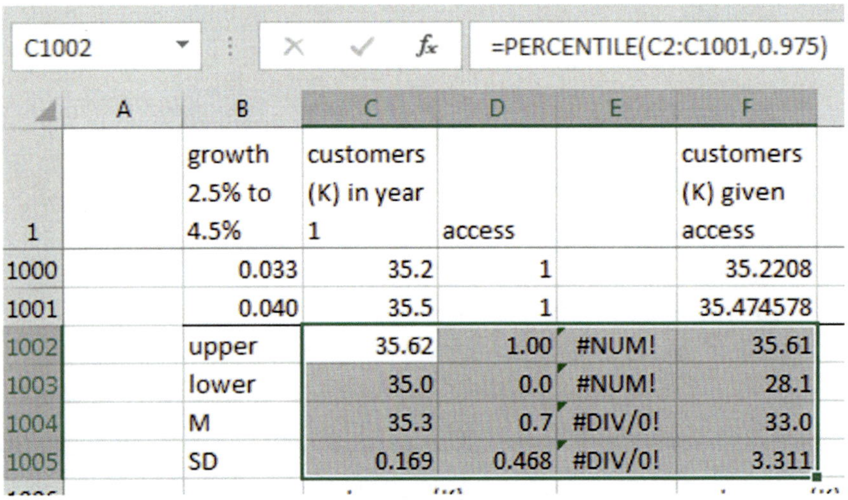
C1002 =PERCENTILE(C2:C1001,0.975)

	A	B	C	D	E	F
1		growth 2.5% to 4.5%	customers (K) in year 1	access		customers (K) given access
1000		0.033	35.2	1		35.2208
1001		0.040	35.5	1		35.474578
1002		upper	35.62	1.00	#NUM!	35.61
1003		lower	35.0	0.0	#NUM!	28.1
1004		M	35.3	0.7	#DIV/0!	33.0
1005		SD	0.169	0.468	#DIV/0!	3.311

Create histogram bins, beginning with 28, ending with 35.75, at increments of .25, which will produce 32 bins and a continuous distribution. (The bin width ought to be an easy to read increment, not too small, but small enough to reveal the distribution shape.) Request a histogram of *customers (K) given access* in F1040.

In F1006,
Customers (K) given access
In F1007,
28

Convert frequencies to probabilities, move probabilities to columns G, adjacent to customers given access in column F, select customers given access and probabilities to create a column chart of the distribution of customers given access.

In F1008,
=F1007+.25
In F1008,
Shift+down to F1038
Cntl+D
Alt AY*n* H
F1:f1001 tab F1006:F1038 tab O F1040

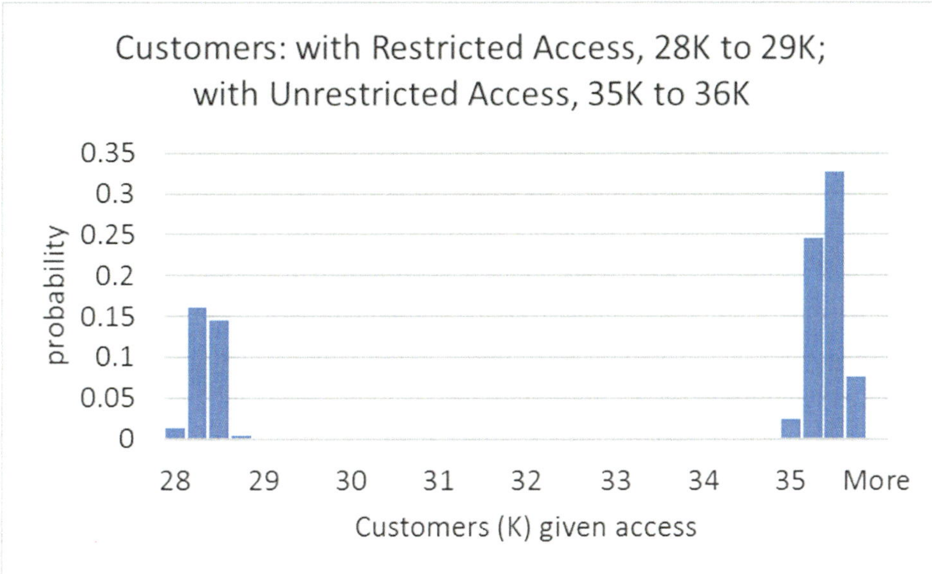

Price. The Team assumes that full access would translate to higher volume and discounted unit costs, enabling a lower competitive price of $1.25 per bottle to be charged, instead of the parity price of $1.50.

Find *price ($)* in column G from each *access* outcome in column D.

In G1,
Price ($)
In G2,
= 1.50 – .25 *D2
In G2,
Double click the lower right corner to fill the column

G2 =1.5-0.25*D2

	A	B	C	D	E	F	G
1		market growth 2.5% to 4.5%	customers (K) in year 1	access		customers	price ($)
2		0.041	35.5	1		35.50214	1.25
3		0.039	35.4	1		35.43486	1.25
4		0.033	35.2	1		35.23801	1.25

Find the Standard Deviation, Given Assumptions. At the lower price, $1.25, with unrestricted access, the Team assumes that at least 20%, and as much as 50%, of potential customers will be captured, with the most likely share equal to 35%, with share possibilities distributed Normal. It is assumed that there is only a 2.5% chance that share could be less than 20%, and only a 2.5% chance that share could be greater than 50%. Therefore, there is a 95% chance that share will fall within the range 20 to 50%, making the standard deviation approximately equal to one quarter of this range, by the Empirical rule:

SD = (50% − 20%) / 4 = 7.5%

Generate a sample of 100 Normally distributed *unrestricted access share* in column H, with mean .35 and standard deviation .075.

Alt AY*n* R
1 tab 1000 tab N tab tab tab .35 tab .075 tab tab tab H2

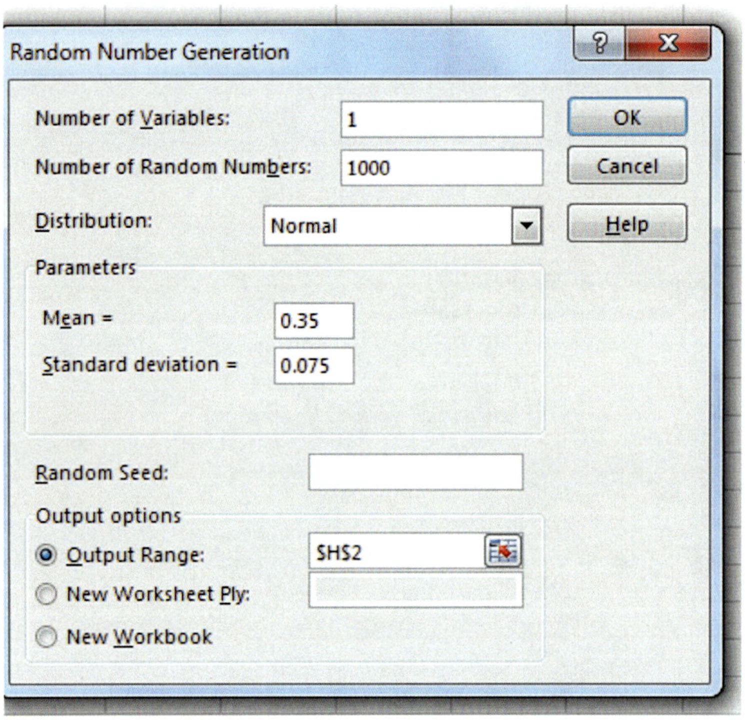

With restricted access and the higher, parity price, $1.50, the Team assumes that at least 10%, and as much as 20%, of customers with access could be captured, with the most likely share in the middle at 15%.

It is assumed that there is only a 2.5% chance that *share given restricted access, higher price*, would be less than 10%, and that there is only a 2.5% chance that share could be greater than 20%. Therefore, from the Empirical Rule, it is assumed that the share standard deviation is one quarter of this range:

SD = (20% − 10%) / 4 = 2.5%

Insert a new column I labeled *restricted access share* and simulate a Normally distributed sample of 1000 values with mean .15 and standard deviation .025.

J2 =D2*H2+(1-D2)*I2

	D	E	F	G	H	I	J
1	access		customers (K) given access	price ($)	unrestricted access share	restricted access share	share given access
2	1		35.502136	1.25	0.309040508	0.14795426	0.3090405
3	1		35.434859	1.25	0.314996618	0.13284912	0.3149966
4	1		35.238009	1.25	0.375835419	0.13763667	0.3758354
5	0		28.210374	1.5	0.218654186	0.13192254	0.1319225

To see share possibilities, add a new column J of *share given access* from access, unrestricted access share, and restricted access share.

In J1,
Share given access
In J2,
=D2*H2+(1-D2)*I2
In J2,
Double click lower right corner to fill column

Find the 95% confidence interval, mean and standard deviation by filling right those formulas in rows 1002 through 1005.

F1002 =PERCENTILE(F2:F1001,0.975)

	F	G	H	I	J
1	customers (K) given access	price ($)	unrestricted access share	restricted access share	share given access
1000	35.2208	1.25	0.406912683	0.17663523	0.4069127
1001	35.474578	1.25	0.299213191	0.18010041	0.2992132
1002	35.61	1.50	0.51	0.20	0.50
1003	28.1	1.3	0.2	0.1	0.1
1004	33.0	1.3	0.3	0.1	0.3
1005	3.311	0.117	0.077	0.025	0.114
1006	customers (K) given acce share				

Find the median share given access in J1006.

In i1006,
Median
In J1006
=median(j2:j1001)

J1006 =MEDIAN(J2:J1001)

	F	G	H	I	J
1	customers (K) given access	price ($)	unrestricted access share	restricted access share	share given access
1001	35.474578	1.25	0.299213191	0.180100409	0.2992132
1002	35.61	1.50	0.51	0.20	0.502
1003	28.1	1.3	0.2	0.1	0.11
1004	33.0	1.3	0.3	0.1	0.3
1005	3.311	0.117	0.077	0.025	0.114
1006	customers (K) given access			median	0.298

Create a histogram of share possibilities given access. Choose bin width .05, which is smaller than the standard deviation for unrestricted access shares and larger than the standard deviation for restricted access shares. In J1008, set the first share bin at .10, the lower 95% confidence interval bound for *share given access*, and add .05 to create bins through the maximum share of .55, the bin in which the upper 95% confidence interval bound would fall. Request a histogram of share given access in J1019. Create probabilities of shares given access in L1020 through L1030 from frequencies in K1020 through K1030.

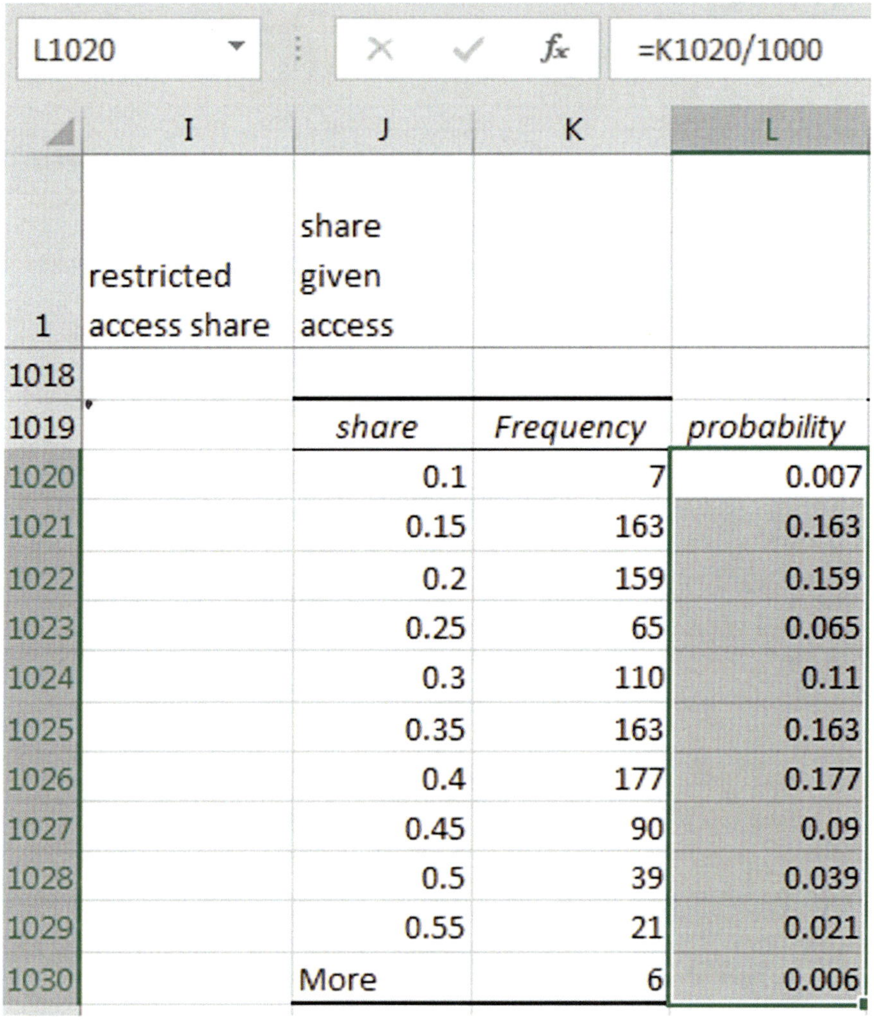

Select and cut probabilities in L1019 through L1030 and insert in K1019 next to share bins in J1019. Then, select the bins and the probabilities in J1019 through K1030 and request a column chart.

From L1019,
Cntl+shift+down
Shift+right
Cntl+X
In K1019,
Alt HIE
From J1019,
Cntl+shift+down
Shift+right
Shift+right
Alt NC

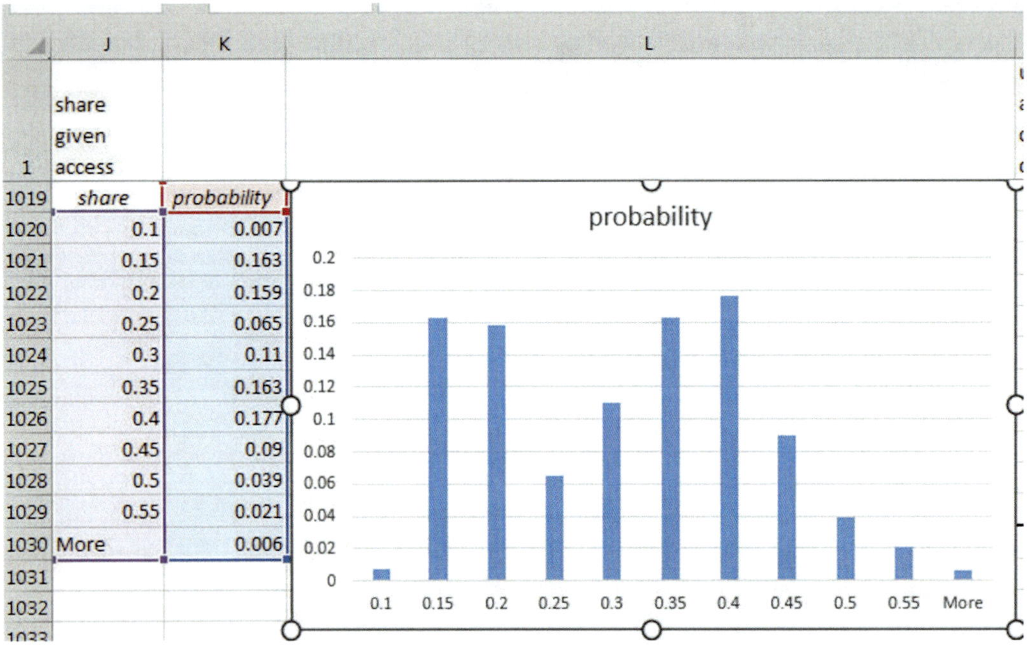

Convert shares to percents in J1020 through J1029.	From J1020, **Cntl+shift+down** **Shift+up** **Alt HP**

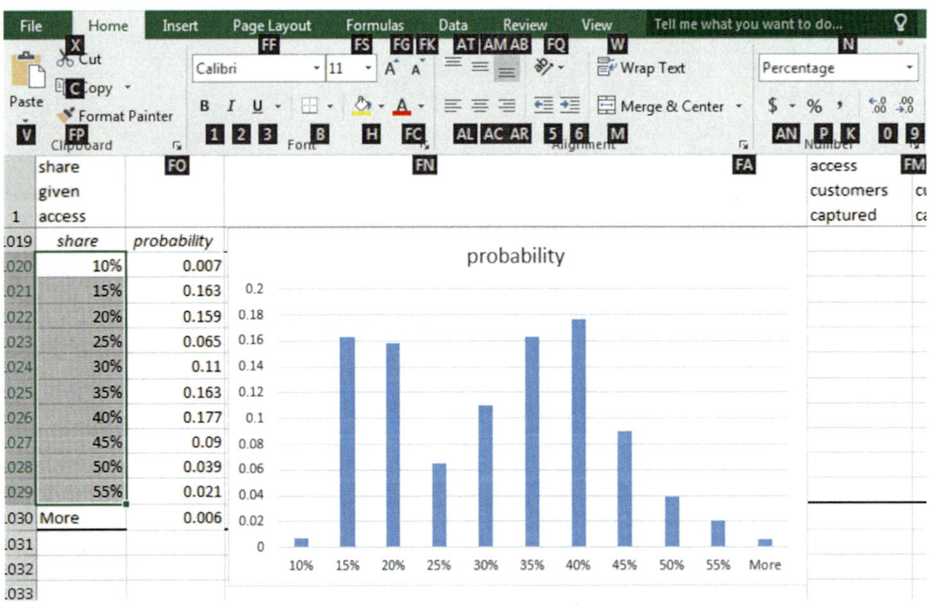

Add axis titles, type in a stand alone chart title, reduce gap widths and set fontsize to 12.

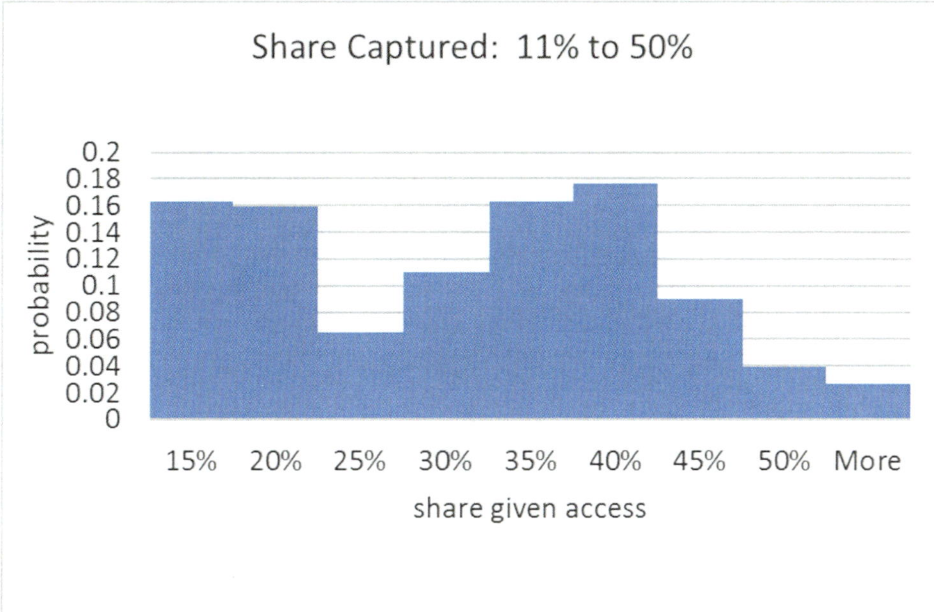

Insert new column M, *customers captured given access,* with values for *customers given access* in F and *share given access* in J.

In M1,
Customers captured given access
In M2,
=f2*j2
From M2,
Shift+right right
Double click lower right corner to fill columns

M2 =J2*F2

	F	G	H	I	J	K	L	M
1	customers (K) given access	price ($)	unrestricted access share	restricted access share	share given access			customers captured given access
2	35.502136	1.25	0.309040508	0.147954	0.309			10.9715981
3	35.434859	1.25	0.314996618	0.132849	0.315			11.1618608
4	35.238009	1.25	0.375835419	0.137637	0.3758			13.243692
5	28.210374	1.5	0.218654186	0.131923	0.1319			3.7215842
6	35.241546	1.25	0.379248156	0.159787	0.3792			13.3652912
7	35.173676	1.25	0.291241242	0.163607	0.2912			10.2440251
8	35.425718	1.25	0.27967264	0.149175	0.2797			9.90760394

Right fill in rows 1002 through 1006 to see 95% confidence interval, mean, standard deviation, and median of customers captured given access.

	M
1	customers captured given access
1002	17.81
1003	3.2
1004	9.7
1005	4.440
1006	10.484

Create bins and a histogram of *customers captured (K) given access*:

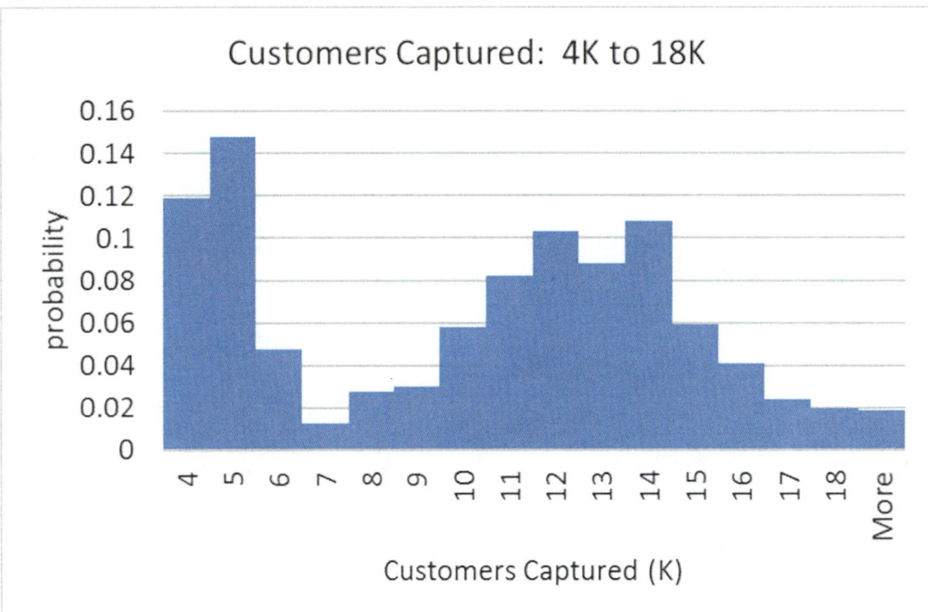

Demand per customer. From earlier market research, the Team believes that average bottles demanded per customer per week falls within the range 8 to 12, with 95% confidence. The Team assumes that demand is Normally distributed, with standard deviation equal to one quarter of the 95% confidence interval:
SD = (12−8)/4 = 1. Generate a sample of 1000 Normally distributed levels for *bottles per customer per week* in column N with mean 10 and standard deviation 1.

Bottles sold(M). Insert a new column O *bottles sold (M)* in millions from *customers captured given access* in M and *bottles per customer per week* in N, assuming that the vending units will be stocked during the 38 weeks in which classes are in session.

In O1,
Bottles sold (M)
In O2,
=38*N2*M2/1000
In O2,
Double click lower right to fill column

Find the *95% confidence interval, mean, standard deviation* and *median* by filling right in rows 1002 through 1006.

Create bins and a histogram for possible volumes of *bottles sold in year 1:*

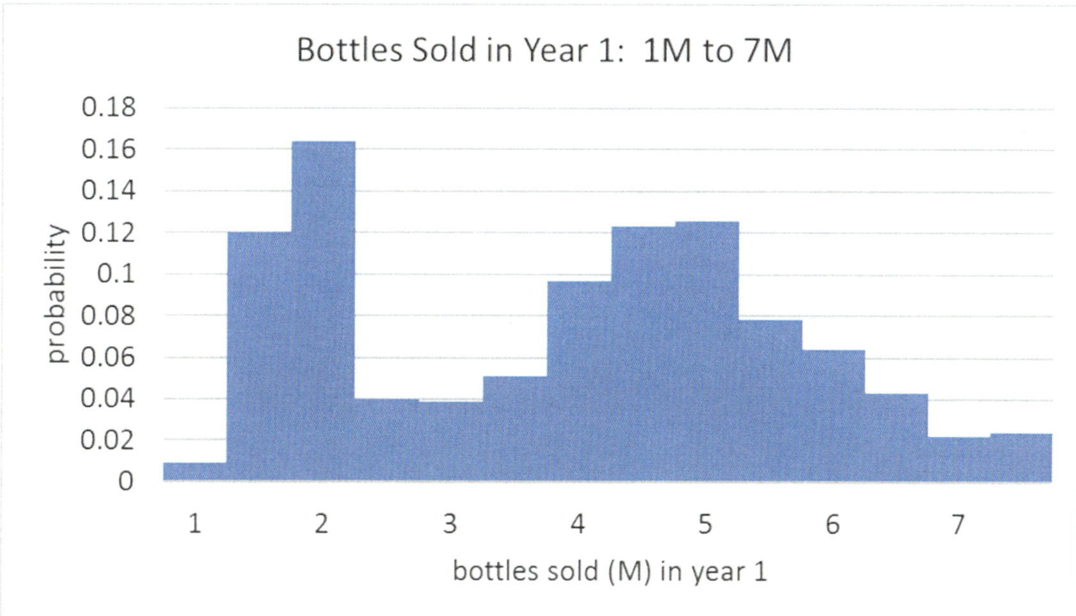

Insert a new column P *revenues ($M)*, from the simulated samples of *price ($) given access* in G and *bottles sold (M)* in O.

In P1,
Revenue ($M) in year 1
In P2,
=g2*o2
In P2
Double click lower right corner to down fill

Find the *95% prediction interval, mean, standard deviation,* and *median* in rows 1002 through 1006.

Create bins and a histogram of *revenues ($M)* in year 1:

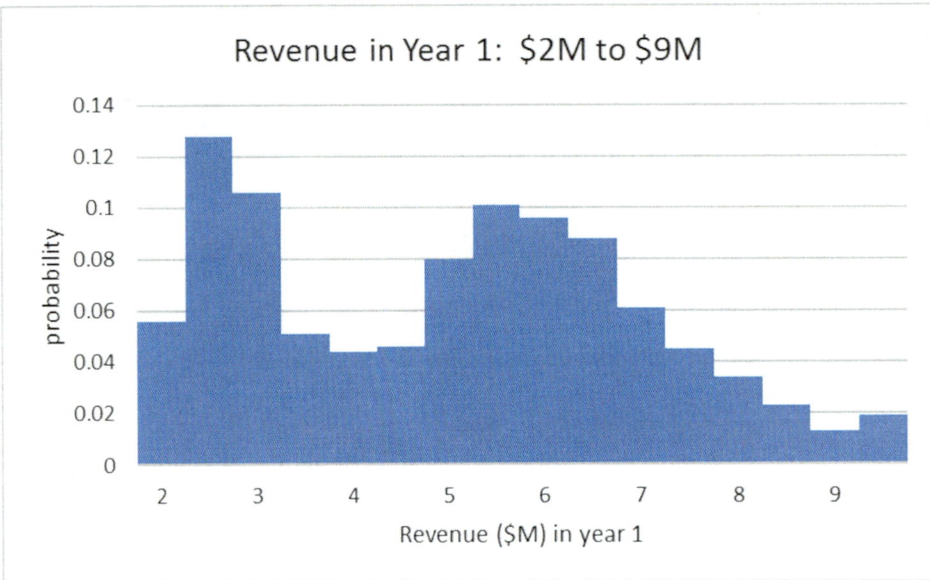
Revenue in Year 1: $2M to $9M
probability
0.14
0.12
0.1
0.08
0.06
0.04
0.02
0
2
3
4
5
6
7
8
9
Revenue ($M) in year 1

Lab 4 Inference: Dell Android Smartphone Plans

Managers at Dell are considering a joint venture with a Chinese firm to launch a new Android platform smartphone.

I. Find the Distribution of Potential Dell Smartphone Shares

Apple and Samsung control 43% of the market, though 42% of the market remains to be captured by a major vendor. The third, fourth and fifth place vendors each control about 5% of the market.

Dell managers agree that 5% is a reasonable estimate of the new smartphone's potential share in 2020, though share could be as low as 3%, or possibly as high as 6%.

Vendor	2015Q1 shipments	2014Q1 shipments	Year-over-year change (2015Q1 vs 2014Q1)	2015Q1 market share	2014Q1 market share
Samsung	82.4	88.5	−6.9%	24.5%	30.7%
Apple	61.2	43.7	40.0%	18.2%	15.2%
Huawei	17	13.5	25.9%	5.1%	4.7%
LG	15.4	12.3	25.2%	4.6%	4.3%
Lenovo	18.8	18.9	−0.5%	5.6%	6.6%
Others	141.7	111.4	27.2%	42.1%	38.6%
Total	**336.5**	**288.3**	**16.7%**	100.0%	100.0%

Use Excel to create a sample of 1000 *Dell Smartphone share* proportions from a Normal distribution with mean equal to management's best guess of 5%.

To choose a standard deviation for your simulation, estimate the standard deviation of *Dell Smartphone share* by dividing the likely range by 4, following the Empirical Rule: ___

II. Simulate Dell Shipments to Assess Potential, Given Assumptions

Managers want to estimate potential *shipments of Dell smartphones* in 2020Q1 with 95% confidence. *Shipments of Dell smartphones* would be driven by *world smartphone shipments, annual growth in world shipments*, and *Dell smartphone share.*

Management assumes that annual growth rates over past year will average 15% in each of the next five years, though annual growth could be as low as 10%, and possibly as high as 20%, given growth in developing markets.

1. Use Excel to draw a random samples of 1000 Normally distributed *growth rates* for 2016Q1, 2017Q1, 2018Q1, 2019Q1 and 2020Q1.

Find the 2020Q1 distribution of possible *world smartphone shipments* from 2015Q1*world smartphone shipments* and the sample of simulated *annual growth rates* in 2016Q1 through 2020Q1.

Make a histogram of *world smartphone shipments* in 2020Q1.

2. Use your samples of simulated *Dell smartphone share* and *World smartphone shipments* to find the distribution of possible *Dell smartphone shipments* in 2020Q1.

   Make a histogram of *Dell smartphone shipments* in 2020Q1.

3. What is the 95% confidence interval for *Dell smartphone shipments* in 2020Q1?

   __________ to ____________

4. Several conservative managers worry that *Dell smartphone share* could be less than 3% (the lower bound of their 95% confidence interval) and annual *growth in world shipments of smartphones* could slow to 10%, producing a forecast of *Dell smartphone shipments* of just 16M n 2020.

   What is the chance that this unfortunate "worst case" could occur? _________

Case 4.1 American Girl in Starbucks

Mattel and Warner Brothers are considering a partnership with Starbucks to promote their new American Girl movie. Starbucks previously backed Lionsgate's "Akeelah and the Bee," which earned $19 million. In exchange for $8 million, Starbucks would install signage and stickers in 6800 of its stores, print American Girl branded cup sleeves, and sell the picture's soundtrack for a 3 month period. Materials for the movie would also appear on the company's website during that period. Starbucks claims 44 million customers in the 6800 stores.

In a pretest of the promotion during 1 week in one Starbucks store, 184 of the 924**, or 20%** of Fast Card customers served that week agreed that they had heard of the movie when surveyed by phone the following week.

Mattel managers believe that roughly 35% of those who are aware of the movie will buy tickets**.**

Managers assumed that there is only a 2.5% that the percent buying tickets would be less than 5%, and there is only a 2.5% change that the percent buying tickets will be more than 9%.

The distribution of *tickets purchased* depends on the number of guests that ticket purchasers bring. 95% of movie-goers are expected to bring 1 to 3 family members or friends**.**

Mattel would earn $1 royalty from each ticket.

1. Illustrate the distributions of *percent buying tickets* and *tickets purchased,* given assumptions.

2. Illustrate the distribution royalties, given assumptions.

3. A conservative manager advises that in the "worst case," royalties could be as low as $4.4M, in which case Mattel would lose $3.6M (= *payment to Starbucks − royalties).*

   What is the chance that this "worst case" could occur, given assumptions?

Case 4.2 Can Whole Foods Hold On?

Organic food sales were roughly $33B in 2010, and had grown to **$39.2B** in 2011, annual growth of 19%. **Analysts forecast annual growth rates between 14 and 16% for years 2014 through 2018.**

In 2013, WFM added 32 new stores. WFM 2013 revenue grew to $12.9B from 362 stores, with same store sales annual growth rates of 6 to 9% in 2010 through 2013. **Managers assume that annual same store sales growth is likely to average 7.5% in years 2014 through 2018, with rates possibly as low as 5% and possibly as high as 10%.**

WFM *share* of organic food sales in a given year depends on *organic food sales,* the number of *WFM stores,* and *sales per WFM store:*

$$WFM\ share_t = WFM\ stores_t \times sales\ per\ WFM\ store_t / organic\ food\ sales_t$$

In 2018, *organic food sales* will depend on the *annual organic food growth rates*:

$$\begin{aligned} Organic\ food\ sales_{2018} = &(1 + organic\ food\ growth\ rate_{2012}) \\ &\times (1 + organic\ food\ growth\ rate_{2013}) \\ &\times (1 + organic\ food\ growth\ rate_{2014}) \\ &\times (1 + organic\ food\ growth\ rate_{2015}) \\ &\times (1 + organic\ food\ growth\ rate_{2016}) \\ &\times (1 + organic\ food\ growth\ rate_{2017}) \\ &\times (1 + organic\ food\ growth\ rate_{2018}) \times organic\ food\ sales_{2011} \end{aligned}$$

***2018 WFM stores* depend on the *annual WFM growth in number of stores,* 4 to 8% in recent years. WFM expects to continue aggressively adding stores, with an expected rate of 6% per year.**

$$\begin{aligned} WFM\ stores_{2018} = &(1 + WFM\ nstore\ growth\ \mathrm{rate}_{2014}) \times (1 + WFM\ nstore\ growth\ \mathrm{rate}_{2015}) \\ &\times (1 + WFM\ nstore\ growth\ \mathrm{rate}_{2016}) \times (1 + WFM\ nstore\ growth\ \mathrm{rate}_{2017}) \\ &\times (1 + WFM\ nstore\ growth\ \mathrm{rate}_{2018}) \times WFM\ stores_{2013} \end{aligned}$$

***2018 Sales per WFM store* depend on *annual growth in same store sales*, 5 to 10% considered possible:**

$$\begin{aligned} Sales\ per\ WFM\ store_{2018} = &(1 + same\ store\ sales\ growth_{2014}) \\ &\times (1 + same\ store\ sales\ growth_{2014}) \\ &\times (1 + same\ store\ sales\ growth_{2016}) \\ &\times (1 + same\ store\ sales\ growth_{2017}) \\ &\times (1 + same\ store\ sales\ growth_{2018}) \times sales\ per\ WFM\ store_{2013} \end{aligned}$$

Can WFM hold on?

Forecast WFM's market share of the organic food sales market in 2018:

1. Present histograms of $organic\ food\ sales_{2018}$, $number\ of\ WFM\ stores_{2018}$, $sales\ per\ WFM\ store_{2018}$ and $WFM\ share_{2018}$, based on your assumptions.

2. What is the 95% confidence interval for $WFM\ share_{2018}$?

3. A conservative manager forecasts that, in the "worst case," $WFM\ share_{2018}$ could be as low as 13%. Based on your assumptions, how likely is such a "worst case?"

Case 4.3 Chipotle's Ambitions to Triple Share of Top 100 Chain Sales in the Recession Rebound

The largest U.S. restaurant brands put the Great Recession behind them as new-store openings and unit-level sales gains in their latest fiscal years pushed their aggregate sales growth into pre-downturn territory, according to 2013 Nations Restaurant News Top 100 research.

The 2013 Top 100 chains collectively pulled in nearly $213.7 billion, a 5.3-percent increase over last year's aggregate sales. Chipotle managers expect annual sales growth of U.S. sales by the Top 100 chains to resemble recent growth, within the 4 to 6 percent range through 2020.

Chipotle is currently ranked #21 in the 2013 Top 100 chains, with 1.3% share of Top 100 sales. Growth in both Chipotle's locations and ESPU (sales per unit) have outpaced the industry. Annual growth in Chipotle locations has ranged from 12 to 18%, from 2010, with 1398 locations in 2013. Annual growth in Chipotle ESPU has ranged from 5 to 9%, from 2010, with 2013 ESPU of $2.9M.

Management expects to continue to grow Chipotle, through addition of locations and through increased ESPU at annual growth resembling rates achieved from 2010.

Management has asked you to forecast Chipotle share of Top 100 chain sales in 2020. They are specifically interested in learning whether or not it is likely that Chipotle can triple its share, from 1.3%, to at least 3.8% by 2020, if their assumptions are correct.

1. What is your forecast for Top 100 chain sales in 2020, if management's assumptions are correct?

2. What is your forecast for Chipotle sales in 2020, if management's assumptions are correct?

3. What is your forecast for Chipotle share of Top 100 chain sales in 2020, if management's assumptions are correct?

4. Illustrate your 2020 Chipotle share forecast, given management's assumptions. Use share bins with width of .005 (.5%):

5. A consultant from a competing firm has told Chipotle executives that Chipotle share of Top 100 chain sales could be as low as 2.8% in 2020, falling short of the goal to triple share. Top 100 chain sales could grow at an annual rate of 6%, Chipotle's annual unit growth could be just 12%, and annual growth in Chipotle's ESPU could be just 5%, all seemingly consistent with management's assumptions. If Chipotle executives' assumptions are correct, what are the chances that 2020 share will be 2.8% or less?

Chapter 5
Simple Regression for Long Range Forecasts

Regression analysis is a powerful tool for quantifying the influence of continuous, *independent, drivers X* on a continuous *dependent, performance* variable *Y*. Often we are interested in both explaining how an independent decision variable *X* drives a dependent performance variable *Y* and also in predicting performance *Y* to compare the impact of alternate decision variable *X* values. *X* is also called a *predictor*, since from *X* we can predict *Y*. Regression allows us to do both: quantify the nature and extent of influence of a performance driver and predict performance or response *Y* from knowledge of the driver *X*.

With regression analysis, we can statistically address these questions:

- Is variation in a dependent, performance, response variable *Y* influenced by variation in an independent variable *X*?

If *X* is a driver of *Y*, with regression, we can answer these questions:

- What percent of variation in performance *Y* can be accounted for with variation in driver *X*?
- If driver *X* changes by one unit, what range of response can we expect in performance *Y*?
- At a specified level of the driver *X*, what range of performance levels *Y* are expected?

Regression analysis is used with both cross sectional data, to explain differences, or variation within a population and to identify drivers, and time series data, to explain changes in a population over time, identify drivers and/or to forecast future values. In this chapter, simple regression based on stable, linear trend is introduced, as the means to make long range forecasts.

In some circumstances, managers want to estimate the *trend*, or stable level of growth in performance, in order to produce a longer term forecast. Regression allows estimation of trend, using the time period as the independent variable. Forecasting based on trend is *naïve,* because the focus is not on understanding why performance varies across time periods, but only on forecasting what future performance would look like if the future performance resembled past performance.

Naïve forecasting based on trend is quick and easy and provides a view of the future, under the status quo. In Chapter 10, multiple regression models with cross sectional data will be introduced for cases in which explanation of variation and identification of drivers is desired. In Chapter 12, more sophisticated forecasting, which includes explanation of variation across time periods, will be introduced. In addition to the introduction of simple regression, this chapter also includes exploration of the link between correlation and simple linear regression, since the two are closely related.

C. Fraser, *Business Statistics for Competitive Advantage with Excel 2016*,
DOI 10.1007/978-3-319-32185-1_5

5.1 The Simple Linear Regression Equation Describes the Line Relating an Independent Variable to Performance

Regression produces an equation for the line which best relates changes or differences in a continuous, dependent performance variable *Y* to changes or differences in a continuous, independent driver *X*. This line comes closest to each of the points in a scatterplot of *Y* and *X*:

$$\hat{y} = b_0 + b_1 \times X$$

Where $\hat{y}$ is the expected value of the dependent performance, or response, variable, called "*y hat*",

X is the value of an independent variable, decision variable, or driver,

b_0 is the *intercept* estimate, which is the expected value of *y* when X is zero,

b_1 is the estimated *slope* of the regression line, which indicates the expected change in performance $\hat{y}$ in response to a unit change from the driver's average $\overline{X}$.

In the context of a naïve model based on stable, linear trend, *X* is a measure of time, such as quarter, *q*, or year, *t*. The slope, b_1, estimates average change per time period. A naïve simple regression model to estimate trend and to forecast is written with subscripts to identify time periods:

$$\hat{y}_t = b_0 + b_1 \times X_t$$

Example 5.1 Concha y Toro. Concha y Toro vineyards, a Chilean global multinational, is the largest producer of wines in Latin America. The firm produces and exports wine to 135 markets in Europe, North, Central and South American, Asia, and Africa. Concha y Toro has been successful in developing wine varieties that appeal to the unique tastes of the various global segments.

In 2010, the Board of Directors was reviewing performance. While the firm's wines had earned multiple distinctions, export revenues had grown only 4% since 2009, which was a startling slow down, relative to revenue growth as high 43% in recent years. Some conservative managers were convinced that revenue growth over the next five years could be as little as 1%. The Board needed a forecast of export revenues. The modeling team elected to estimate the trends in volume in the four largest export regions, Europe, the U.S, Asia and Latin America.

5.2 Hide the Two Most Recent Datapoints to Validate a Time Series Model

Before a time series model is used to forecast future performance, it is validated:

- the two most recent observations are hidden while the model is built,
- the model equation is used to forecast performance in those two most recent periods,

- model prediction intervals are compared with actual performance values in those two most recent periods, and if the prediction intervals contain actual performance values, this is evidence that the model has *predictive validity* and can be reliably used to forecast unknown performance in future periods.

The process for naïve time series models is shown in Figure 5.1.

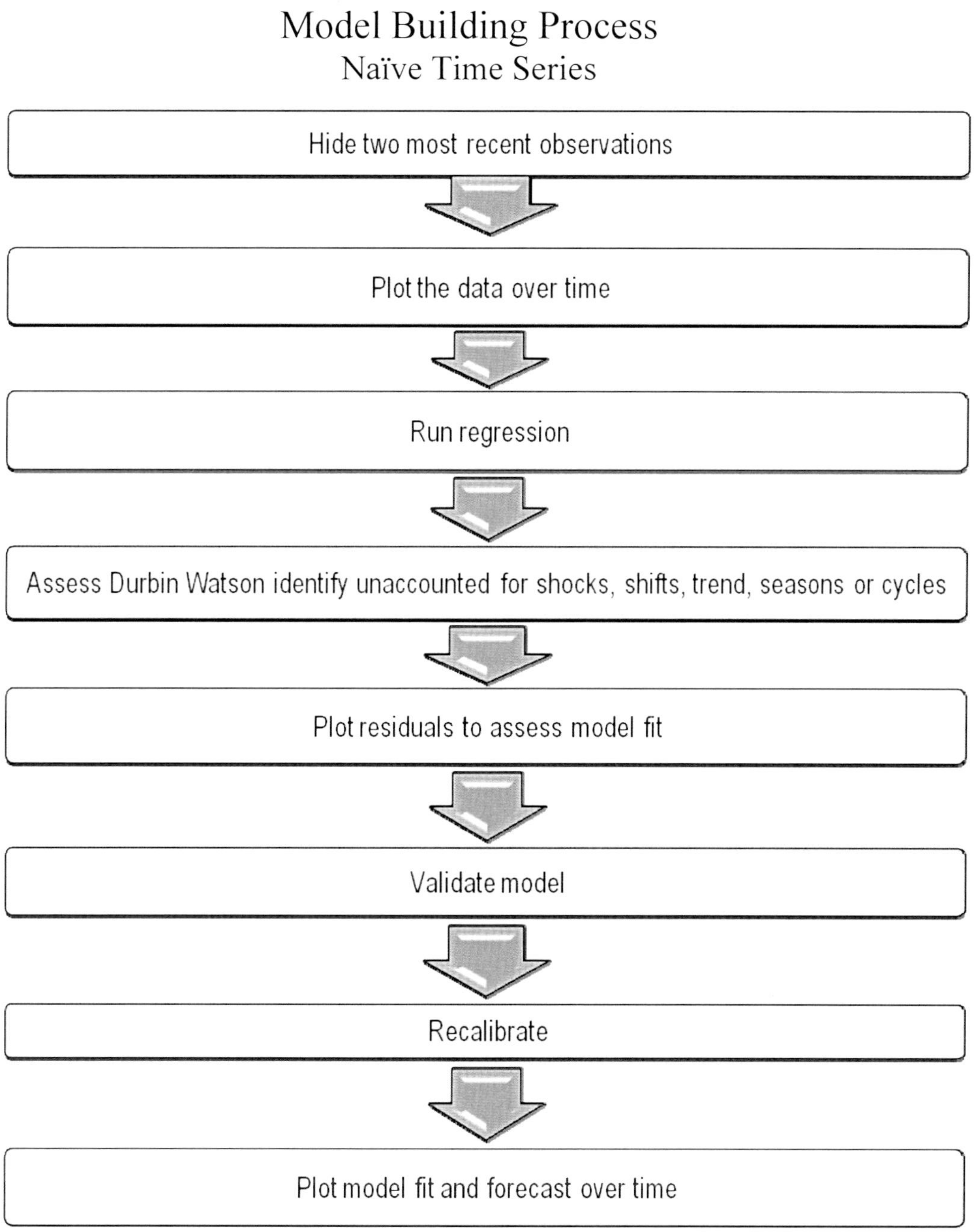

Figure 5.1 Model building processes with a naïve model and time series data

In the U.S. market, volume had grown consistently, but at a much slower rate than volume growth in other global markets, as Figure 5.2 illustrates. While volume growth elsewhere was celebrated, there was concern that growth in the U.S. had fallen short of that desired. Wine consumption in the U.S. was growing faster than in other global regions, presenting an opportunity that could not be ignored.

In order to inform future discussions, regressions were used to estimate the annual trends in sales volume for the four major export regions. Trends were estimated with the following model:

$$sales\ vo\hat{l}ume(ML)_{i,t} = b_{0_i} + b_{1_i}(\frac{ML}{t}) \times t$$

where i is the i'th global region, and t is the year.

The slope estimates, b_{1_i} will provide estimates of average annual changes in sales volume.

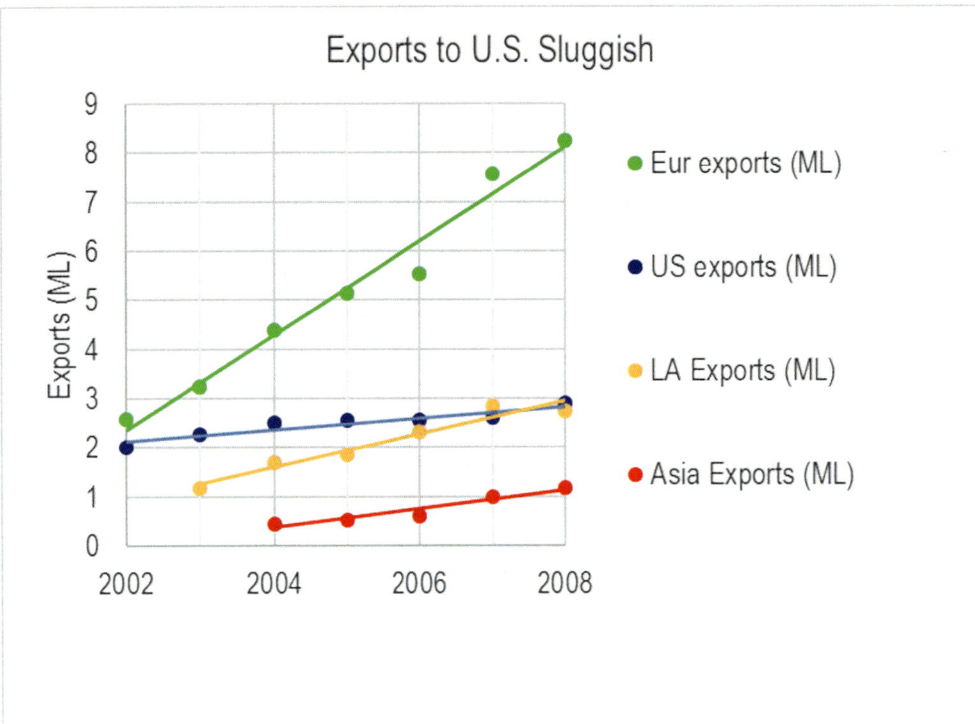

Figure 5.2 Growth in sales volumes across global regions

Since sluggish exports to the U.S. market were of greatest concern, management focused initial analysis on those. The null and alternate hypotheses which the modelling team wanted to test are:

H_0: Average annual growth in export volume β_1 is less than or equal to zero.
H_1: Average annual growth in export volume β_1 is greater than zero.

Regression results, shown in Table 5.1, suggest that sales volume in the U.S. is increasing annually, since the slope estimate (coefficient for year t) is positive.

Table 5.1 Regression of U.S. export volume by year

Regression Statistics: U.S. sales (ML)	
R Square	.88
Standard Error	.104
Observations	7

ANOVA

	df	*SS*	*MS*	*F*	*Significance F*
Regression	1	.391	.391	36.1	.0018
Residual	5	.054	.011		
Total	6	.445			

	Coefficients	*Standard Error*	*t Stat*	*p value*	*Lower 95%*	*Upper 95%*
Intercept	−234	39	−5.9	.0019	−336	−133
T	.118	.020	6.0	.0018	.068	.169

The average annual increase in export volume to the U.S. is .13(ML) per year. Expected annual exports to the U.S. $\hat{y}$ increase at a constant rate of .118ML per year.

Because variation in revenues y is related linearly to variation in titles X, the linear regression line is a good summary of the data:

$$ex\hat{p}orts(ML)_t = -234(ML) + .118(ML/t) \times t$$

5.3 Test and Infer the Slope

Because the true trend, or average annual change β_1 in exports y is unknown, this *slope*, or *coefficient*, is estimated from a sample. This estimate b_1 and its sample standard error s_{b_1} are also used to test the hypothesis that the trend drives variation in y:

H_0: The regression slope is less than or equal to zero. $\beta_1 \leq 0$.

Alternatively,

H_1: The regression slope is not zero: $\beta_1 > 0$

In many instances, including this example, from experience or logic, we know the likely direction of influence. In those instances, the alternate hypothesis requires a *one tail* test. This one sided alternate hypothesis describes an upward slope. A similar alternate hypothesis could be used when logic or experience suggests a downward slope.

Sample slopes b_1 are Normally distributed around the population slope β_1, which is less than or equal to zero, under the null hypothesis. Whether the sample slope is consistent with the null hypothesis depends on its distance from zero and dispersion of the sample slope distribution. Figure 5.3 illustrates two possibilities. The sample slope to the left is "close" to the hypothetical population slope of zero. The sample slope on the right is "far" from the hypothetical population slope, providing evidence that the population slope is unlikely to be zero.

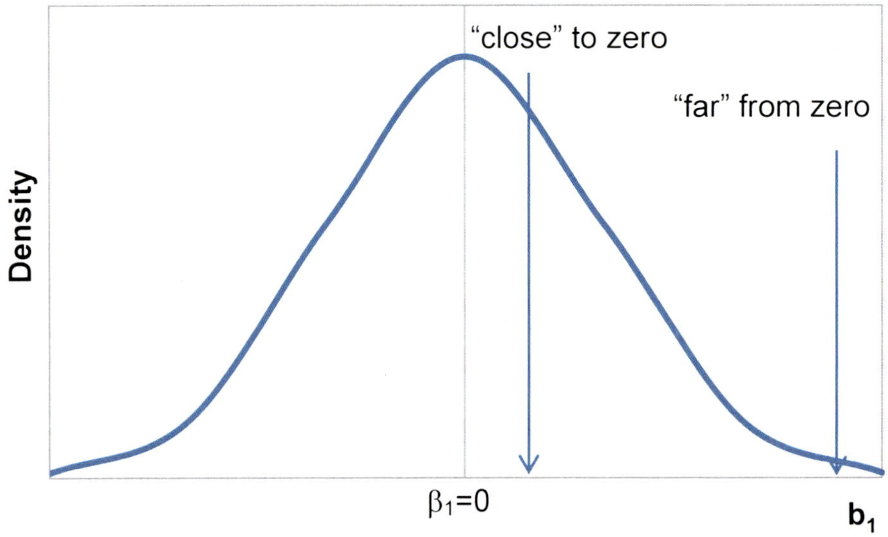

Figure 5.3 *Distribution of sample slope under the null hypothesis*

To judge whether a sample slope is close or far from zero, the *standard error of the slope* is needed. The slope standard error depends on unexplained variation, the sample size, and population dispersion, and is equal to .118 for the trend in U.S. export volume:

$$s_{b_1} = \sqrt{\frac{\Sigma(y-\hat{y})^2/(N-2)}{\Sigma(X-\bar{X})^2}} = .020$$

To form a conclusion about the significance of the slope, calculate the number of standard errors which separate the slope estimate b_1 from zero.

$$t_{N-2} = b_1/s_{b_1} = .118\ /\ .020 = 6.0$$

The slope estimate is six standard errors from zero.

From both experience and logic, managers had a good idea that the trend in U.S. export volume had been positive, so a *one tail* test is appropriate, corresponding to the alternate hypothesis is that the slope is positive. Figure 5.4 illustrates *p values* for *t* statistics for a sample of 7. A *t* value of 6.0 is far from zero, with corresponding *p value* .0092.

There is a very small chance that we would observe the sample data were the trend not positive. From our sample evidence, we reject the null hypothesis of a flat slope and accept the alternate hypothesis of a positive slope.

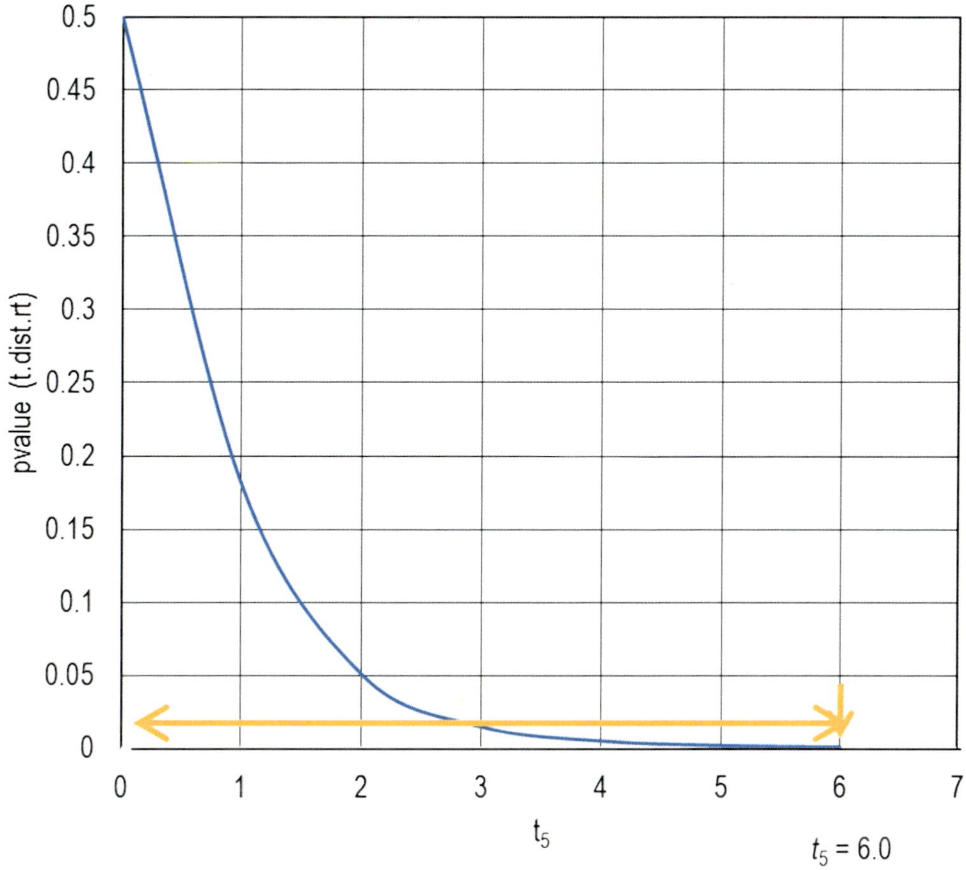

Figure 5.4 *p values for a one tail t test with 5 degrees of freedom*

Excel does these calculations for us. The slope and intercept estimates are labeled *Coefficients* in Excel, shown in Table 5.1, on the left. To the right of the coefficient estimates are their standard errors, *t* statistics, and *p values,* as well as 95% confidence intervals for the population intercept and slope.

Excel assumes that we have no prior information concerning the direction of driver influence, and so Excel provides a *two tail p value.* Divide the Excel *p value* by 2 to find the *one tail p value.*

There is a 95% chance that the true population slope will fall within $t_{.05,(N-2)}$ standard errors of the slope estimate:

$$b_1 - t_{.05,50} \times s_{b_1} < \beta_1 < b_1 + t_{.05,50} \times s_{b_1}$$

$$.118 - 2.57 \times (.020) < \beta_1 < .118 + 2.57 \times .020$$

$$.068 < \beta_1 < .169$$

The expected annual increase in U.S. export volume is .068 (ML) to .169 (ML)

5.4 The Regression Standard Error Reflects Model Precision

Using the regression formula, we can predict the expected volume $\hat{y}$ for in a given year *t*. The differences between expected and actual revenue are the *residuals* or errors. Errors from these four years are shown in Table 5.2 and Figure 5.5.

Table 5.2 Residuals from the regression line

Year	*Actual exports(ML)*	*Expected exports (ML)*	*Residual (ML)*
T	*y*	$\hat{y}$	$e = y - \hat{y}$
2005	2.55	2.48	.063
2006	2.55	2.60	−.061
2007	2.60	2.72	−.118
2008	2.89	2.84	.050

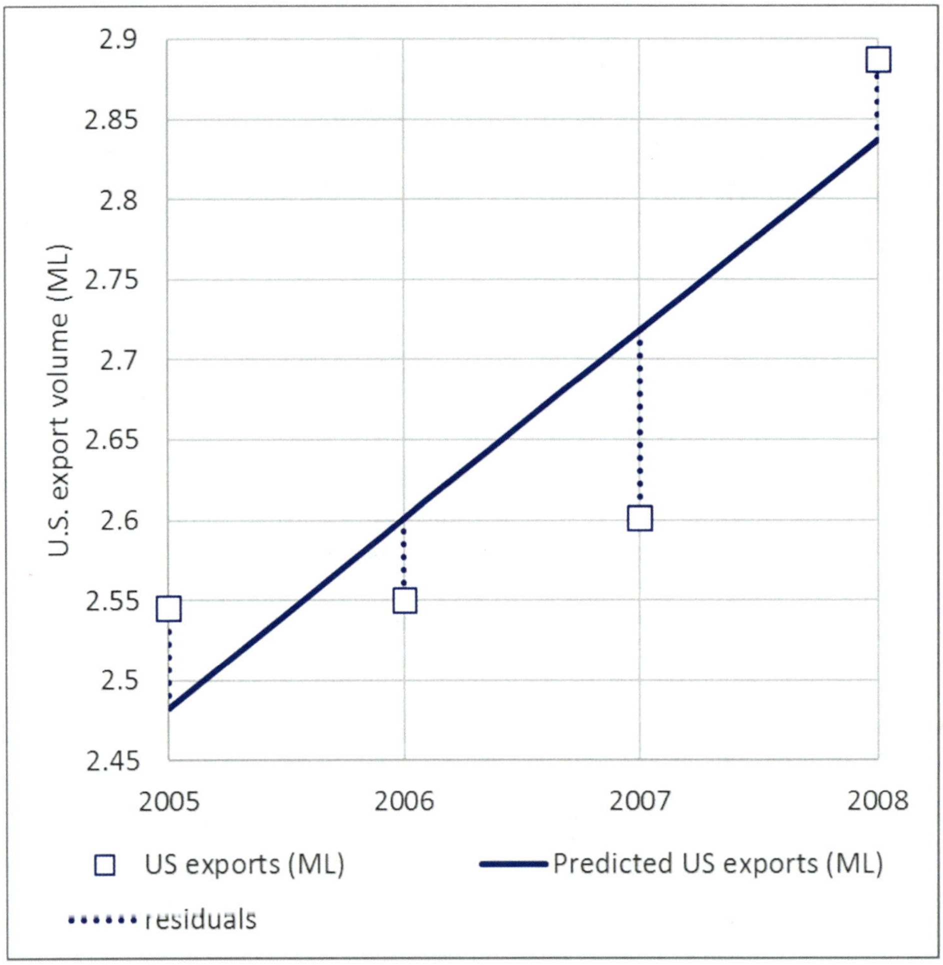

Figure 5.5 Four residuals from the regression line

The *Sum of Squared Errors* in a regression,

$$SSE = \sum e_i^2 = \sum (y_i - \hat{y})^2 = \sum (y_i - b_0 - b_1 \times X_i)^2$$
$$= .054$$

is the portion of total variation in the dependent variable, *SST*, which remains unexplained after accounting for the impact of variation in *X*. In this case, *SSE* is .054. The *Least Squares* regression line is the line with the smallest SSE of all possible lines relating *X* to *Y*.

The regression *standard error*, equal to the square root of *Mean Square Error*, reflects the precision of the regression equation.

$$s_{\hat{y}} = \sqrt{SSE/(N-2)} = \sqrt{MSE}$$

In the Concha y Toro U.S. export volume regression, the standard error is .104(ML):

$$s_{\hat{y}} = \sqrt{.054/5} \qquad = \sqrt{.011} \qquad = .104(\text{ML})$$

The forecast *margin of error* is approximately twice the standard error:

$$me = t_{.05,residual\ df} \times s_{\hat{y}}$$

where the residual degrees of freedom are N − 2 for a regression model with one driver (and the intercept).

The margin of error is .267(ML) in the Concha y Toro U.S. export volume regression:

$$me = 2.57 \times .104(\text{ML}) = .267(\text{ML})$$

We expect forecasts to be no further from actual performance than the margin of error 95% of the time.

5.5 Prediction Intervals Estimate Average Population Response

95% prediction intervals

$$\hat{y} \pm me$$

for U.S. export volumes by year are shown in Table 5.3 and Figure 5.6. Note that though data from 2009 and 2010 were not used to fit the model, the 95% prediction intervals correctly forecast actual exports in 2009 and 2010. We have evidence that the model has predictive validity.

Table 5.3 Individual 95% prediction intervals

Year t	*expected exports (ML)* $\hat{y}$	*standard error* $s_{\hat{y}}$	*margin of error me* $t_{.05,5} \times s_{\hat{y}}$	*95% prediction interval* $\hat{y} \pm me$	
2008	2.84	.104	.267	2.57	3.10
2009	2.95	.104	.267	2.69	3.22
2010	3.07	.104	.267	2.81	3.34

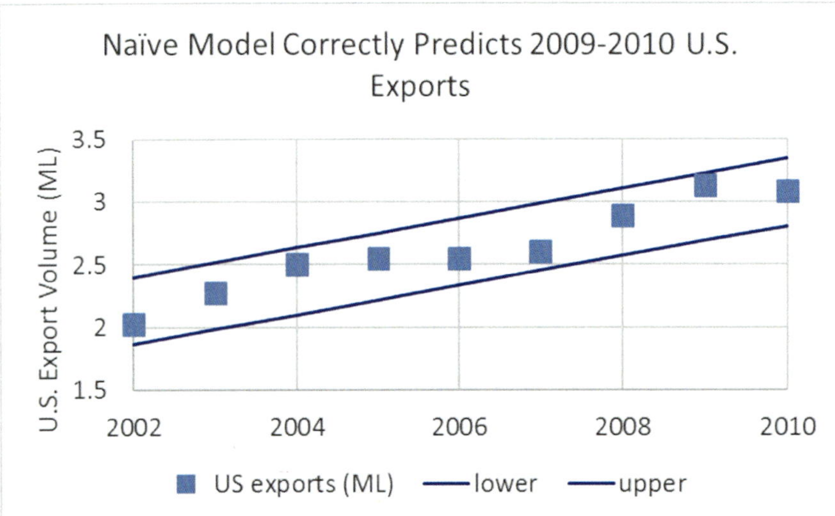

Figure 5.6 95% prediction intervals for annual U.S. export volume

5.6 *Rsquare* Summarizes Strength of the Hypothesized Linear Relationship and *F* Tests Its Significance

ANOVA, an acronym for *Analysis of Variance,* focuses on explained and unexplained variation. The difference, SST − SSE, called the *Regression Sum of Squares, SSR,* or *Model Sum of Squares,* is the portion of total variation in y influenced by variation in X.

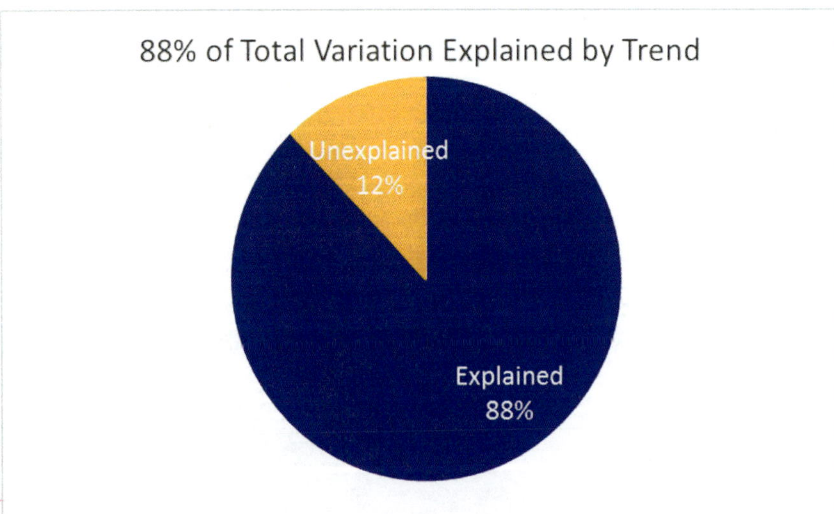

Figure 5.7 ANOVA showing explained variation, SSR, and unexplained variation, SSE

RSquare is the ratio of explained to unexplained variation:

$$RSquare = SSR/SST$$

RSquare reflects the power of the driver *X* in explaining variation in performance *y*. At the extremes, *RSquare* is zero if no variation is explained and one, if all of the variation is explained. In the Concha y Toro example of U.S. export volumne, *RSquare* is .88, or 88%.

$$RSquare = .391\ /\ .445 = .88$$

Trend in U.S. export volume accounts for 88% of the variation. Other factors account for the remaining 12%.

A test of the null hypothesis that the independent variable does not influence the dependent variable in the population is equivalent to a test that RSquare is equal to zero, and no variation in the dependent variable is explained by variation in the independent variable:

H_0: Trend *t* does not drive variation in *y, U.S. export volume,*

OR H_0: *RSquare* = 0

Versus

H_1: Trend *t* does drive variation in *y*, U.S. export volume,

OR H_1: *RSquare* > 0

Adding independent variables to a model adds explanatory power. Explained variation, *SSR,* is divided by the number of independent variables for the hypothesis test on variation explained per independent variable. Unexplained variation, *SSE,* is divided by the sample size, less the number of variables in the model (including the intercept), for the relevant comparison of variation explained per independent variable, *MSR,* to unexplained variation for a model of given size and sample size, *MSE*. This ratio of mean squares is distributed as an *F,* and the particular *F* distribution is indexed by model size and sample size. The numerator degrees of freedom is the number of predictors and the denominator degrees of freedom is the sample size, less the number of variables in the model (including the intercept).

$$F_{regression\ df, residual\ df} = \frac{SSR/regression\ df}{SSE/residual\ df} = \frac{MSR}{MSE}$$

The *F* statistic can also be determined with *RSquare:*

$$F_{regression\ df, residual\ df} = \frac{RSquare/regression\ df}{(1 - RSquare)/residual\ df}$$

In Concha y Toro U.S. exports volumes, a relatively large proportion of variation is explained by just one driver, the year *t*, making the *F* statistic large:

$$F_{1,50} = \frac{.391/1}{.054/5} = \frac{.391}{.011} \cong 36.1$$

OR

$$= \frac{.878/1}{(1-.878)/5} = \frac{.878}{.024} \cong 36.1$$

F distributions are skewed with minimum value of zero. *p values* for *F* statistics with 1 and 5 degrees of freedom are shown in Figure 5.8. In the Concha y Toro U.S. export volume regression, the $F_{1,5}$ equal to 36.1 has a *p value* < .0018, providing evidence that the sample data would not be observed if *RSquare* were zero.

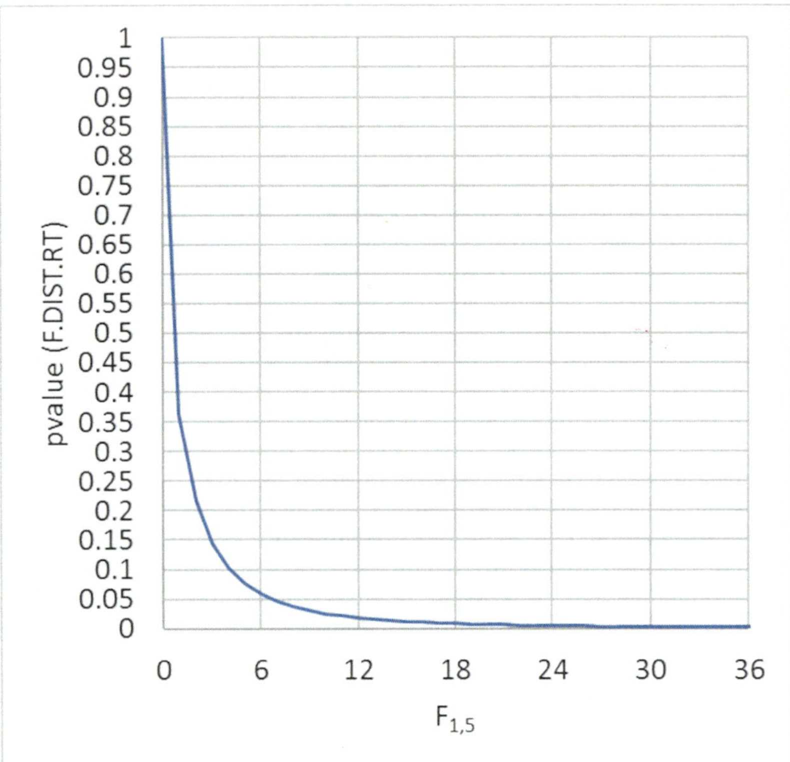

Figure 5.8 *p values* from *F* tests with one independent variable and sample size 5

In Excel, regression model fit and ANOVA statistics appear in two tables. *RSquare* and the *standard error* appear in SUMMARY OUTPUT, which is followed by the ANOVA table with Regression, Residual, and Total *SS* (*SSR, SSE,* and *SST*), Regression and Residual MS (*MSR* and *MSE*), *F*, and *significance F* (*p value* for the *F* test). The SUMMARY OUTPUT and ANOVA tables from Excel for the Concha y Toro regression are shown in Table 5.4.

Table 5.4 Model summary of fit and ANOVA table

SUMMARY OUTPUT					
Regression Statistics					
RSquare	.88				
Standard Error	.104				
Observations	7				
ANOVA	*df*	*SS*	*MS*	*F*	*Significance F*
Regression	1	.391	.391	36.1	.0018
Residual	5	.054	.011		
Total	6	.445			

5.7 Assess Residuals to Learn Whether Assumptions Are Met

The Durbin Watson (*DW*) statistic incorporates correlation between residuals across adjacent time periods which allows assessment of the presence of unaccounted for trend or cycles in the residuals. If there is an unaccounted for cycle or trend, higher residuals are likely to be followed by similar higher residuals, and lower residuals are likely to be followed by similar lower residuals.

In the Concha y Toro model, the trend accounts for much of the variation in export volumes to the U.S., however some unexplained variation remains.

DW indicates *positive autocorrelation,* the correlation of residuals over time, which signals that a shift, shock or cycle has been ignored. The Durbin Watson statistic compares the sum of squared differences between pairs of adjacent residuals with the sum of squared residuals:

$$DW = \frac{\sum_2^N (e_q - e_{q-1})^2}{\sum_1^N e_q^2}.$$

If all of the trend, shifts, shocks and cycles in the data have been accounted for *DW* will be "high." Exactly how high depends on the length of time series, which is the number of observations used in the regression, and the number of independent variables, including the intercept. *DW* critical values are available online at stanford.edu/~clint/bench/dwcrit.htm, found by googling "Durbin Watson critical values." (In this online table, sample size is indexed by *T*, and the number of independent variables, plus intercept, is indexed by *K*.)

There are two relevant critical values, a lower value and an upper value, dL and dU.

> DW below the lower critical value, dL, indicates presence of positive autocorrelation from unaccounted for trend, cycle, or seasons which we would then attempt to identify and incorporate into the model.
>
> DW above the upper critical value, dU, indicates lack of autocorrelation and freedom from unaccounted for trend, cycle or seasons, which is the goal.

DW between dL and dU is the gray area, indicative of possible autocorrelation and presence of unaccounted for trend, cycle or seasons. When DW is in the gray area, we look for pattern in the residuals from unaccounted for trend, cycle or seasons, knowing that there is a reasonable chance that pattern may not be identified.

Figure 5.9 illustrates critical values for several sample and model sizes:

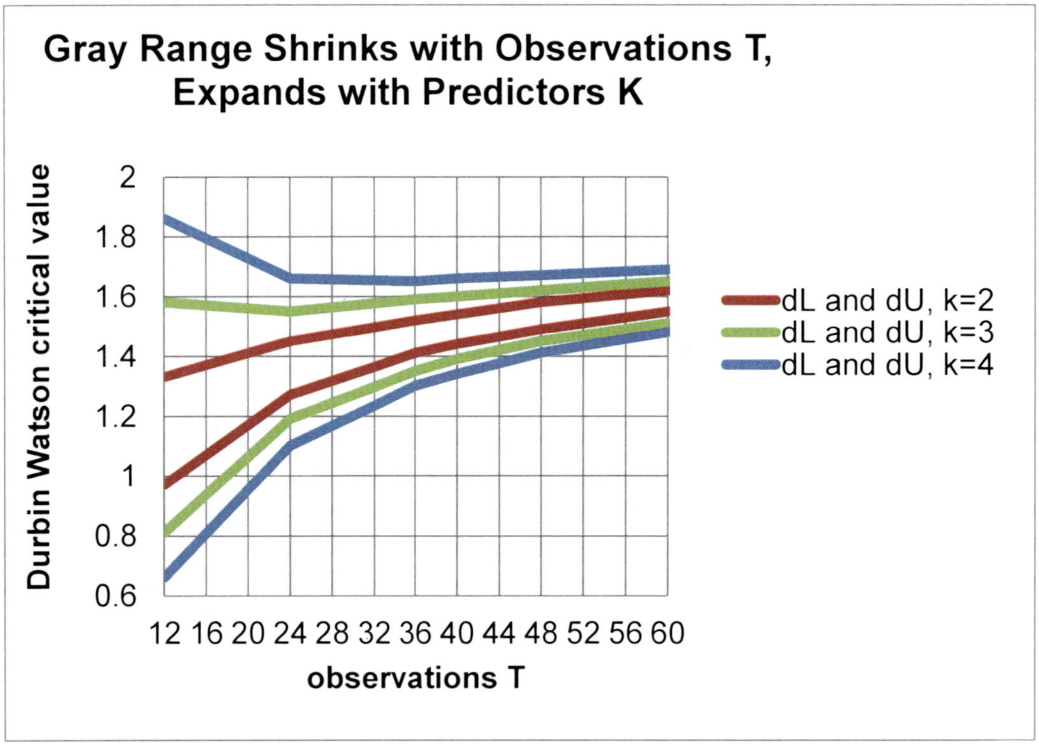

Figure 5.9 Durbin Watson critical values by sample size T and predictors k

Notice that the gray area, dL to dU, shrinks as sample size increases, but expands as the number of predictors increases.

The Concha y Toro model, with one driver, plus intercept, and a sample size of 7, has *DW* critical values of *dL*=.70 and *dU*=1.36 The model *DW* statistic is 1.51, leading to the conclusion that the residuals are free of positive autocorrelation. The residuals are pattern free, and no important shifts, shocks or cycles have been ignored.

A plot of the residuals by predicted values, Figure 5.10, with gridlines set to the standard error, suggests that all are within two standard errors, and that fit is equally good in earlier and later years. All predictions are within the margin of error, .267ML, of actual export volumes. There is some indication, however, that exports are cyclical.

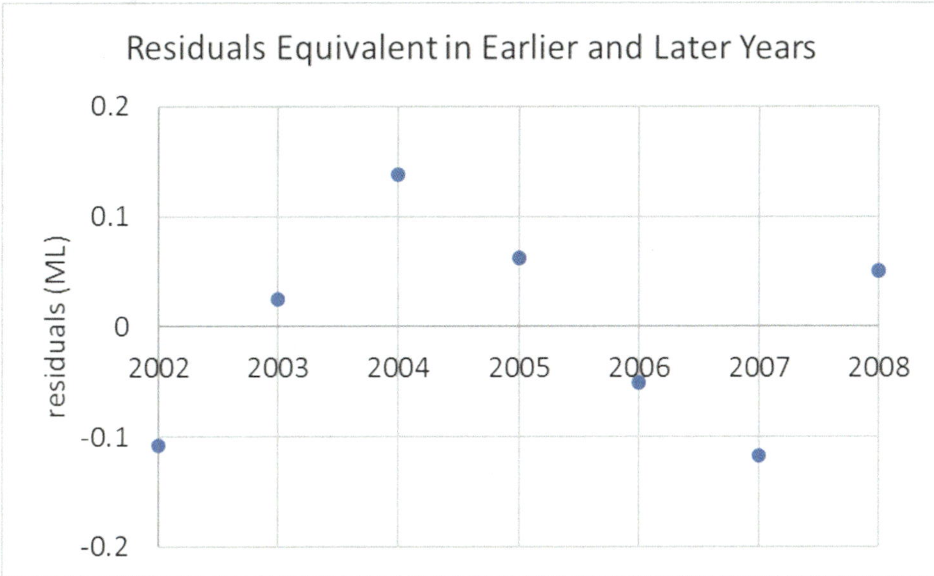

Figure 5.10 Residuals by year

Linear regression assumes that the residuals are *Normally* distributed. The distribution of residuals has skewness of 0.00, confirming that residuals are symmetric.

5.8 Recalibrate to Update a Valid Model

The modelling team has created a valid model that correctly predicts the two most recent datapoints. Next, the model will be recalibrated by including those two datapoints. Those two most recent points are likely to resemble future export volumes more than earlier data.

The recalibrated model shows a slightly stronger trend:

$$\widehat{exports}(ML)_t = -253(ML) + .127(ML/t) \times t$$

To illustrate both the model fit to past data and the forecast, the data are plotted with the lower and upper 95% prediction intervals, shown in Figure 5.11.

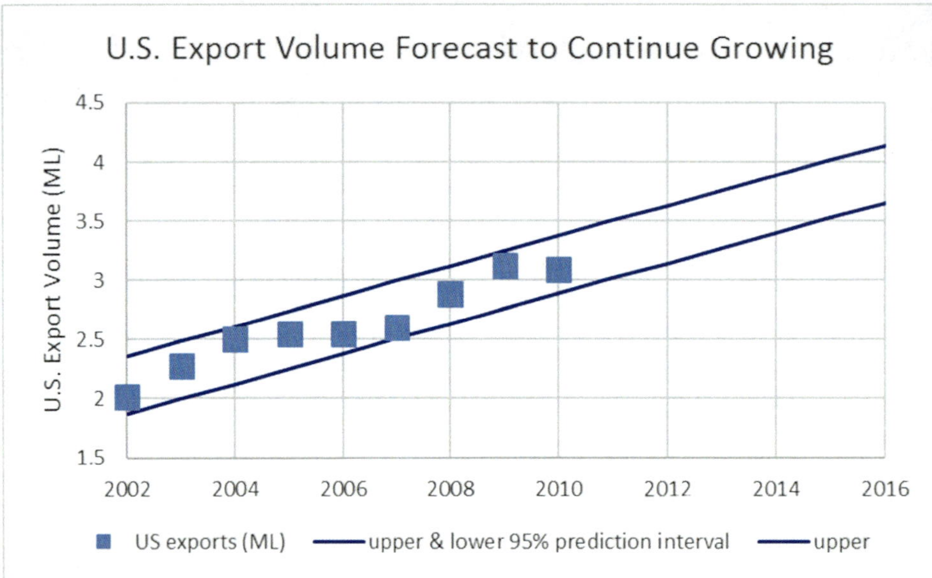

Figure 5.11 Fit and forecast of exports to the U.S.

Not all models are valid. It is not uncommon for a shift, such as an acquisition, entry into new markets, entry of exit of competitors, or a shock, such as global recession or hurricane, has occurred in the two most recent time periods. Without those two most recent time periods, a regression model would ignore recent driver changes and may not successfully forecast those two most recent periods. Should such a model, lacking predictive validity be recalibrated? Given the choice of information from a model lacking predictive validity and no information, the model, though without predictive validity, is still valuable and provides the best information available. Results from such a model would be presented with the caveat that the model ought to be updated in the near future, since recent changes are expected to influence future performance.

From the regression analysis, the modelling team could

- conclude that the trend in U.S. exports is positive,
- estimate the trend, and
- predict export volumes in future years.

In the presentation of results to management, the team could conclude:

"Sample evidence suggests that the trend in U.S. export volume is positive.

The positive trend accounts for 93% of the variation in annual export volume to the U.S. within the series of nine previous years.

Future export volumes can be estimated with a margin of error of 244K liters.

Each year a volume increase of 96K to 159K liters can be expected.

The forecast for 2016 U.S. export volume is 3.65K to 4.14K liters, representing annual growth of three to five percent."

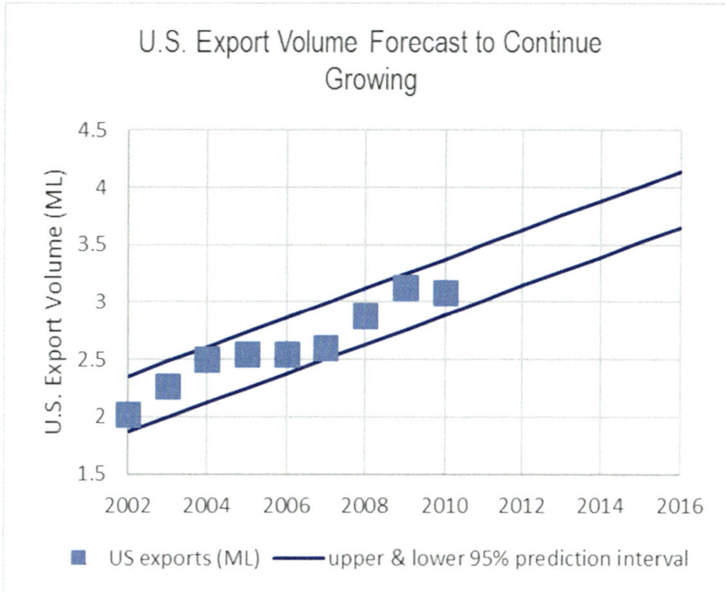

$$\hat{exports}(ML)_t = -253^a(ML) + .127^a\left(\frac{ML}{t}\right) \times t$$

RSquare: .93

[a]Significant at .0001

5.9 Present Regression Results in Concise Format

The regression equation is presented in a standard format, with the dependent variable on the left, RSquare below the equation, and significance levels of the model and parameter estimates indicated with superscripts:

$$\hat{y} = b_0^a + b_1^a \times X$$

RSquare=___[a]

[a]Significant at ___,

where the variable names and units are specified.

Significance is reported at two levels, .05 and .01.

p values greater than .01, but less than or equal to .05 are reported as significant at .05.
p values less than or equal to .01 are reported as significant at .01, shown in Table 5.5:

Table 5.5 Significance levels from p values

	Reported as significant at
$.01 < p\ value \leq .05$	.05
$p\ value \leq .01$	.01

For the general business audience, the verbal description with graphical illustration conveys all of the important information. The three additional lines provide the information that statistically savvy readers will want in order to assess how well the model fits and which parameter estimates are significant.

5.10 Assumptions We Make When We Use Linear Regression

Often a group of independent variables together jointly influence a dependent variable. If we attempt to explain or predict a dependent variable with an independent variable, but omit a third (or fourth) important influence, results will be misleading. If just one from the group is included in a regression, it may seem to be responsible for the joint impact of the group. It will seem that the independent variable chosen is more important than it actually is. Chapters 10 and 11 introduce diagnosis of *multicollinearity,* the situation in which predictors are correlated and jointly influence a dependent variable.

Linear regression assumes that the dependent variable, which is often a performance variable, is related linearly to the independent variable, often a decision variable. In reality, few relationships are linear. More often, performance increases or decreases in response to increases in a decision variable or time period, but at a diminishing rate. In these cases, linear regression doesn't fit the data perfectly. Extrapolation beyond the range of values within a sample can be risky if we assume constant response when response is actually diminishing or increasing. Though often not perfect reflections of reality, linear relationships can be useful approximations. In Chapter 12, we will explore simple remedies to improve linear models of nonlinear relationships by simply rescaling to square roots, logarithms or squares.

5.11 Correlation Reflects Linear Association

A correlation coefficient ρ_{xy} is a simple measure of the strength of the linear relationship between two continuous variables, *X* and *y*. The sample estimate of the population correlation coefficient ρ_{xy} is calculated by summing the product of differences from the sample means $\overline{\mathrm{X}}$ and $\overline{\mathrm{Y}}$, standardized by the standard deviations s_x and s_y:

$$r_{xy} = \frac{1}{(N-1)}\sum \frac{(x_i-\overline{\mathrm{X}})}{s_x}\frac{(y_i-\overline{\mathrm{Y}})}{s_y},$$

where x_i is the value of *X* for the *i*th sample element, and

y_i is the value of *Y* for the *i*th sample element.

When *x* and *y* move together, they are positively correlated. When they move in opposite directions, they are negatively correlated.

Table 5.6 contains years t and U.S. export volumes y from a sample of nine years:

Table 5.6 Year and U.S. export volumes

Year T	*U.S. export volume (ML)* Y
2002	2.02
2003	2.27
2004	2.50
2005	2.55
2006	2.55
2007	2.60
2008	2.89
2009	3.31
2010	3.08

A scatterplot in Figure 5.12 reveals that export volumes are higher in later years.

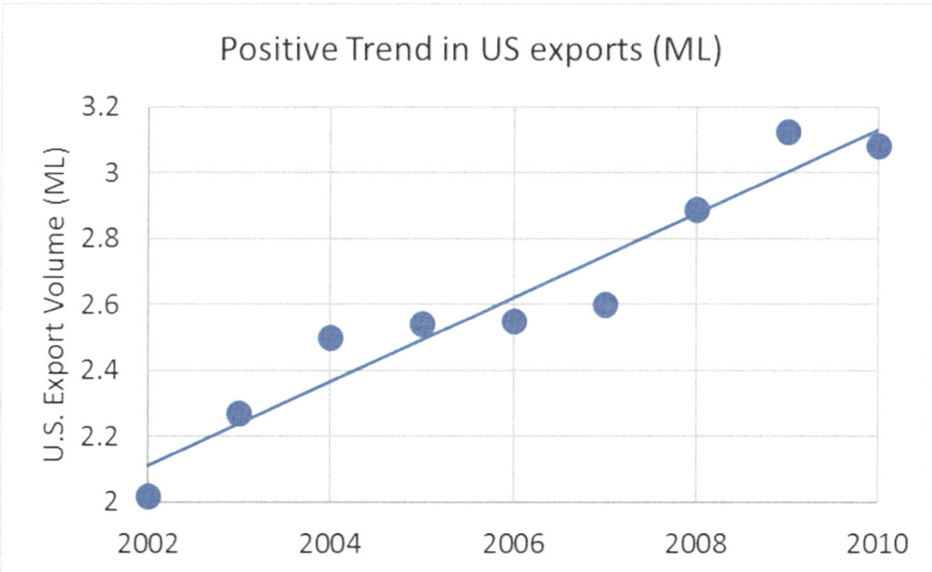

Figure 5.12 Positive trend in U.S. export volume

Differences from the sample means and their products are shown in Table 5.7.

Table 5.7 Differences from sample means and crossproducts

	Year			*Export volume*		
x_i	$\overline{X}$	$x_i - \overline{X}$	y_i	$\overline{Y}$	$y_i - \overline{Y}$	$(x_i - \overline{X}) \times (y_i - \overline{Y})$
2002	2006	−4	2.02	2.62	−0.60	2.40
2003	2006	−3	2.27	2.62	−0.35	1.05
2004	2006	−2	2.50	2.62	−0.12	0.24
2005	2006	−1	2.55	2.62	−0.08	0.08
2006	2006	0	2.55	2.62	−0.07	0.00
2007	2006	1	2.60	2.62	−0.02	−0.02
2008	2006	2	2.89	2.62	0.27	0.53
2009	2006	3	3.13	2.62	0.50	1.51
2010	2006	4	3.08	2.62	0.46	1.85

The sample standard deviations are $s_x = 2.74$ years and $s_y = .362$(ML).

The correlation coefficient is:

$$r_{xy} = \frac{1}{(9-1)}\left[\frac{2.40 + 1.05 + .24 + .08 + .00 - .02 + .53 + 1.51 + 1.85}{(2.74)(.362)}\right]$$

$$= \frac{1}{8}\left[\frac{7.64}{.991}\right]$$

$$= .964$$

A correlation coefficient can be as large in absolute value as 1.00, if two variables were perfectly correlated. All of the points in the scatterplot lie on the regression line in that case. RSquare, which is the squared correlation in a simple regression, would be 1.00, whether the correlation coefficient were −1.00 or +1.00.

In the Concha y Toro example above, *RSquare* is

$$RSquare = r_{xy}^2 = .964^2 = .929$$

In some cases, *x* and *y* are not related *linearly*, though they *are* strongly related. There are situations, for example, where more is better up to a point and improves performance, then, *saturation* occurs and, beyond this point, response deteriorates.

- Without enough advertising, customers will be not aware of a new product. Spending more increases awareness and improves performance. Beyond some saturation point, customers grow weary of the advertising, decide that the company must be desperate to advertise so much, and switch to another brand, reducing performance.

- A factory with too few employees x to man all of the assembly positions would benefit from hiring. Adding employees increases productivity y up to a point. Beyond some point, too many employees would crowd the facility and interfere with each other, reducing performance.

5.12 Correlation Coefficients Are Key Components of Regression Slopes

Correlation coefficients are closely related to regression slopes. From the correlation between x and y, as well as their sample standard deviations s_x and s_y, the regression slope estimate can be calculated:

$$b_1 = r_{xy}\frac{s_y}{s_x}$$

Similarly, from the regression slope estimate and sample standard deviations s_x and s_{xy}, the correlation coefficient can be calculated:

$$r_{xy} = b_1\frac{s_x}{s_y}$$

The t test of hypothesis that a slope is zero

$$\mathrm{H_0}:\ \beta_1 = 0$$

Versus

$$\mathrm{H_1}:\ \beta_1 \neq 0$$

is equivalent to the t test used to test the hypothesis that a correlation is zero:

$$\mathrm{H_0}:\ \rho_{xy} = 0$$

Versus

$$\mathrm{H_1}:\ \rho_{xy} \neq 0:$$

$$t_{N-2} = \frac{b_1}{s_{b_1}} = \sqrt{N-2}\frac{r_{xy}}{\sqrt{1\ \ r_{xy}^2}}$$

In the Concha y Toro example, with the correlation coefficient, $r_{t,exports}$=.963, and the sample standard errors, s_t = 2.74 and $s_{exports}$ = .362, the regression slope estimate can be calculated,

$$b_{titles} = .963\frac{.362}{2.74} = .127$$

as well as the t value to test the hypothesis that the slope is less than or equal to zero or, equivalently, that the correlation is less than or equal to zero:

$$t_7 = \sqrt{7} \times \frac{.963}{\sqrt{(1-.963)}} = 9.57$$

with *p value* < .01.

Based on sample evidence, there is little chance that titles stocked and vending unit revenues are uncorrelated or negatively correlated.

5.13 Correlation Complements Regression

The correlation coefficient summarizes direction and strength of linear association between two continuous variables. Because it is a standardized measure, taking on values between −1 and +1, it is readily interpretable. Unlike regression analysis, it is not necessary to designate a dependent and an independent variable to summarize association with correlation analysis. Later, in the context of multiple regression analysis, the correlations between independent variables will be an important focus in our diagnosis of multicollinearity, introduced in Chapters 9 and 10.

Correlation analysis should be supplemented with visual inspection of data. It would be possible to overlook strong, nonlinear associations with small correlations. Inspection of a scatterplot will reveal whether or not association between two variables is linear.

Correlation is closely related to simple linear regression analysis:

- The squared correlation coefficient is *RSquare*, our measure of percent of variation in a dependent variable accounted for by an independent variable.
- The regression slope estimate is a product of the correlation coefficient and the ratio of the sample standard deviation of the dependent variable to sample standard deviation of the independent variable.
    - Slope estimates from simple linear regression are unstandardized correlation coefficients.
    - Correlation coefficients are standardized simple linear regression slope estimates.

5.14 Linear Regression Is Doubly Useful

Linear regression handles two modeling jobs, quantification of a driver's influence and forecasting. Regression models quantify the direction and nature of influence of a driver on a response or performance variable. Regression models also enable forecasts and to compare decision alternatives. In this chapter, focus was on naïve models of trend, which provide long range forecasts, but without explanation. In Chapter 9, regression models to identify drivers with cross sectional data are introduced, and in Chapter 10, regression models to both identify drivers and to forecast short term with time series data are introduced.

Excel 5.1 Build a Simple Linear Regression Model

Impact of Trend on Concha y Toro Export Volume to the U.S. Use regression analysis to explore the linear influence of trend on *U.S. export volume* differences across a time series sample of nine years.

Open **Excel 5.1 Concha y Toro US exports.**

Use shortcuts to run regression. Be sure to exclude the two most recent datapoints from 2009 and 2010, so that you can later validate your model.	**Alt AY3, R, down, down** **B1:b8 tab a1:a8 tab LR**

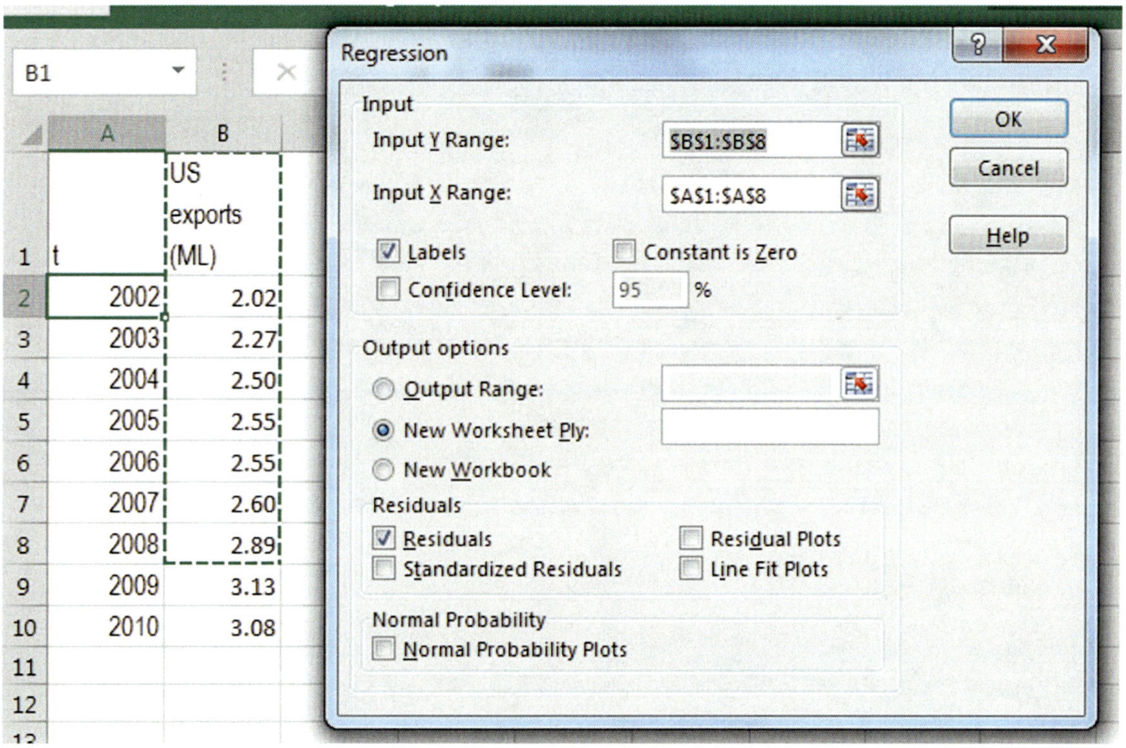

	A	B	C	D	E	F	G
1	SUMMARY OUTPUT						
2							
3	*Regression Statistics*						
4	Multiple F	0.937184					
5	R Square	0.878315					
6	Adjusted I	0.853978					
7	Standard E	0.104047					
8	Observati	7					
9							
10	ANOVA						
11		*df*	*SS*	*MS*	*F*	*gnificance F*	
12	Regressio	1	0.390695	0.390695	36.08958	0.001836	
13	Residual	5	0.054128	0.010826			
14	Total	6	0.444823				
15							
16		*Coefficients*	*andard Err*	*t Stat*	*P-value*	*Lower 95%*	*Upper 95%*
17	Intercept	-234.357	39.42424	-5.94449	0.001924	-335.7	-133.014
18	t	0.118124	0.019663	6.007461	0.001836	0.067579	0.16867
19							
20							
21							
22	RESIDUAL OUTPUT						
23							
24	*bservatio*	*d US expo*	*Residuals*				
25	1	2.12806	-0.1076				
26	2	2.246184	0.024816				

Excel 5.2 Assess Residuals

Assess the residuals by finding the Durbin Watson statistic to test presence of positive autocorrelation, unaccounted for trend, shifts, shocks or cycles.	In D24, DW In D25, **=sumxmy2(c25:c30,c26:c31)/sumsq(c25:c31)**

Compare DW with the lower and upper critical values on the Stanford website, T=7, K=2:

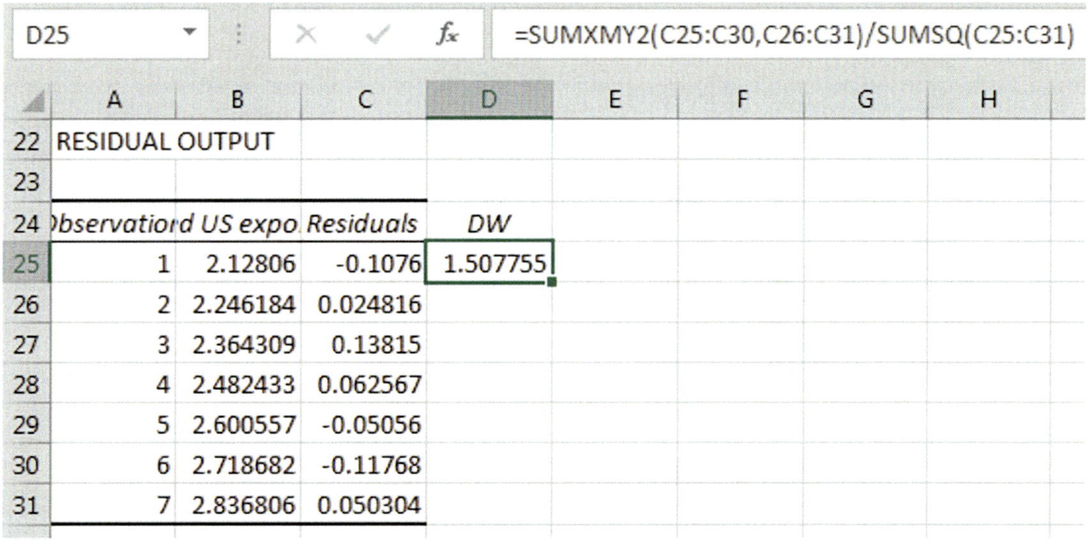

D25 =SUMXMY2(C25:C30,C26:C31)/SUMSQ(C25:C31)

	A	B	C	D
22	RESIDUAL OUTPUT			
23				
24	*Observation*	*d US expo*	*Residuals*	*DW*
25	1	2.12806	-0.1076	1.507755
26	2	2.246184	0.024816	
27	3	2.364309	0.13815	
28	4	2.482433	0.062567	
29	5	2.600557	-0.05056	
30	6	2.718682	-0.11768	
31	7	2.836806	0.050304	

To assess Normality of the residuals, find the residual skew.

In C32, **=skew(c25:c31)**

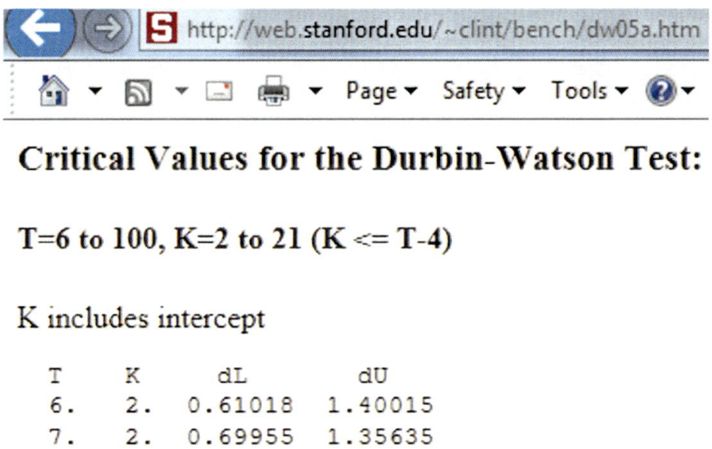

Critical Values for the Durbin-Watson Test:

T=6 to 100, K=2 to 21 (K <= T-4)

K includes intercept

T	K	dL	dU
6.	2.	0.61018	1.40015
7.	2.	0.69955	1.35635

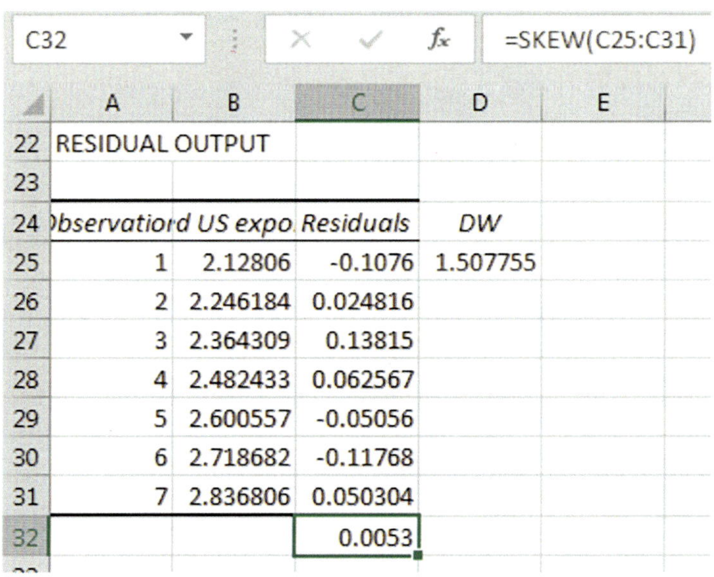

C32 =SKEW(C25:C31)

	A	B	C	D
22	RESIDUAL OUTPUT			
23				
24	*Observation*	*d US expo*	*Residuals*	*DW*
25	1	2.12806	-0.1076	1.507755
26	2	2.246184	0.024816	
27	3	2.364309	0.13815	
28	4	2.482433	0.062567	
29	5	2.600557	-0.05056	
30	6	2.718682	-0.11768	
31	7	2.836806	0.050304	
32			0.0053	

Excel 5.3 Construct Prediction Intervals to Validate

Copy the data and paste next to residuals, and then use the regression equation with the coefficients (always in B17 and B18) to find predicted exports. (Use **f4,** function 4, to lock cell references so that your equation will use the coefficients to make predicted values in each row.)	**Cntl+Page down** In A1, **Cntl+shift+down** **Shift+right** **Cntl+C** **Cntl+Page up** In E24, **Cntl+V** In G24, Predicted exports (ML) In G25, **=b17 f4 +f18 f4 *e24** In G25, Double click the lower right corner to down fill

G25 =$B$17+$B$18*E25

	A	B	C	D	E	F	G
16		*Coefficients*	*andard Err*	*t Stat*	*P-value*	*Lower 95%*	*Upper 95%*
17	Intercept	-234.357	39.42424	-5.94449	0.001924	-335.7	-133.014
18	t	0.118124	0.019663	6.007461	0.001836	0.067579	0.16867
19							
20							
21							
22	RESIDUAL OUTPUT						
23							
24	*bservatio*	*d US expo*	*Residuals*	*DW*	t	US exports (ML)	*predicted exports (ML)*
25	1	2.12806	-0.1076	1.507755	2002	2.02	2.12806
26	2	2.246184	0.024816		2003	2.27	2.246184
27	3	2.364309	0.13815		2004	2.50	2.364309
28	4	2.482433	0.062567		2005	2.55	2.482433
29	5	2.600557	-0.05056		2006	2.55	2.600557
30	6	2.718682	-0.11768		2007	2.60	2.718682
31	7	2.836806	0.050304		2008	2.89	2.836806
32			0.0053		2009	3.13	2.954931
33					2010	3.08	3.073055
34					2011		3.19118
35					2012		3.309304

Find the critical t value and the margin of error from the critical t for residual degrees of freedom (always in B13) and the standard error (always in B7).

In C6,
Critical t
In C7,
=t.inv.2t(.05,b13)
In D6,
ME
In D7,
=b7*c7

D7 | f_x | =B7*C7

	A	B	C	D	E
3	*Regression Statistics*				
4	Multiple F	0.937184			
5	R Square	0.878315			
6	Adjusted I	0.853978	critical t	me	
7	Standard E	0.104047	2.570582	0.26746	
8	Observati	7			
9					
10	ANOVA				
11		*df*	*SS*	*MS*	*F*
12	Regressio	1	0.390695	0.390695	36.08
13	Residual	5	0.054128	0.010826	

To test the predictive validity of the model, find the *lower 95%* and *upper 95%* prediction interval bounds by adding and subtracting the margin of error (always in D7) to and from *predicted revenues,* locking the *margin of error* cell reference with **f4**.

In H24,
Lower 95% prediction interval bound
In I24,
Upper 95% prediction interval bound
In H25,
=g25-d7 f4
In I25,
=g25+d7 f4
In H25,
Shift+right
Double click lower right corner to fill in columns

H25 =G25-$D$7

	D	E	F	G	H	I
6	me					
7	0.26746					
8						
9						
10						
11	*MS*	*F*	*gnificance F*			
12	0.390695	36.08958	0.001836			
13	0.010826					
14						
15						
16	*t Stat*	*P-value*	*Lower 95%*	*Upper 95%*	*ower 95.0%*	*pper 95.0%*
17	-5.94449	0.001924	-335.7	-133.014	-335.7	-133.014
18	6.007461	0.001836	0.067579	0.16867	0.067579	0.16867
19						
20						
21						
22						
23						
24	*DW*	t	US exports (ML)	*predicted exports (ML)*	lower 95% pi bound	upper 95% pi bound
25	1.507755	2002	2.02	2.12806	1.860599	2.39552
26		2003	2.27	2.246184	1.978724	2.513644
27		2004	2.50	2.364309	2.096848	2.631769
28		2005	2.55	2.482433	2.214973	2.749893

Compare actual exports for 2009 and 20010, in F32 and F33, with 95% prediction interval bounds in H32, I32, H33 and I33.

	D	E	F	G	H	I
24	*DW*	t	US exports (ML)	*predicted exports (ML)*	lower 95% pi bound	upper 95% pi bound
25	1.507755	2002	2.02	2.12806	1.86	2.40
26		2003	2.27	2.246184	1.98	2.51
27		2004	2.50	2.364309	2.10	2.63
28		2005	2.55	2.482433	2.21	2.75
29		2006	2.55	2.600557	2.33	2.87
30		2007	2.60	2.718682	2.45	2.99
31		2008	2.89	2.836806	2.57	3.10
32		2009	3.13	2.954931	2.69	3.22
33		2010	3.08	3.073055	2.81	3.34

Excel 5.4 Recalibrate and Present Fit and Forecast in a Scatterplot

Recalibrate to update your fit and forecast. Run regression again, this time including the two most recent datapoints.

	A	B	C	D	E	F	G
1	SUMMARY OUTPUT						
2							
3	*Regression Statistics*						
4	Multiple F	0.963858					
5	R Square	0.929023					
6	Adjusted I	0.918883					
7	Standard E	0.103082					
8	Observati	9					
9							
10	ANOVA						
11		*df*	*SS*	*MS*	*F*	*gnificance F*	
12	Regressio	1	0.973579	0.973579	91.62281	2.85E-05	
13	Residual	7	0.074382	0.010626			
14	Total	8	1.04796				
15							
16		*Coefficient*	*andard Err*	*t Stat*	*P-value*	*Lower 95%*	*Upper 95%*
17	Intercept	-252.909	26.69558	-9.4738	3.05E-05	-316.034	-189.784
18	t	0.127383	0.013308	9.571981	2.85E-05	0.095914	0.158851

Update your fit and forecast by reusing formulas for the critical t, margin of error, predicted exports and the lower and upper 95% prediction intervals from the first regression.

Cntl+Page up
From C6,
Shift+down
Shift+right
Cntl+C
Cntl+page down
In C6,
Cntl+V
Cntl+page up
From E24,
Cntl+shift+down
Cntl+shift+right
Cntl+C
Cntl+page down
From E24,
Cntl+V

	C	D	E	F	G	H	I
6	critical t	me					
7	2.364624	0.243751					
8							
9							
10							
11	*SS*	*MS*	*F*	*gnificance F*			
12	0.973579	0.973579	91.62281	2.85E-05			
13	0.074382	0.010626					
14	1.04796						
15							
16	*andard Err*	*t Stat*	*P-value*	*Lower 95%*	*Upper 95%*	*ower 95.0%*	*pper 95.0%*
17	26.69558	-9.4738	3.05E-05	-316.034	-189.784	-316.034	-189.784
18	0.013308	9.571981	2.85E-05	0.095914	0.158851	0.095914	0.158851
19							
20							
21							
22							
23							
24	*Residuals*		t	US exports (ML)	*predicted exports (ML)*	lower 95% pi bound	upper 95% pi bound
25	-0.09072		2002	2.02	2.111182	1.87	2.35
26	0.032436		2003	2.27	2.238564	1.99	2.48
27	0.136512		2004	2.50	2.365947	2.12	2.61

To make a scatterplot of the fit and forecast, first move *predicted revenues* to the right of the *95% prediction intervals*. Select year *t*, *predicted exports* and the *lower and upper prediction interval* cells and request a scatterplot.

From G24,
Cntl+Shift+down
Cntl+X
From J24,
Alt HIE
From E24,
Cntl+Shift+down
Cntl+Shift+right right right
Alt ND

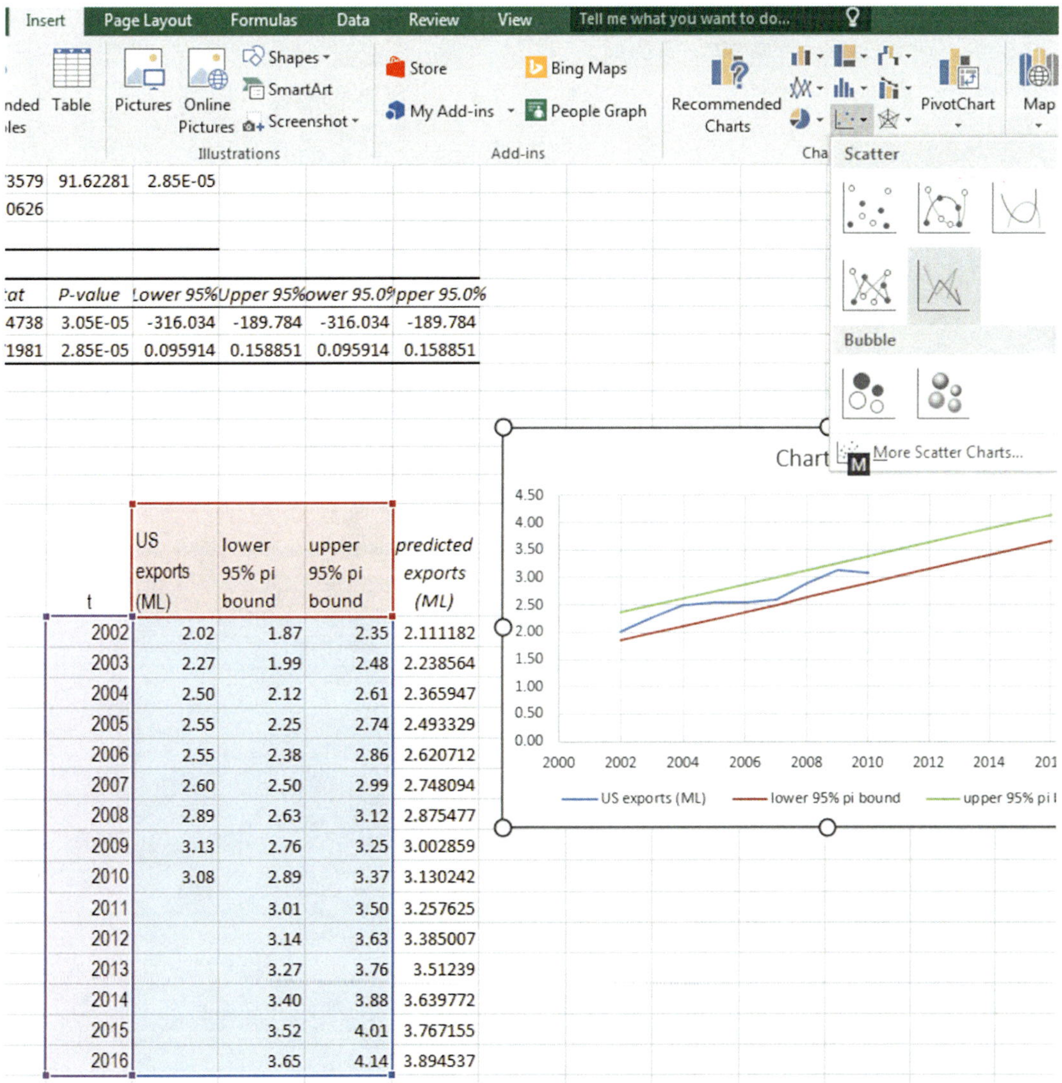

t	US exports (ML)	lower 95% pi bound	upper 95% pi bound	predicted exports (ML)
2002	2.02	1.87	2.35	2.111182
2003	2.27	1.99	2.48	2.238564
2004	2.50	2.12	2.61	2.365947
2005	2.55	2.25	2.74	2.493329
2006	2.55	2.38	2.86	2.620712
2007	2.60	2.50	2.99	2.748094
2008	2.89	2.63	3.12	2.875477
2009	3.13	2.76	3.25	3.002859
2010	3.08	2.89	3.37	3.130242
2011		3.01	3.50	3.257625
2012		3.14	3.63	3.385007
2013		3.27	3.76	3.51239
2014		3.40	3.88	3.639772
2015		3.52	4.01	3.767155
2016		3.65	4.14	3.894537

Change the actual exports from a line to markers. Select the Series U.S. Exports.

Alt JAE down to Series U.S. Exports

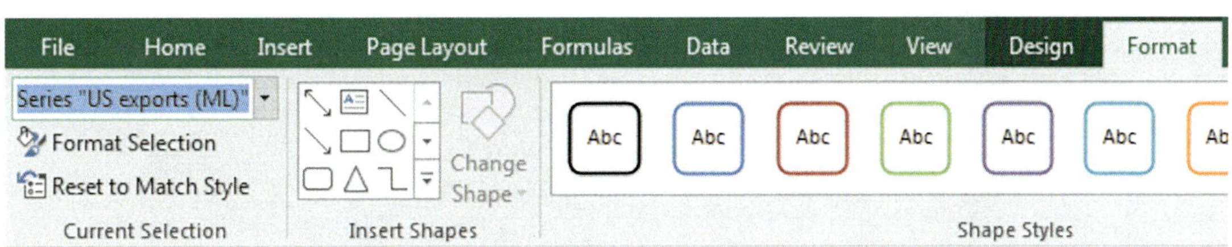

Format the series.
Alt JAM

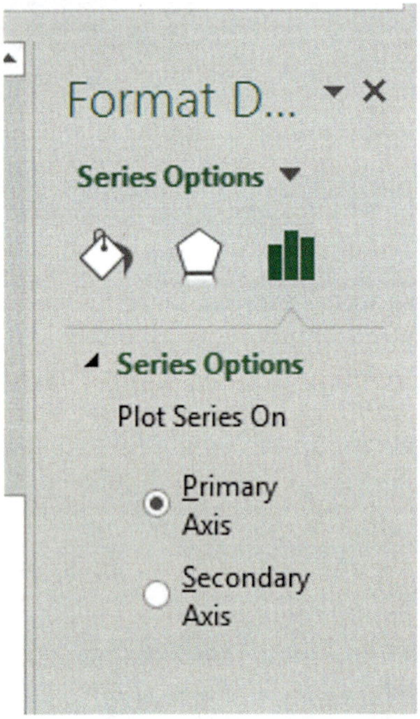

Click the tent like icon on the left and choose No Line.

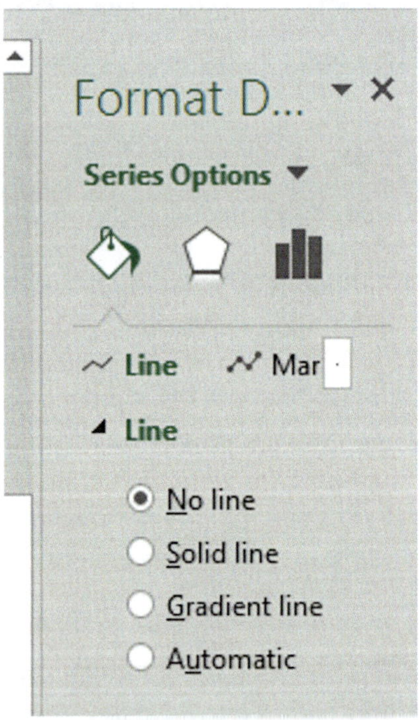

Click Marker, Marker Options, and Built In.

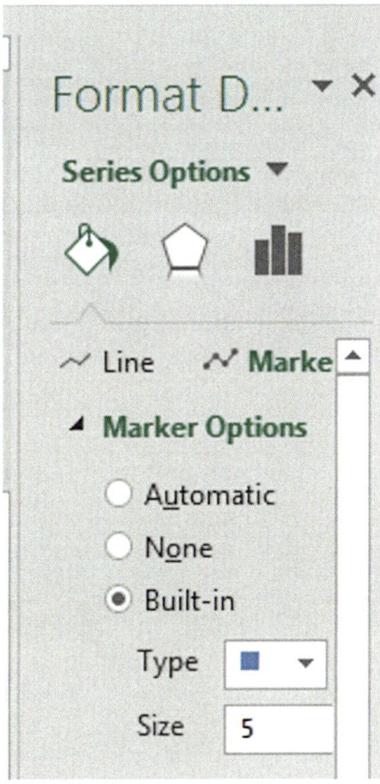

Change the color of one of the prediction interval lines to match the other.
Alt JAE down to Series lower 95% prediction interval bound
Alt JASO
Use arrow keys to choose color

Add a vertical axis title and chart title. Adjust the axes to make better use of white space and set fontsize to 12.

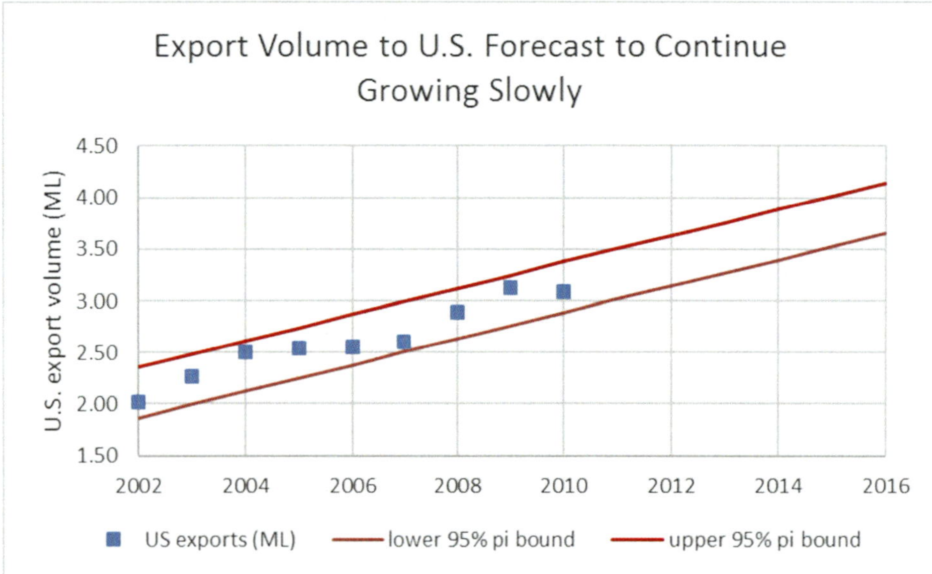

Excel 5.5 Find Correlations Between Variable Pairs

Find the correlation between year *t* and *U.S. export volume.*
In A17,
Correlation
In B17,
=correl(a2:a10,b2:b10)

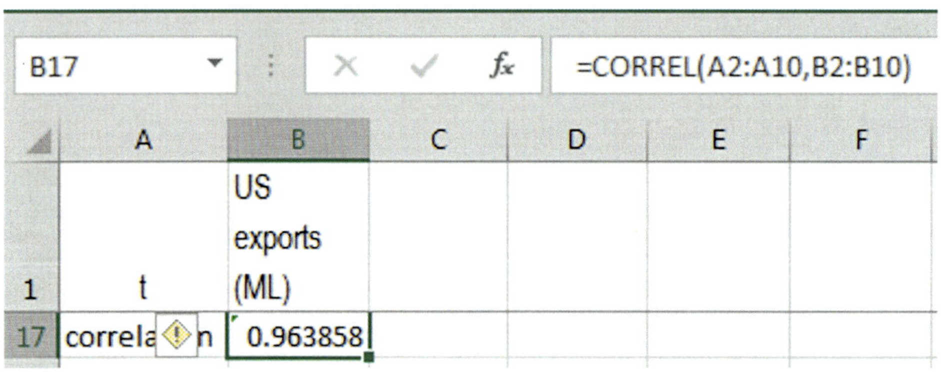

Lab 5 Forecast Concha y Toro Exports to Latin America

With forecasts for the U.S. market segment, executives are convinced that forecasts for the remaining export segments would provide invaluable information. Build naïve forecasts for export volume in Latin America.

The time series of annual export volume in Latin America segments is in **Concha y Toro exports LA.**

1. Is there a positive trend in exports to Latin America? Y or N pvalue: ______
2. How powerful is your naïve model of trend? RSquare: _____
3. Are residuals free from unaccounted for trend or cycles? Y or N DW: __________
4. How precise will your forecasts be? Margin of error: ________
5. Is your model valid for forecasting? Y or N
6. Present your final equation for trend in exports to Latin America:
7. What is your forecast for exports to Latin America in 2016? ______ to ______
8. Plot your fit and forecast of exports to Latin America:

Assignment 5.1 Forecast Concha y Toro Exports to Europe and Asia

With forecasts for the U.S.and Latin American market segments executives are convinced that forecasts for the remaining export segments would provide invaluable information. Build naïve forecasts for export segments in Asia and Europe.

The time series of annual export volumes to Asia and Europe are in **Concha y Toro exports Europe Asia.**

I. Asia

1. Is there a positive trend in exports to Asia? Y or N pvalue: ______
2. How powerful is your naïve model of trend in exports to Asia? RSquare: _____
3. Are residuals free from unaccounted for trend or cycles? Y or N DW: __________
4. How precise will your forecasts of exports to Asia be? Margin of error: ________
5. Is your model valid for forecasting? Y or N
6. Present your final equation for trend in exports to Asia:
7. What is your forecast for exports to Asia in 2016? ______ to ______

II. Europe

1. Is there a positive trend in exports to Europe? Y or N pvalue: ______
2. How powerful is your naïve model of trend in exports to Europe? RSquare: _____
3. Are residuals free from unaccounted for trend or cycles? Y or N DW: __________
4. How precise will your forecasts of exports to Europe be? Margin of error: ________
5. Is your model valid for forecasting? Y or N
6. Present your final equation for trend in exports to Europe:
7. What is your forecast for exports to Europe in 2016? ______ to ______

III. Plot your Fits and Forecasts

Plot your fits and forecasts of exports to Asia and Europe in a single scatterplot and paste in here:

Chapter 6
Consolidating Multiple Naïve Forecasts with Monte Carlo

The modeling team compared export market forecasts for the U.S. and Latin America, shown in Figure 6.1. Exports to the U.S. were much more consistent, stable and predictable. The trend accounted for 93% of the variation in U.S. exports, and the margin of error in forecasts was small, .24 ML. However, forecast growth was low, 4 to 5% per year. Exports to Latin America differed. While the trend accounted for 92% of the variation in Latin American exports, the margin of error was larger, .51ML. Yet future growth was higher, 5 to 9% per year.

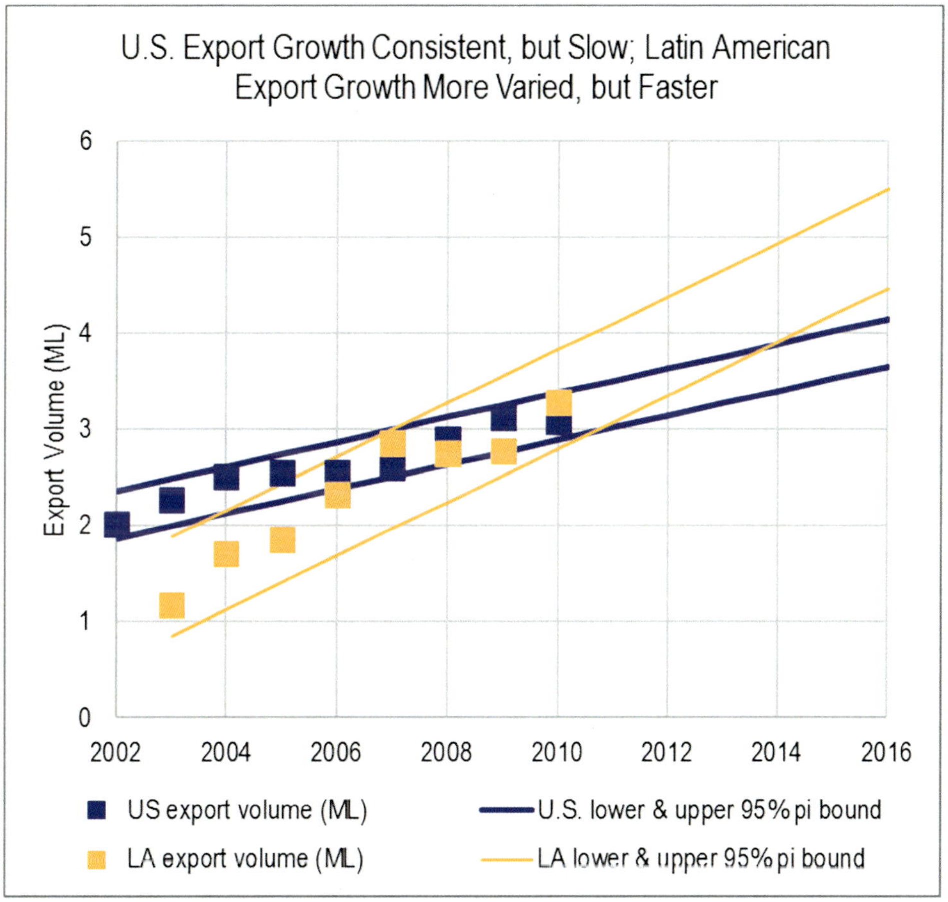

Figure 6.1 Fits and forecasts of export volumes to the U.S. and Latin America

The management had asked for comparison of "best" and "worst" scenarios. If exports in both New World markets slowed, how low might the 2016 consolidated forecast volume be? If growth in New World markets quickened, how high might the 2016 consolidated forecast volume be? The modelling team combined the two forecasts to produce a consolidated forecast of export volume in the New World:

C. Fraser, *Business Statistics for Competitive Advantage with Excel 2016*,
DOI 10.1007/978-3-319-32185-1_6

$$New\ World\ ex\hat{p}ort\ volume\ (ML)_t = \sum_i ex\hat{p}ort\ volume\ (ML)_{it}$$

These consolidated New World export volume forecasts are shown in Figure 6.2.

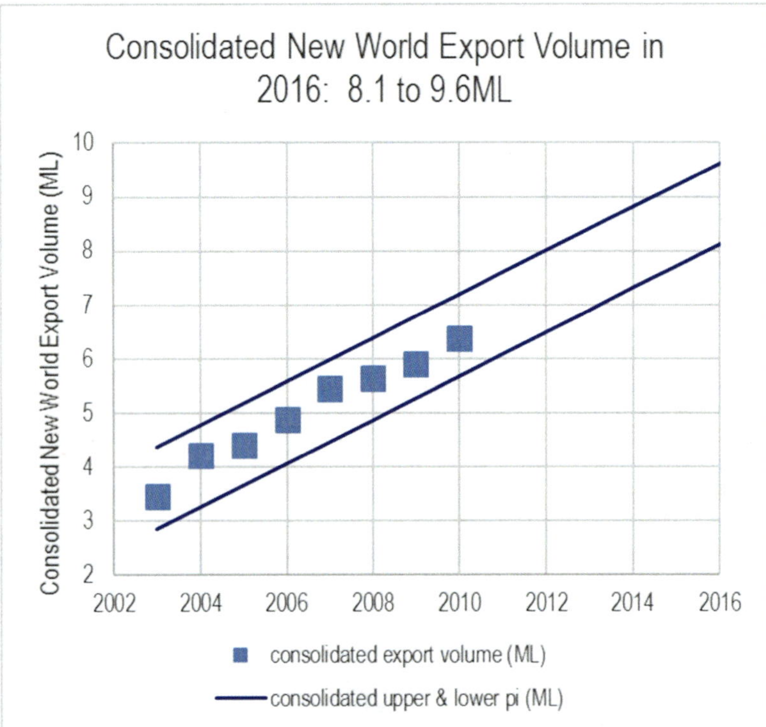

Figure 6.2 Consolidated New World export volume forecast

The consolidated forecasts aggregated the two global segments. If volumes were at the upper 95% confidence interval bound, consolidated export volume in the two major segments would experience annual growth as high as 7% per year.

If, on the other hand, volumes were at the lower 95% confidence interval bound, consolidated major export market revenues could fall below 2010 volume in 2011 and experience growth of just 4% per year.

The range of possibilities, from worst case to best case, was large. In 2016, for example, the worst case forecast was 8.1ML, and the best case forecast was 9.6ML, a range of 1.5ML. However, neither of these extreme outcomes was likely.

The chance that export volume would be at the lower 95% interval boundary in either of the two global regions is 2.5%. The joint probability that worst case outcomes for both volumes would occur is less than one tenth of one percent, .06% (= 2.5% × 2.5%). The best case outcome is equally unlikely, making the interval from worst to best a 99.875% prediction interval.

6.1 Use Monte Carlo to Integrate Multiple Uncertain Naïve Forecasts

In order to be useful to management, a 95% prediction interval was needed for consolidated volumes. Using expected 2016 volumes and the corresponding standard errors from the naïve models, the modeling team used Monte Carlo simulation to find the 95% prediction interval for consolidated New World export market volumes in 2016, shown in Figure 6.3.

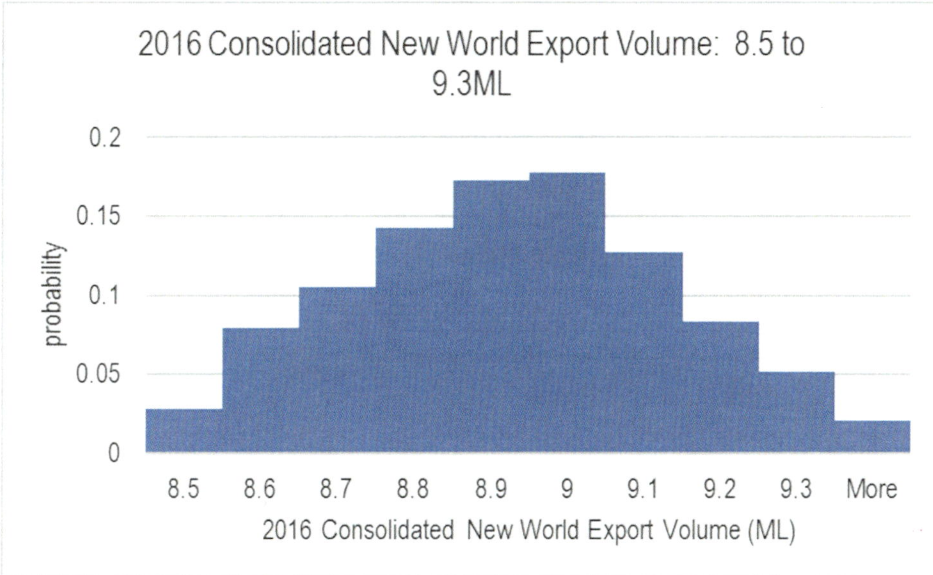

Figure 6.3 Simulated samples of 2015 export volumes

Based on the assumption that export volume trends would continue, export volume would continue to grow at about 5 to 6% annually, to at least 8.5ML by 2016.

Compared with the unlikely best and worst case forecasts, the 99.875% prediction interval of 1.5ML, management now had a much better picture of 2016 possibilities, the 95% prediction interval of .8ML.

6.2 Monte Carlo Offers Likely Possibilities from Consolidated Multiple Naïve Forecasts

It is not uncommon for management to have interest in multiple markets or multiple product lines as long term consolidated performance is forecast and assessed. Naïve models, estimating systematic trend and the amount of systematic variation in performance (through the standard error) are characterized by margins of error. Consolidating multiple forecasts produces an unlikely range of possible outcomes, complicating decision making. With Monte Carlo, a clearer picture of likely outcomes is possible.

Excel 6.1 Use Monte Carlo to Produce a 95% Prediction Interval of Consolidated Possibilities from Multiple Naïve Forecasts

Use the modelling team's naïve forecasts for U.S. and Latin American export volumes to find 95% prediction intervals and likely outcomes for consolidated Concha y Toro New World exports in 2016. **Concha y Toro 2016 New World forecasts** contains data and these regression analyses and forecasts.

From each of the recalibrated regressions, copy the predicted value for 2016 exports and the standard error, and then paste into the data sheet.

In E2,
M
In E3,
SD
In F1,
U.S.
In G1,
Latin America
Cntl+page up page up page up
From I39,
Cntl+C
Cntl+page down page down page down
In F2,
Alt HVSU
Cntl+page up page up page up
In B7,
Cntl+C
Cntl+page down page down page down
In F3,
Alt HVSU
Cntl+page up
In I38,
Cntl+C
Cntl+page down
In G2,
Alt HVSU
Cntl+page up
In B7,
Cntl+C
Cntl+page down
In G3,
Alt HVSU

	E	F	G
1		US	Latin America
2	M	3.89	4.98
3	SD	0.103	0.210

With the mean and standard deviation from naïve forecasts, request random samples of 1000 Normally distributed possible outcomes for exports to the U.S. and to Latin America in columns H and I.

In H1,
2016 U.S. exports (ML)
In I1,
2016 LA exports (ML)
Alt AY*n* R
1 tab 1000 tab N tab tab tab 3.89 tab .103 tab tab tab H2
Alt AY*n* R
1 tab 1000 tab N tab tab tab 4.98 tab .210 tab tab tab I2

	E	F	G	H	I
1		US	Latin Ameri	*2016 U.S. exports (ML)*	*2016 LA exports (ML)*
2	M	3.89	4.98	3.937764	4.78388
3	SD	0.103	0.210	3.964173	5.148152
4				3.887206	5.205358

Consolidate exports volumes in column J.

In J1,
Consolidated New World exports (ML)
In J2,
=h2+i2
In J2,
Double click lower right corner to fill column

J2 =H2+I2

	H	I	J	K	L
1	*2016 U.S. exports (ML)*	*2016 LA exports (ML)*	consolidated 2016 New World exports (ML)		
2	3.937764	4.78388	8.721644		
3	3.964173	5.148152	9.112325		
4	3.887206	5.205358	9.092564		

Find the 95% upper and lower prediction interval bounds.

In I1002,
Upper 95% prediction interval bound
In I1003,
Lower 95% prediction interval bound
In J1002,
=percentile(j2:j1001,.975)
In J1003,
=percentile(j2:j1001,.025)

J1003 =PERCENTILE(J2:J1001,0.025)

	H	I	J	K	L	M
1	*2016 U.S. exports (ML)*	*2016 LA exports (ML)*	consolidated 2016 New World exports (ML)			
1000	3.800514	4.839429454	8.639944			
1001	3.783758	4.412625918	8.196384			
1002		upper 95% prediction interval bound	9.353686			
1003		lower 95% prediction interval bound	8.378635			

Create consolidated export bins beginning at the 95% lower prediction interval bound, 8.4, and then adding .1 to each bin and ending with 9.4. Request a histogram, find probabilities for each bin, and plot probabilities by consolidated exports:

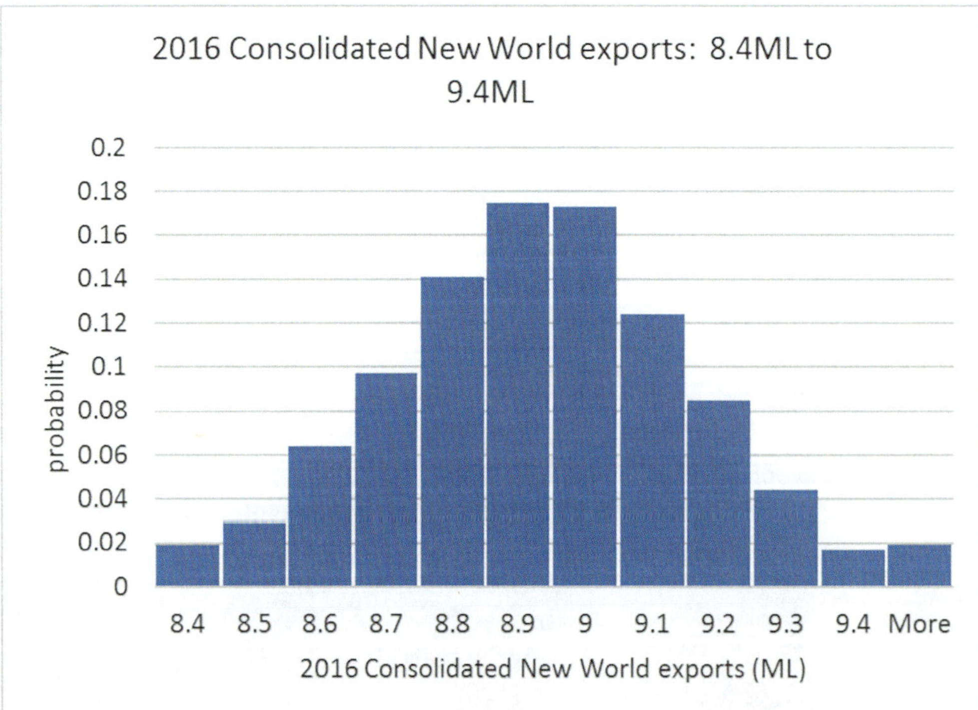

Lab 6 Forecast Concha y Toro Consolidated Exports to the New World

Management acknowledges that both forecasts for New World exports to the U.S. and Latin America are uncertain. Consolidate the 2016 forecasts and find the 95% prediction interval for the consolidated 2016 volume using Monte Carlo.

The naïve trend regressions and prediction intervals for the U.S. and Latin American are in **Concha y Toro New World forecasts.**

1. If annual growth in export volume is stable, what volume will Concha y Toro export to the U.S. in 2016?

   ___________ to _____________

2. What are the margin of error and standard error in this forecast? me:______ se:______

3. If annual growth in export volume is stable, what volume will Concha y Toro export to Latin America in 2016?

   ___________ to ______________

4. What are the margin of error and standard error in this forecast? me:______ se:______

5. What is the worst case for 2016 consolidated export markets in Asia and Latin America?

   ____

6. What is the probability that consolidated exports to New World markets will be as low as the worst case in 2016?

   __________ OR 1 in ______________

7. What is the best case for 2016 consolidated exports in Asia and Latin America?

   ____

8. What is the probability that consolidated exports to New World markets will be as high as the best case in 2016?

   __________ OR 1 in ______________

Use expected 2016 export volume and the standard error from the export volume regression to simulate a sample of 1000 hypothetical 2016 export volumes to the U.S. and to Latin America.

Use the two samples of hypothetical 2016 exports to the U.S. and Latin America to find a sample of hypothetical consolidated 2016 exports to New World segments.

9. What is the 95% prediction interval for consolidated 2016 export volume to New World segments?

   ___________ to ______________

10. Make a column chart to illustrate 2016 export possibilities, given the assumption of stable growth in export volume to the U.S. and Latin America.

Assignment 6 Forecast Concha y Toro Consolidated Exports Worldwide

Management acknowledges that both forecasts for each of the major segments, U.S., Latin America, Europe and Asia, are uncertain. Consolidate the 2016 forecasts and find the 95% prediction interval for the consolidated Worldwide 2016 volume using Monte Carlo.

The naïve trend regressions and prediction intervals for the four major export markets are in **Concha y Toro Worldwide forecasts.**

1. What is the worst case for 2016 consolidated exports Worldwide?

   ____

2. What is the probability that consolidated export Worldwide will be as low as the worst case in 2016?

   __________ OR 1 in ______________

3. What is the best case for 2016 consolidated exports Worldwide?

   ____

4. What is the probability that consolidated exports Worldwide will be as high as the best case in 2016?

   __________ OR 1 in ______________

Use expected 2016 export volume and the standard error from the export volume regression to simulate a sample of 1000 hypothetical 2015 sales volumes for exports in each of the four major global regions.

Use the four samples of hypothetical 2016 exports to the major global regions to find a sample of hypothetical consolidated 2016 exports Worldwide.

5. What is the 95% prediction interval for consolidated 2016 export volume Worldwide?

   ___________ to ______________

6. Make a column chart to illustrate 2016 export possibilities, given the assumption of stable growth in export volume to the four major export regions.

Case 6 Can Arcos Dorados Hold On?

Arcos Dorados ("golden arches," in Spanish), headquartered in Argentina, is the largest franchiser of McDonalds restaurants in the World. Arcos Dorados operates restaurants in Brazil, the Caribbean, North and Central Latin America (NOLAD), and South America (SLAD). (Arcos Dorados listed its shares on the New York Stock Exchange in April, 2011, making its CEO, Colombian Woods Staton, a billionaire.)

With increasing wealth in Latin America, more families are electing to dine out, and fast food revenues, as well as McDonalds' revenues, have been growing at a steady clip. Arcos Dorados claimed the largest share of the fast food business in Latin America in 2009, 12.4%. Though maintaining the largest share, Arcos Dorados' share slipped to 10.4% in 2010. Mr. Staton is concerned that the business may be losing ground.

Arcos Dorados contains annual revenues per restaurant and number of restaurants in the four global Latin business segments, Brazil, Caribbean, NOLAD, and SLAD, for years 2007 through 2011. Data from earlier years is not publicly available, since Arcos Dorados issued shares recently; therefore, "valid" models of revenues per restaurant and number of restaurants cannot be built. However, naïve models can be built, to forecast future performance based on historical trends.

1. In which global segments is the trend in revenue per restaurant increasing?

   ___ Brazil ___ Caribbean ___ NOLAD ___ SLAD

2. Illustrate your fits and forecasts for revenue per restaurant in the four global segments on a single graph and embed below:

3. In which global segments is the trend in number of restaurants increasing?

   ___ Brazil ___ Caribbean ___ NOLAD ___ SLAD

4. Illustrate your first and forecasts for revenue per restaurant in the four global segments in a single graph and embed below:

5. Forecast 2016 consolidated revenues, reporting the 95% prediction interval:

Fast Food Market and Growth in Latin America. The fast food industry in Latin America and the Caribbean was valued at $29.0B in 2010. The industry has been growing at an average annual rate of 9.6%, though industry experts have suggested that growth over the next six years could be as high as 11.4%, though there is only a 2.5% chance that growth would exceed 11.4%. Management agrees that the chance of growth less than 9.6% could occur with only a 2.5% chance. Average annual growth of 10.5% is expected over the next six years.

6. Determine the 95% prediction interval for Arcos Dorados' market share in 2016:

7. Illustrate the distribution of possible Arcos Dorados' market shares in 2016 with a column chart and embed, below:

8. What is the chance that Arcos Dorados' 2016 market share will exceed the 2010 level of 10.4%?

9. Some Arcos Dorados executives worry that 2016 market share could be as low as 7.7%, if revenues per restaurant and restaurant expansion slow, and if the fast food industry in Latin America and the Caribbean grows as fast as 11.4% each year. What is the probability that 2016 market share could be as low as 7.7%, based on management's assumptions?

Chapter 7
Presenting Statistical Analysis Results to Management

While it is important to be able to conduct the correct statistical analyses, it is equally important that results of analyses are packaged into easily understood PowerPoint presentations and Memos which translate those results to management and decision makers.

Results are often presented with a PowerPoint slide deck or Memo to management. In this chapter, guidelines for clear and compelling presentations are offered.

7.1 Use PowerPoints to Present Statistical Results for Competitive Advantage

PowerPoint presentations are a powerful tool that can greatly enhance your presentation of the results of your analysis. They are your powerful sidekick. Tonto to your Lone Ranger. PowerPoints help your audience member key points and statistics and make available graphics to illustrate and enhance the story you are telling.

The key to effective use of PowerPoints for presenting your results for competitive advantage is to be sure that they are not competing with you. PowerPoints with too much text draw audience attention away from you. Cliff Atkinson, in his 2008 book, beyond bullet points, (Microsoft Press) explains clearly how audience members process information during PowerPoint presentations and why you should move beyond bullet points in the design of your PowerPoints. Much of the material that follows reflects Mr. Atkinson's wisdom, and his book is a recommended investment.

Audience Brains Are Designed to Process and Remember Information

Our brains are ingeniously designed to filter and process large amounts of information, selecting the most relevant to be stored in long term memory. Only a small portion of incoming information gains admission into working memory, and only some portion of information processing in working memory survives and is stored in long term memory, as shown in Figure 7.1.

C. Fraser, *Business Statistics for Competitive Advantage with Excel 2016*,
DOI 10.1007/978-3-319-32185-1_7

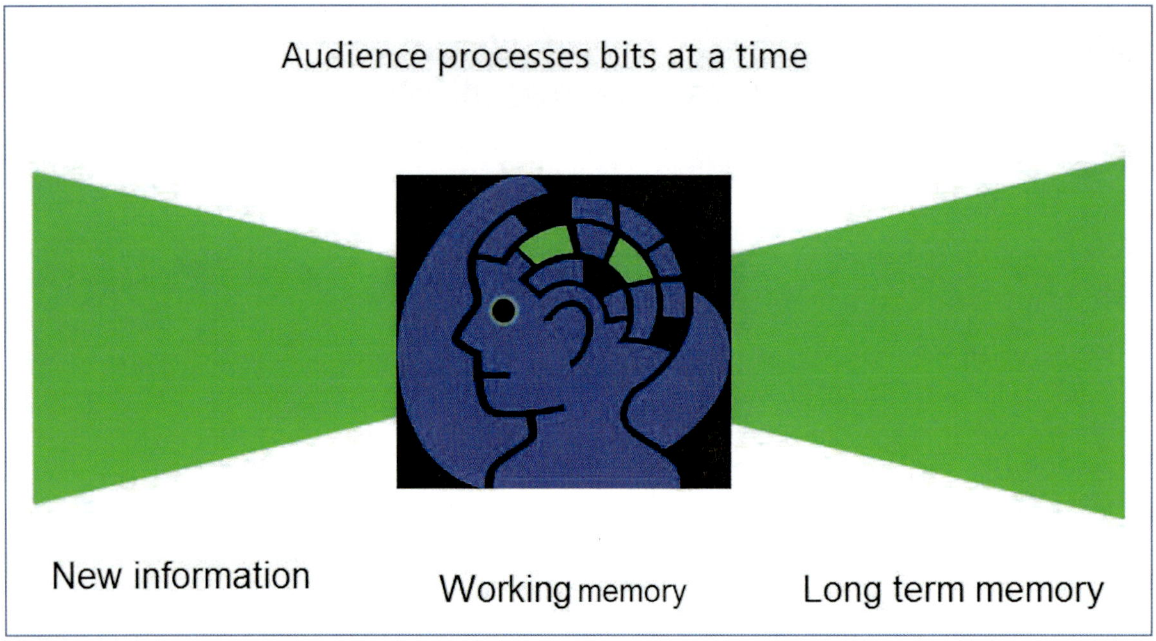

Figure 7.1 Information processing in a given moment

The goal is to help your audience filter information, direct their attention to your key results and interpretation, so that your message will be remembered.

Limit Text to a Single Complete Sentence per Slide

Since brains process only a few select bits of information in a given moment, increase the chance that the key points in your presentation become those select elements, illustrated in Figure 7.2. If your slides are loaded with text, the critical point has a small chance of being processed in working memory.

Figure 7.2 Limited processing in working memory

Work to design your slides so that each slide presents a single idea. Use only one complete sentence per slide, as illustrated in Figure 7.3.

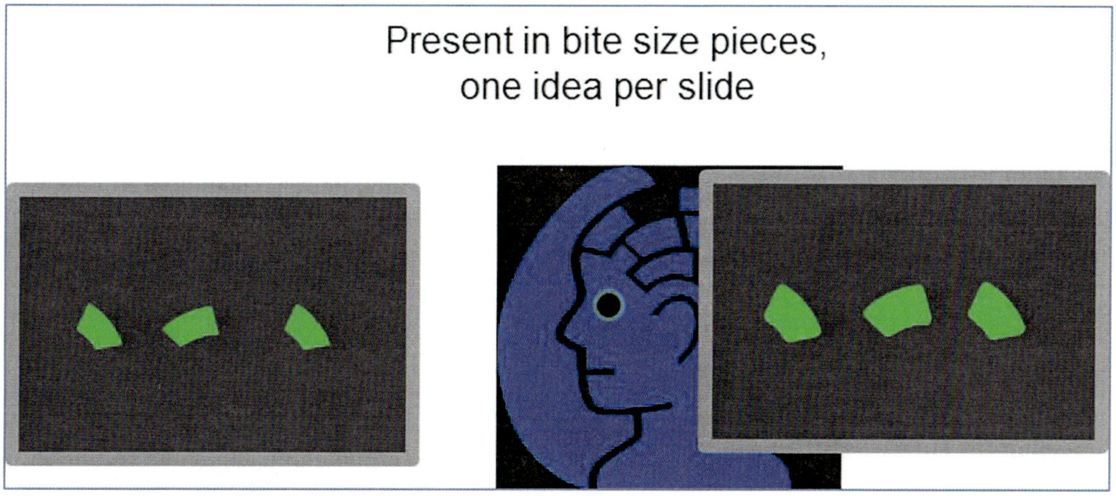

Figure 7.3 Present one idea at a time

Pause to Avoid Competing with Your Slides

Brains process a single channel at a time, as shown in Figure 7.4. Attention is directed toward either visuals or audio in any given moment.

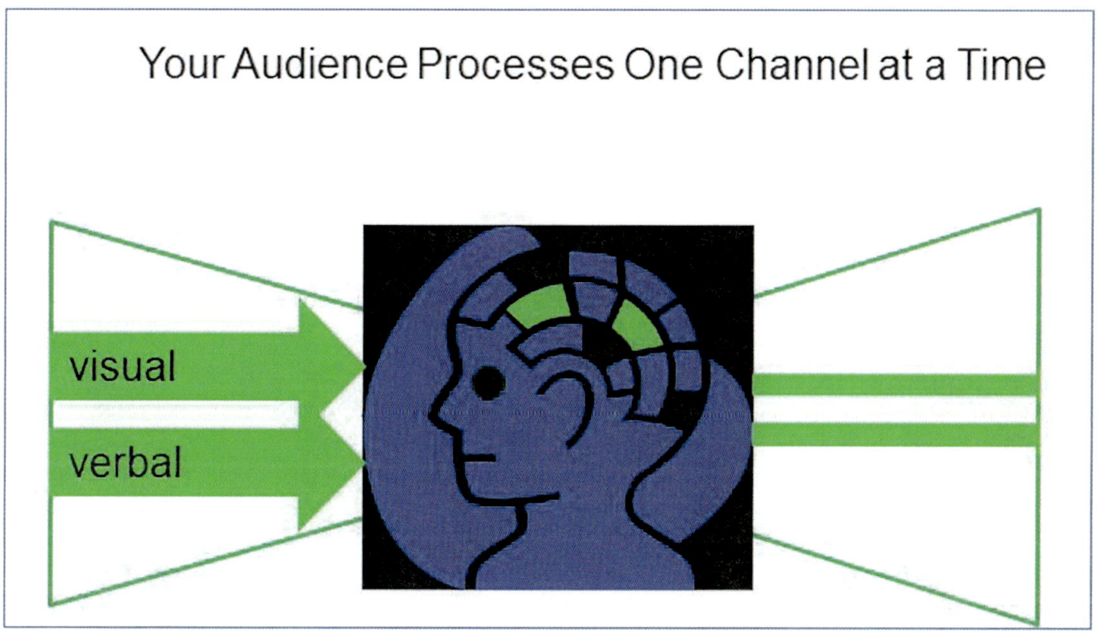

Figure 7.4 Processing of a single channel in a given moment

Your PowerPoints should complement the story that you are delivering. Your PowerPoints should not compete with you for attention. You are the star in the focus of attention. Your PowerPoints should play a supporting role. Pause to allow time for the audience to process that single idea, and then elaborate and explain. This will avoid competition between your slides and you.

Illustrate Results with Graphs Instead of Tables

Tables are effective elements in reports which convey a lot of information for readers to refer to and ponder. Tables are not processed in seconds, which is the time available to process each of your PowerPoint slides. Figure 7.5 illustrates information overload that tables create.

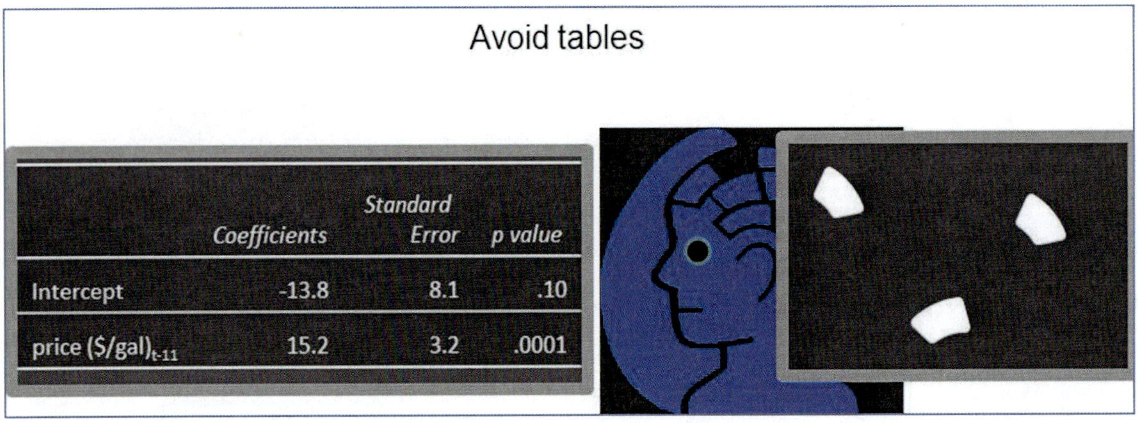

Figure 7.5 Information overload from tables

Synthesize the results in your tables into graphs. Graphs organize your results and illustrate key take aways. Well designed graphs can be processed in seconds, allowing audience attention to flow from a slide back to you, the speaker and, ideally, the focus of attention, as shown in Figure 7.6.

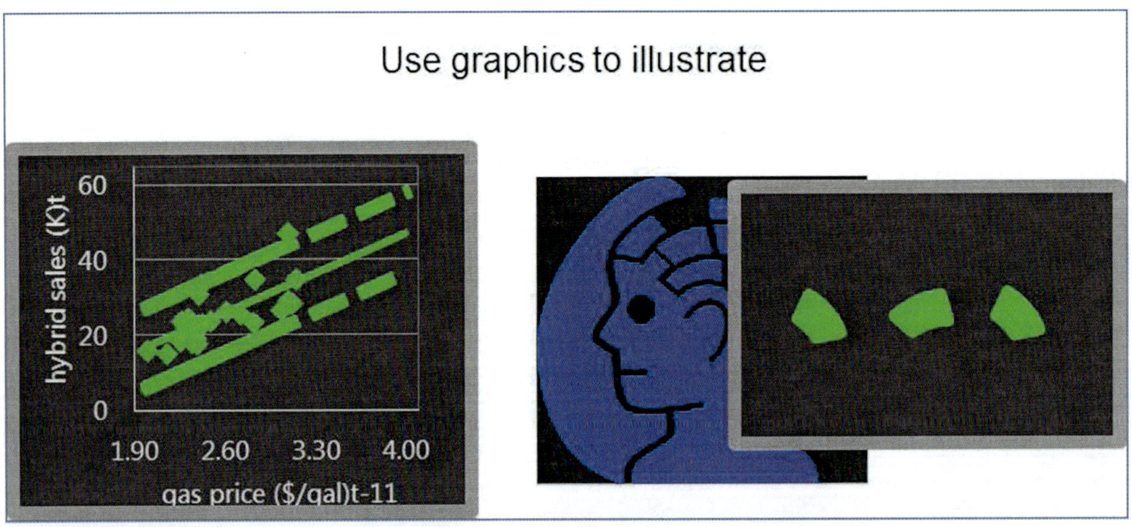

Figure 7.6 Effective presentation of results with graphs

An effective slide contains a single complete sentence, the *headline,* and a graph to illustrate.

Start PowerPoint Design in Slide Sorter

Insure that your PowerPoints are organized effectively. Build your deck by beginning in Slide Sorter view, shown in Figure 7.7. Choose the main points that you want the audience to remember.

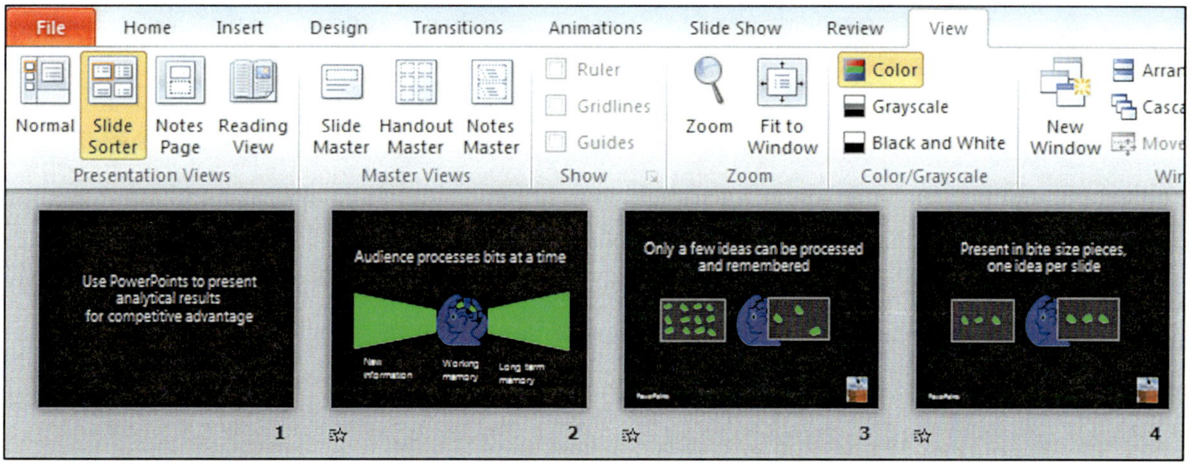

Figure 7.7 Slide sorter view

Next, add slides with supporting information the main point slides, illustrated in Figure 7.8.

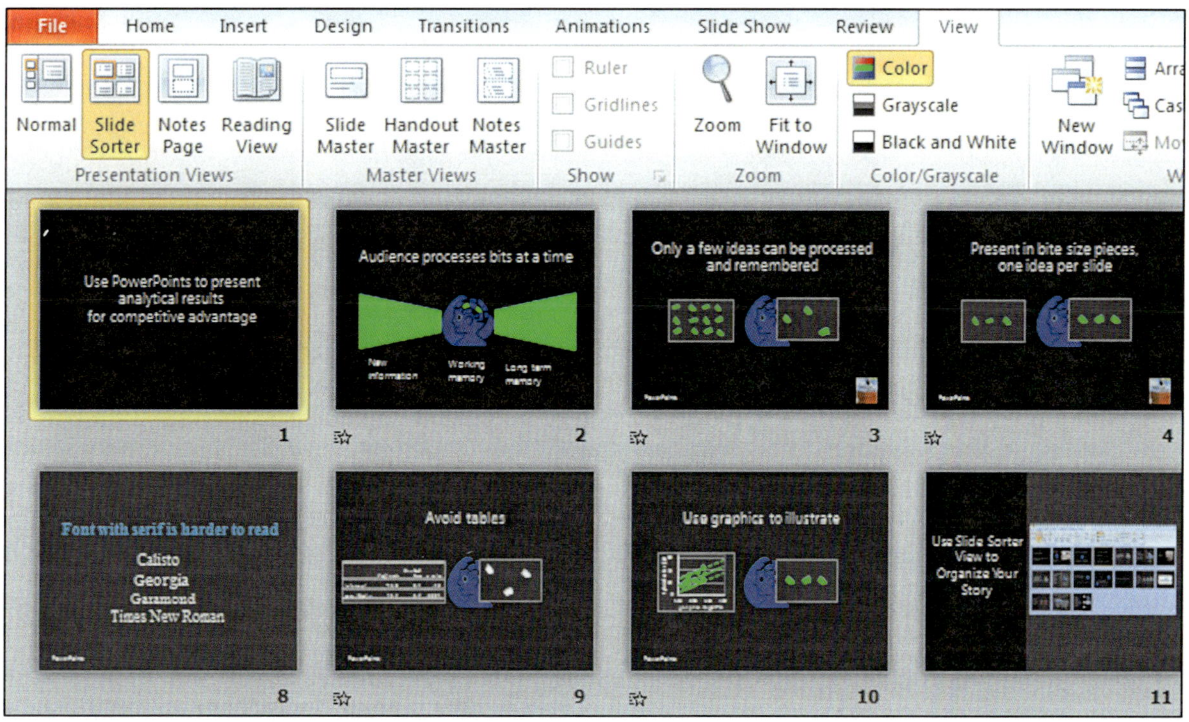

Figure 7.8 Slide sorter view of main and supporting ideas

Put Supporting Text in Slide Notes

Presenters sometimes worry that they will forget the story. For insurance, they include all of the text to be delivered in their slides. You can guess the consequence. Audience members attempt to read and process all of the text in the slides. To do this, they must ignore the presenter. In the few seconds that a slide appears, there is much too little time to read and process all of the text. As a result, the audience processes only remnants of the story. Audience members are frustrated, because at the end of the presentation, they have incomplete information that doesn't make sense.

In addition to supporting your presentation, your PowerPoints deliver an impression. Slides filled with text deliver the impression that the presenter lacks confidence. When audience members fail to process slides laden with text and tables, the natural conclusion is that the speaker is ineffective. "She spoke for 15 minutes, but I can't remember what she said. Made no sense."

Audience members can reach a second, unfortunate conclusion in cases where a presenter has simply converted report pages into slides. Slides converted from reports are crammed with text and tables, and too often look like report pages, with white backgrounds, black text. This sort of unimaginative PowerPoint deck delivers the impression that the speaker is lazy.

In contrast, slides with a single, complete sentence headline and graph deliver the impression that the presenter is confident. After easily processing the slides and then focusing on explanation and elaboration delivered by the presenter, audience members understand and remember the story.

If you present a single idea in each slide, you will remember what you want to say to explain the idea and add elaboration. The audience will focus on your presentation, since you will provide the missing links.

You can have the best of both worlds. You can include your explanation and elaboration of the main points in the slide Notes. The Notes are not seen during your presentation, but they are available later. Provide handouts at the end of your presentation from the Notes view of your slides, as shown in Figure 7.9.

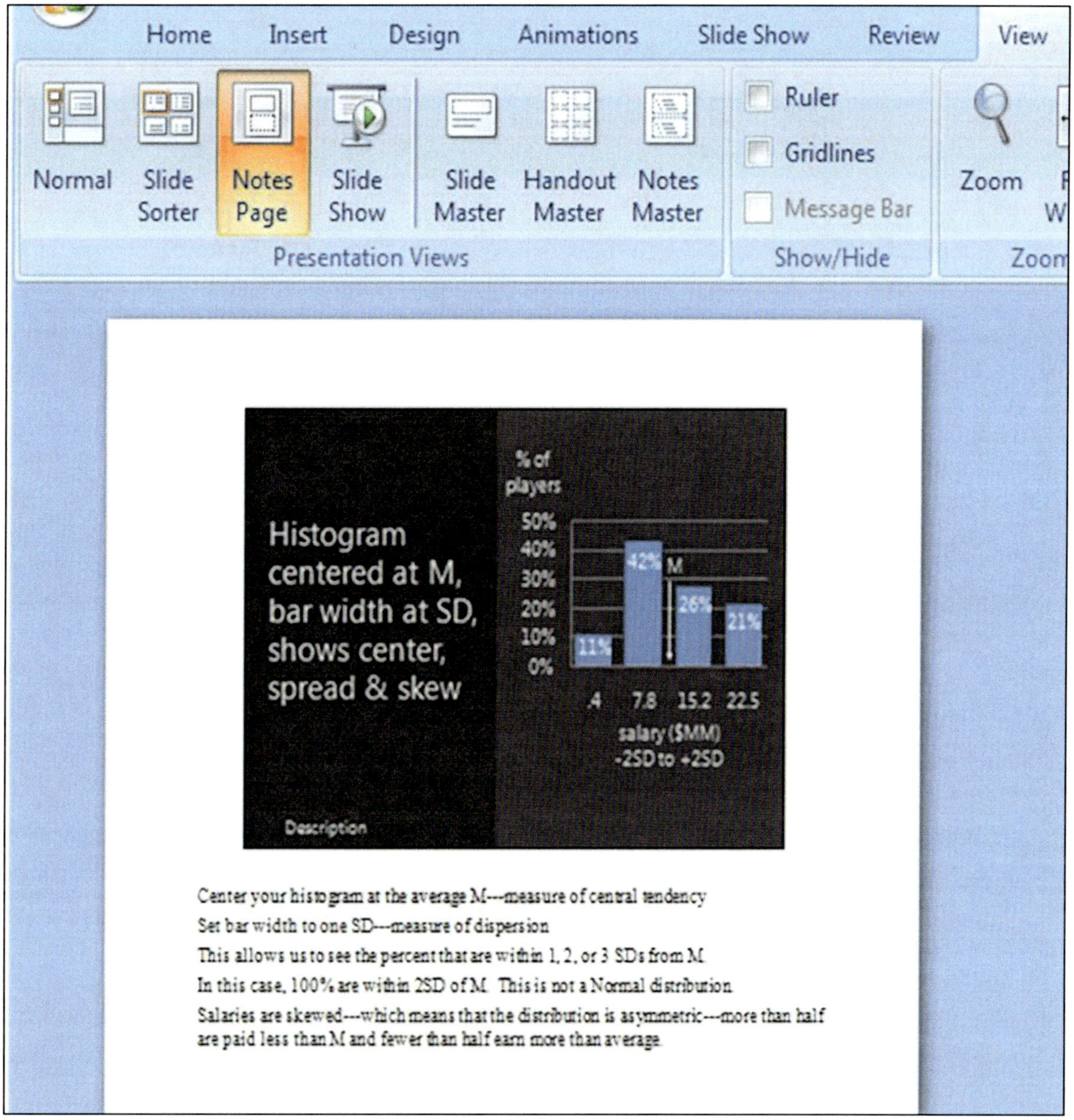

Figure 7.9 Notes view

Choose a Slide Design That Reduces Distraction

Design your slides so that elements are minimally distracting. You want the audience to be able to quickly and easily see the idea in each slide and then focus on you for explanation and elaboration.

Use a Font That Can Be Easily Read

Use *at least 24 pt* font so that audience members can easily read your headline, numbers, and labels. If you include numbers in your graphs, in the axes or as data labels, they must be easily read. (Be sure to round your numbers to two or three significant digits, also.) Axes labels, and other text must also be easily read. Any font smaller than 24 pt will challenge easy reading.

Choose a *sans serif* font (Ariel, Lucida, or Garamond). *Sans serif* fonts, without "feet" are easier to read in PowerPoints. *San serif* characters, without extra lines, are clearer in slides. (The opposite is true for reports, where the *serif* enhances reading ease.) If you have any doubts about readability, test your slides in a room similar in size and shape to the presentation location.

Choose Complementary Colors and Limit the Number

In cases where the slides will be presented in a darkened room, the background should be darker than the title and key words. Choose a medium or darker background, with complementary, contrasting, lighter text color. PowerPoints in this setting are more like television, movies, and internet media, and less like books or reports, and should feature darker backgrounds like those you see in movie credits.

When presentations are in well lit rooms, backgrounds can be lighter than title and key words. In a light setting, PowerPoints resemble text pages, with lighter backgrounds and darker text colors.

If we see more than 5 colors on a slide (including text), our brains overload and we have difficulty processing the message and remembering it. Limit the number of distinct colors in each slide.

7.2 Write Memos that Encourage Your Audience to Read and Use Results

Memos are the standard for communication in business. They are short and concise, which encourages the intended audience to read them right away. Memos which present statistical analysis to decision makers

- feature the bottom line in the subject line,
- quantify how the bottom line result influences decisions,
- are ideally confined to one single spaced page,
- include an attractive, embedded graphic which illustrates the key result.

Many novice analysts copy and paste pages of output. The output is for consumption by analysts, whose job it is to condense and translate output into general business language for decision makers. Decision makers need to be able to easily find the bottom line results without referring to a statistics textbook to interpret results. It is our job to explain in easily understood language how the bottom line result influences decisions. For the quantitative members of the audience, key statistics are included.

On the following page is an example of a memo which might have been written by the quantitative analysis team at Procter & Gamble to present a key result of a concept test of Pampers Preemies to brand management.

Notice that

- the subject line contains the bottom line result,
- results are illustrated,
- results are described in general business English.

Description of the concept test and results are condensed and translated. Brand management learns from reading the memo what was done, who was involved, what results were, and what implications are for decision making.

MEMO

Re: Worldwide exports forecast to grow modestly through 2016
To: Concha y Toro Management
From: Concha y Toro Quantitative Analysis Team
Date: July 2011

Summary

Consolidation of naïve models of trend in exports to the U.S., Latin American, European and Asian global regions through Monte Carlo simulation suggest model growth in export volumes through 2016.

Sample and method

The Consolidated Simulated Samples. From 2016 naïve model forecasts of export volumes to each of the four global regions, a simulated sample of likely export volumes was drawn. Those four samples were consolidated to identify likely Worldwide export volumes in 2016.

Results

Worldwide exports in 2016: 26.5 to 28.2ML

Given the assumption that exports in each of the four major global export markets will continue to exhibit stable, linear growth, 2016 Worldwide export volume is forecast to fall between 26.5ML and 28.2ML. This forecasts suggests annual growth in Worldwide export volume between 7 and 8 percent. The forecast has a margin of error of .85ML.

Europe and Asia will contribute slightly larger percentages to total exports, with Europe dominating 2016 export volumes (shown in Exhibit 1). U.S. exports will continue to be an important region, though the U.S. percentage of total worldwide exports will decline over the next six years. Global region shares of worldwide export volumes are shown in Exhibit 2.

Conclusions

Modest, stable growth forecast

Growth worldwide exports over the next six years is forecast to slow to 7 to 8 percent, down from annual growth of 11 percent the past six years. Europe and Asian markets are increasingly more important, and exports to the U.S. are forecast to grow less.

Additional considerations

Latin American exports increasing, but at a declining rate

In the Latin American market, exports have slowed in recent years. There, exports are growing at a decreasing rate, and a nonlinear naïve trend model would be more appropriate. Consequently, the forecast Latin American export contribution to consolidated Worldwide exports may overestimate the 2016 contribution and, consequently, the 2016 consolidated total.

Exhibit 1. Forecast 2016 Exports by Global Region

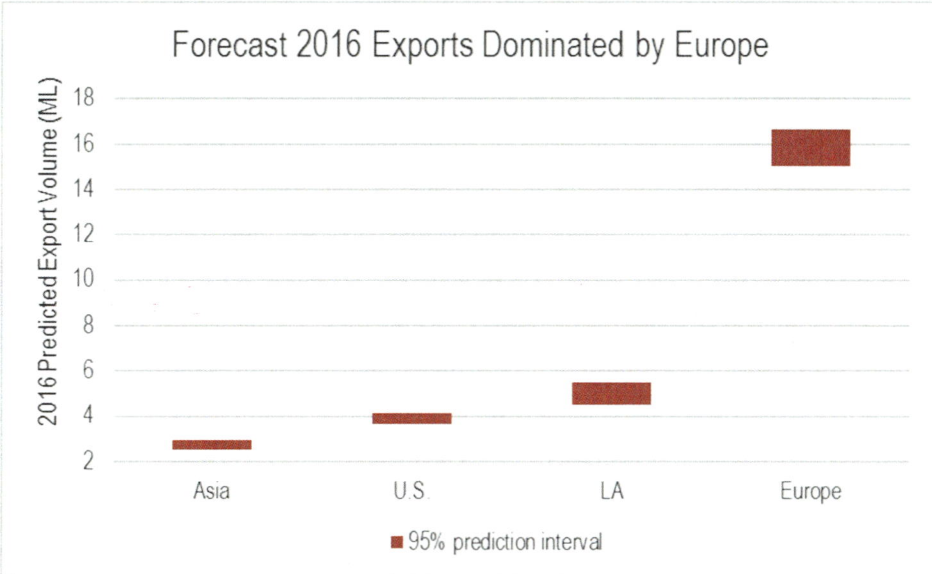

Exhibit 2. Worldwide Export Volume Shares by Global Region

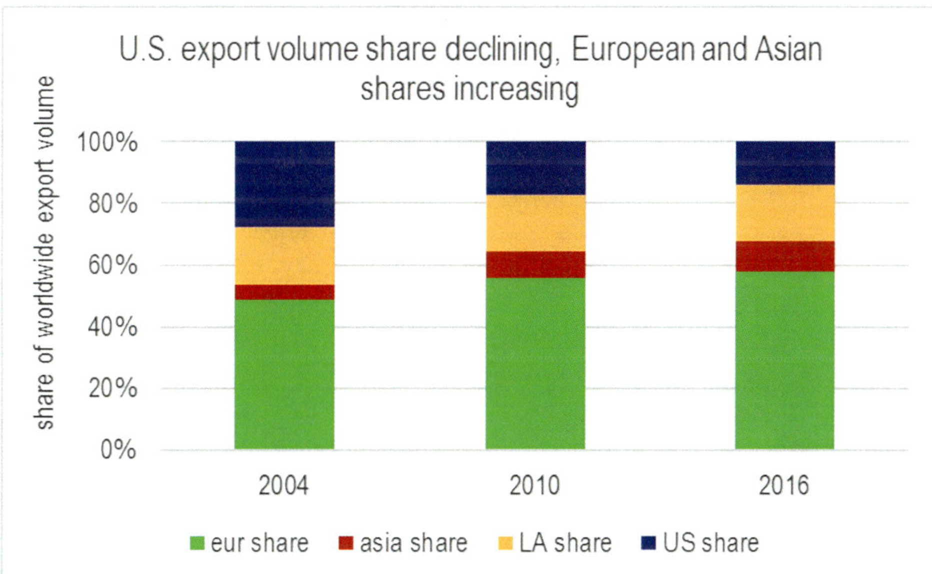

CASE 7 Segmentation of the Market for Preemie Diapers

Deb Henretta is about to commit substantial resources to launch *Pampers Preemies*. The following article from the *Wall Street Journal* describes Procter & Gamble's involvement in the preemie diaper market:

New York, N.Y.

P&G Targets the 'Very Pre-Term' Market

Wall Street Journal

THE TARGET MARKET for Procter & Gamble Co.'s newest diaper is small. Very small.

Of the nearly half a million infants born prematurely in the U.S. each year, roughly one in eight are deemed "very pre-term," and usually weigh between 500 grams and 1,500 grams (one to three pounds). Their skin is tissue-paper-thin, so any sharp edge or sticky surface can damage it, increasing the chance of infection. Their muscles are weak, and unlike full-term newborns, excessive handling can add more stress that in turn could endanger their health.

Tiny as they are, the number of premature infants is increasing—partly because of improved neonatal care: From 1985 to 2000, infant mortality rates for premature babies fell 45%, says the National Center for Health Statistics. Increasingly, such babies are being born to older or more affluent women, often users of fertility drugs, which have stimulated multiple births.

It's a testament to the competitiveness of the $19 billion global diaper market that a behemoth like Procter & Gamble, a $40 billion consumer-products company, now is focusing on a niche that brought in slightly more than $1 million last year; just 1.6% of all births are very pre-term. But P&G sees birth as a "change point," at which consumers are more likely to try new brands and products. Introducing the brand in hospitals at an important time for parents could bring more Pampers customers, the company reasons.

P&G's Pampers, which is gaining ground on rival Kimberly-Clark, but still trails its Huggies brand, has made diapers for premature infants for years. (P&G introduced its first diaper for "pre-emies" in 1973; Kimberly-Clark in 1988), but neither group had come up with anything that worked well for the very smallest of these preemies.

The company that currently dominates the very-premature market is Children's Medical Ventures, Norwell, Mass., which typically sells about four million diapers a year for about 27 cents each. The unit of Respironics Inc., Murrysville, Pa., has been making its "WeePee" product for more than a decade. But the company, which also makes incubator covers, feeding tubes and extra small bathtubs for preemies, hadn't developed certain features common in mass-market diapers, such as softer fabric coverings.

By contrast, P&G's preemie diapers, which it started distributing to hospitals in August, sell for about 36 cents each; about four cents more than P&G's conventional diapers. P&G's "Preemie Swaddler" fits in the palm of an adult's hand and has no adhesives or hard corners. It closes with mild velcro-like strips and is made of breathable fabric, not plastic. It has an extra layer of fabric close to the infant's skin to avoid irritation.

Children's Medical Ventures is coming out with another size of the WeePee, and plans to introduce velcro-like closures, a development the company says was in the works before P&G came out with a rival diaper. The new diapers won't cost any more, Children's Medical Ventures says.

P&G says the new diaper is the natural extension of its Baby Stages initiative, which took effect in February 2002 when P&G revamped its Pampers brand in the U.S. to cater to various stages of a baby's development. Working with very small preemies helps the company better understand infant development and become "more attuned to new products they might need," says Deb Henretta, president of P&G's global baby-care division.

But the marketing director for Children's Medical Ventures believes the increasing affluence of preemie parents is a greater inducement for big companies to enter the market. In the past, the typical mother of a preemie was poorer, often a teenager, but today more preemie "parents tend to be older, well-educated, and have money for things like fertility treatments," says Cathy Bush, marketing director for Children's Medical Ventures.

> The competition may raise the bar for the quality of diapers for these smallest of preemies. P&G says the parents of premature babies are demanding better products. "They have much higher expectations than they did years ago," Ms. Henretta says.
>
> Neonatal nurses have all sorts of opinions about the relative merits of Preemie Swaddlers and WeePees. Pat Hiniker, a nurse at the Carilion Roanoke Community Hospital in Virginia, says the new Pampers diaper, while absorbent, is too bulky for small infants. Allison Brooks of Alta Bates Hospital in Berkeley, Calif., says P&G's better absorbency made the babies less fidgety when they needed to be changed. "That sounds small, but you don't want them wasting their energy on squirming around," she says. "They need all their energy to grow."
>
> In any case, if health professionals have their way, the very-premature market will shrink, or at least stop growing. The March of Dimes recently launched a $75 million ad campaign aimed at stemming the rise of premature births. P&G is donating 50,000 diapers to the nonprofit organization.
>
>

Before resources are dedicated, Deb wants to confirm that preemie parents are attracted to the *Pampers Preemies* concept of superior comfort and fit. She commissioned a concept test to assess consumers' intentions to try the product. There is evidence that preferences and motivations of preemie mothers may differ and differences may be linked to lifestyle or demographics.

If the concept is attractive to at least one segment, would commercialization produce revenues sufficient to justify the investment? Deb requires a forecast of future revenues in order to make sound decisions.

The Market for Preemie Diapers

The market for preemie diapers is unusual in that the first diapers that a preemie baby wears are chosen by the hospital. Procter & Gamble is banking on positive experiences with *Pampers Preemies* in the hospital and consumer brand loyalty once baby goes home. If parents see *Pampers Preemies* in the hospital, are satisfied with their performance, and find them widely available at the right price, parents may adopt the *Pampers* brand after their infant comes home. Satisfaction and brand loyalty to *Pampers* could then lead to choice of other Pampers products as their baby grows.

Preemie Parent Segments

Based on focus group interviews and market research, Deb's team has learned that there are five broad segments of preterm parents:

- Younger (15 to 19), unemployed mothers who live with their parents. These young mothers are inexperienced and their pregnancies are unplanned. They tend to differ widely in their attitudes and preferences, and so a further breakdown is necessary:
    - ***Younger, Single, Limited Means***. The means of these young mothers are limited, and they are highly responsive to low prices and price promotions. Due to lack of knowledge about prenatal care and lack of access to healthcare, the preterm birth rate is relatively high in this segment, and minorities make up a disproportionately large proportion of this segment.

- ***Younger, Single, with Means***. These young mothers have their parents' resources at their disposal and want the best diapers. They are inexperienced consumers and could be attracted by a premium diaper. Brand name appears to be very important to these young women, and they believe that better mothers rely on name brands seen on television. This segment has access to healthcare and the majority of mothers in this segment are ethnically white.

- ***Young*** (20 to 35) mothers tend to be married and have adequate resources. Their pregnancies tend to be planned and this segment is virtually indistinguishable from the larger segment of disposable diaper users for full-term babies. This group has the fewest preterm births.
- ***Later in Life Moms*** (35 to 39) and ***Latest*** (40+) mothers tend to be wealthier, more highly educated professionals with higher incomes. A large proportion has no other children and has undergone fertility treatment. Multiple preemie births are more likely in this segment. Some of these mothers are single parents. This group is particularly concerned about functional diaper features and wants the best diaper their dollars can buy. They are willing to pay for a premium diaper perceived as the highest quality, offering superior fit and comfort. This segment is predominantly ethnically white.

The Concept Test

A market research agency has conducted a concept test of *Pampers Preemies* to gauge interest among consumers in a variety of potential target markets. 180 mothers with preemies who had been born at four local hospitals were asked to fill out a survey about purchase intensions after trying the product on their babies. If that data supports the launch, Deb will need to know which types of mothers and families to feature in the ads.

Data from the concept test is contained in **Case 7 Pampers Concept test**. Below is an overview of the questions asked in the survey, the manner in which they were coded, and the variable names contained in the dataset (which are in italics).

Trial Likelihood

Participants were asked, "How likely would you be to try Pampers Preemies if they were available in the store where you normally buy diapers and were sold at a price of $X.XX per diaper?"
The interviewer flipped a coin and inserted the "premium" price of $0.36, for heads, and the "value" price of $0.27, for tails.
Responses were coded as follows:

Definitely Would Not Try	= .05
Probably Would Not Try	= .25
Maybe Would Try	= .5
Probably Would Try	= .75
Definitely Would Try	= .95

Demographic Information

Consumers were asked to report their ethnicity (*ethnicity*).

Data Recoding

A new variable, *likely trier,* was created from the intention to try question. "Likely triers" were identified using a "Top two box rule" (i.e., those who indicated that they "Probably" or "Definitely" would try the product). Therefore, for *intent* ≥ .75, *likely trier* = 1; otherwise *likely trier* = 0.

Information Needed

Deb's team needs an estimate of revenue potential, plus additional information on target segments.

I. Revenue Potential

Deb's team has devised a method to estimate potential revenues, based on demographics. Their logic is explained below.

The number of very preterm births in a year is the product of number of births and the chance that a newborn will be very preterm, the very preterm birthrate. Advances in infertility treatments have led to more births by *older, predominantly white*, high risk mothers. Immigration has led to more births by the *youngest, predominantly Hispanic* mothers, many with little information about prenatal care and lack of access to adequate healthcare. The *very preterm* percentage of births is expected to increase in future years. The number of very preterm babies is uncertain and thought to vary by *ethnicity*.

Preterm Diaper Market. *Very preterm diaper sales volume* is the product of the average number of days a very preterm baby remains very preterm, approximately 30, the average number of diapers used per day, approximately 9, and the sum of *very preterm babies* in each segment *i:*

$$\begin{aligned} \textit{Very preterm diaper sales volume}_t &= \textit{30 days per very preterm baby} \\ &\times \textit{9 diapers per very preterm baby per day} \\ &\times \textstyle\sum_i very\ perterm\ babies_{i,t} \end{aligned} \tag{7.1}$$

Procter & Gamble's Preemie Business. From past experience, Procter & Gamble managers have learned that 75% of the proportion of *Likely Triers,* the *trial rate*, become loyal customers in the first year. Managers expect the *trial rate* to depend on price *j, premium* or *value,* and on appeal at price *j* to each ethnic segment *i,* which is uncertain:

$$trial\ rate_{ij} = .75 \times likely\ triers_{ij} \tag{7.2}$$

P&G's *share of diaper purchases* in a given year *t* at each price *j*, would then be the product of *trial rate* for an ethnic segment *i* at price *j* and the number of *very preterm babies* in that ethnic segment *i* in year *t:*

$$P\&G\ share\ of\ diaper\ purchases_{jt} = \sum_i trial\ rate_{ij} \times very\ preterm\ babies_{it} / \sum_i very\ perterm\ babies_{i,t} \tag{7.3}$$

Procter & Gamble consolidated *very preterm diaper sales volume* would depend on *very preterm diaper sales volume* and *P&G share of diaper purchases* at price *j*.

$$P\&G\ very\ preterm\ diaper\ sales\ volume_{j,t} = P\&G\ share\ of\ diaper\ purchases_{jt} \times very\ preterm\ diaper\ sales\ volume_t \tag{7.4}$$

Consolidated *revenue($)* would then be the product of *price j*, and *P&G very preterm diaper sales volume* at *price j:*

$$Revenue(\$)_{j,t} = price(\$)_j\ x\ P\&G\ very\ preterm\ diaper\ sales\ volume_{j,t} \tag{7.5}$$

To effectively evaluate revenue potential, managers believe the uncertainties in the number of *very preterm births* by ethnic segment and *trial rates* by ethnic segments at the alternate prices must be incorporated in analyses.

Case 7 Very Preterm Births contains time series of very preterm births for the three largest ethnic segments.

95% prediction interval for 2016 very preterm births by ethnicity. Build a naïve model of *very preterm births* to forecast expected *very preterm births* in 2016 and to find the standard error for each of the three largest ethnic segments.

1. Illustrate the trends in possible *very preterm births* in 2016 by ethnicity, showing actual births in 2001–2008 and predicted births in 2009–2016 in a stacked area chart.

Use Monte Carlo simulation to find a distribution of possible *very preterm births* in 2016 for each of the three largest ethnic segments, and then find possible 2016 *very preterm diaper sales* (7.1).

2. State the assumptions upon which your forecast of 2016 very preterm diaper sales rely.

3. Illustrate possible 2016 *very preterm diaper sales* with a histogram.

From the concept test sample, use a PivotTable to find the proportions of *Likely Triers* at each of the two prices for each of the three largest ethnic segments.

From the proportions of *Likely Triers,* find the *trial rates (7.2),* and the *conservative confidence interval* for *trial rate* at each of the two alternate prices, *premium* and *value,* for each of the three major ethnic segments. (Set any lower confidence interval bound that is below zero to zero.)

4. Illustrate the *conservative 95% confidence intervals* for *trial rate* for each of the largest ethinic segments at the two alternative prices, side by side, with a column chart. (Make four columns, *price (premium or value), ethnicity (black, Hispanic or white), upper* and *lower* 95% conservative confidence interval bounds. Request a column chart, format data series, setting series overlap to 100%, and changing shape fill of the *lower* to white (or the color of your slide background.)

Use Monte Carlo simulation to find a distributions of possible *trial rates* for the two alternate prices, *premium* and *value* within each of the three major ethnic segments. Use the *conservative standard error* for the standard deviation in each of your simulated samples.

From the simulated samples of possible *trail rates* for *premium* and *value* prices in each of the three major ethnic segments, find the distribution of possible 2016 *P&G shares of diaper purchases* given each of the alternative prices, using (7.3) in your spreadsheet.

5. State the assumptions upon which your forecasts of 2016 *P&G shares of preemie diaper purchases* rely.

6. Illustrate the distribution of possible 2016 *P&G shares of diaper purchases* at the *premium* price, and the distribution of possible 2016 *P&G shares of diaper purchases* at the *value* price, one on top, one on bottom, for easy comparison. Use the same share bins for each.

P&G Revenues at alternate prices. From the sample of possible consolidated *very preterm diaper sales volume* in 2016 (7.1) and the distribution of possible *P&G shares of diaper purchases* (7.3) for the two alternative prices, find the distribution of possible *P&G very preterm diaper sales volumes* in 2016 for *premium* and *value* prices, using (7.4) in your spreadsheet.

From the distributions of possible *P&G preterm diaper sales volumes* in 2016 (7.4) at the two alternative prices, find the distribution of possible consolidated *revenues* in 2016 for *premium* and *value* prices, using (7.5) in your spreadsheet.

7. Illustrate possible 2016 *revenues*, showing the two distributions (at *premium* and at *value* prices), one on top, one on bottom, for easy comparison. Use the same bins for each.

Team Assignment

I. PowerPoint Presentation to Management. Each Team is responsible for the presentation to management of market potential in the three largest ethnic segments at alternate prices, potential P&G *share* and *revenue* forecasts.

To facilitate your presentation, construct ten PowerPoint slides that illustrate your key results, using the guidelines from class.

- Slide 1 introducing your team
- Slides 2 through 8 presenting your assumptions and your results, 1. to 7., above.
- One slide with your conclusions and recommendations

Use graphs, rather than tables. Round to no more than three significant digits. Use fonts no smaller than 24 pt, including text in graphs. Label and adjust axes, including appropriate units. Use a Stand Alone Sense Slide title OR Stand Alone Sense Chart Title.

II. Memo to Management. Each Team is also responsible for creating a single page, single-spaced memo, using 12 pt font, presenting your analysis to P&G management.

Include one embedded figure which illustrates a key result. Include additional pages with exhibits containing graphics *which are referred to* in your memo. Exhibits should contain only graphs, and *only graphs that are referred to in the memo.* Round to no more than three significant digits.

Chapter 8
Finance Application: Portfolio Analysis with a Market Index as a Leading Indicator in Simple Linear Regression

Simple linear regression of stock rates of return with a Market index provides an estimate of *beta*, a measure of risk, which is central to finance investment theory.

Investors are interested in both the mean and the variability in stock price growth rates. Preferred stocks have higher expected growth—expected *rates of return*—shown by larger percentage price increases over time. Preferred stocks also show predictable growth—low variation—which makes them less risky to own. A portfolio of stocks is assembled to diversify risk, and we can use our estimates of portfolio *beta* to estimate risk.

8.1 Rates of Return Reflect Expected Growth of Stock Prices

Example 8.1 Wal-Mart and Apple. Figure 8.1 contains plots of share prices of two well known companies, Wal-Mart and Apple, over a 60 month period, July 2010 to July 2015.

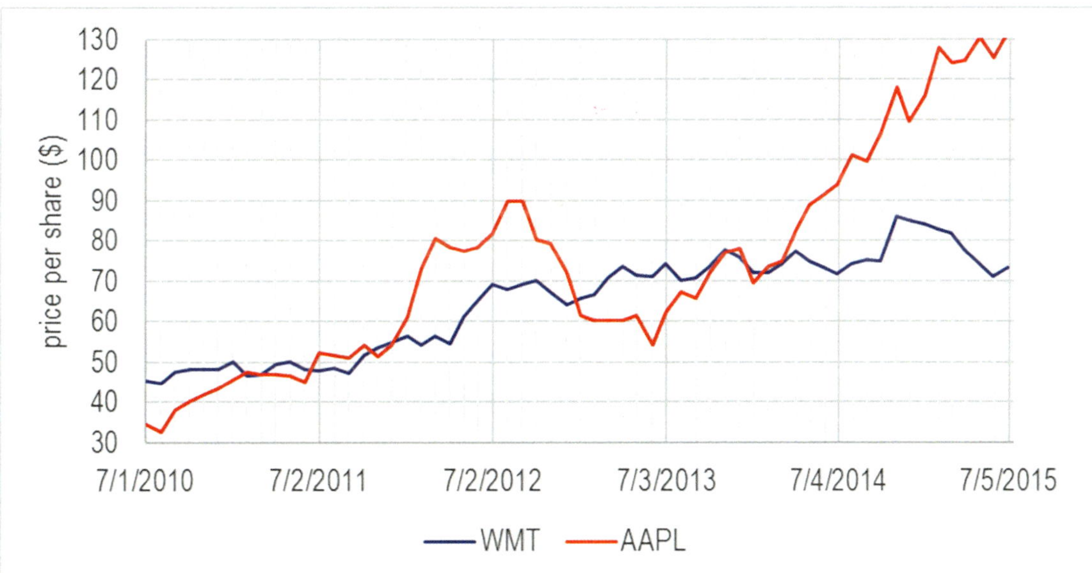

Figure 8.1 Monthly share prices of Wal-Mart and Apple, July 2010 to 2015

It is important to note that although prices in some months were statistical outliers, those unusual months were not excluded. A potential investor would be misled were unusually high or low prices ignored. Extreme values are expected and included, since they influence conclusions about the appeal of each potential investment. The larger the number of unusual months, the greater the dispersion in a stock price, and the riskier the investment.

C. Fraser, *Business Statistics for Competitive Advantage with Excel 2016*,
DOI 10.1007/978-3-319-32185-1_8

To find the growth rate in each of the stock investments, calculate the monthly percent change in price, or *rate of return, RR*:

$$RR_{stock,t} = \frac{(price_{stock,t} - price_{stock,t-1})}{price_{stock,t-1}}$$

where t is month.

Investors seek stocks with higher average rates of return and lower standard deviations. They would prefer to invest in stocks that exhibit higher expected, average growth and less volatility or risk. The standard deviation in the rate of return captures risk. If a stock price shows little variability, it is a less risky investment.

Figure 8.2 illustrates monthly rates of return in Wal-Mart and Apple stocks and a Market index, the S&P 500, over the five year period:

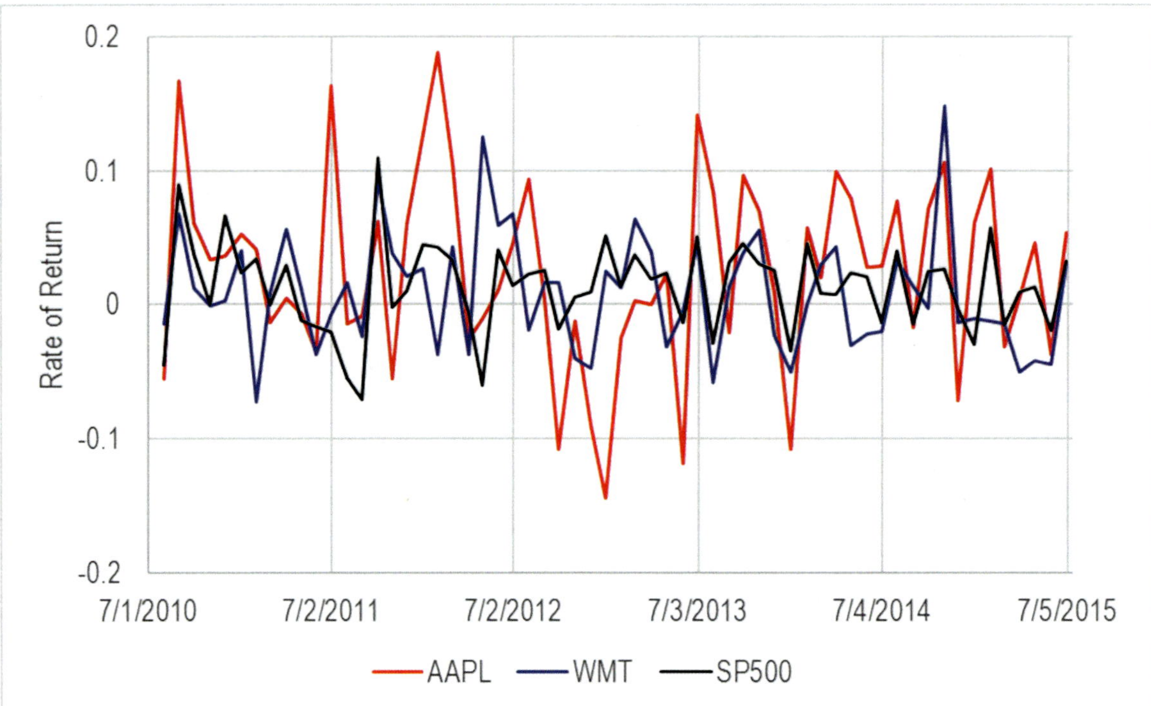

Figure 8.2 Monthly rates of return of Wal-Mart and Apple, July 2010 to July 2015

Table 8.1 Monthly rates of return of Wal-Mart and Apple Stock, July 2010 to July 2015

Monthly Rate of Return					
Wal-Mart WMT		*Apple AAPL*		*S&P 500*	
M	.009	*M*	.025	*M*	.013
SD	.044	*SD*	.071	*SD*	.034
Minimum	−.07	*Minimum*	−.14	*Minimum*	−.07
Maximum	.15	*Maximum*	.19	*Maximum*	.15

From Table 8.1, notice that Apple's mean monthly rate of return of 2.5% exceeds the Market mean monthly rate of return of 1.3%, though Apple's rates of return are more volatile, with a standard deviation of .071, compared with the Market standard deviation of .034. Wal-Mart rates of return are less volatile than Apple's, with standard deviation of .044, but also show a smaller mean, .9%, lower than the Market. The greater expected return from Apple, 2.7%, versus Wal-Mart's .9%, comes at the cost of added risk, with Apple's standard deviation, .071, exceeding Wal-Mart's, .044.

We would report to a potential investor:

- *Over the 60 months examined, Apple offered a greater expected monthly rate of return of 2.5%, relative to the S&P500 Index of the Market, with expected monthly return of 1.3%, but at higher risk with standard deviation in return .071 versus .034 for the Market.*
- *In this five year period, Wal-Mart offered an expected rate of return below the Market, .9%, but with lower risk, reflected in the lower standard deviation in return .044.*
- *Over the five year period, Apple stock delivered a higher rate of return than Wal-Mart stock, though Wal-Mart returns were was less volatile.*

8.2 Investors Trade Off Risk and Return

Investors seek stocks which offer higher expected rates of return *RR* and lower risk. Relative to a Market index, such as the S&P 500, which is a composite of 500 individual stocks, many individual stocks offer higher expected returns, but at greater risk. Market indices are weighted averages of individual stocks. Like other weighted averages, a Market index has an expected rate of return in the middle of the expected returns of the individual stocks making up the index. An investor attempts to choose stocks with higher than average expected returns and lower risk.

8.3 Beta Measures Risk

A Market index reflects the state of the economy. When a time series of an individual stock's rates of return is regressed against a Market index, the simple linear regression slope β indicates the expected percent change in a stock's rate of return in response to a percent change in the Market rate of return. β is estimated with b using a sample of stock prices:

$$\hat{R}R_{stock,t} = a + b \times RR_{Market,t}$$

where $RR_{stocki,t}$ is the estimated rate of return of a stock i in month t, and

$RR_{Market,t}$ is the rate of return of a Market index in month t.

In this specific case, the simple linear regression slope estimate b is called *beta.* Beta captures Market specific risk. If, in response to a percent change in the Market rate of return, the expected change in a stock's rate of return b is greater than one, the stock is more volatile, and exaggerates Market movements. A one percent increase in the Market value is associated with an expected change in the stock's price of more than one percent change. Conversely, if the expected change in a stock's rate of return b is less than one, the stock dampens Market fluctuations and is less risky. A one percent change in the Market's value is associated with an expected change in the stock's price of less than one percent. Beta reflects the amount of risk a stock contributes to a well diversified portfolio.

Recall from Chapter 5 that the sample correlation coefficient between two variables r_{xy} is closely related to the simple regression slope estimate b:

$$b = r_{X,y} \times \frac{s_y}{s_X}$$

In a model of an individual stock's rate of return against a Market index, the estimate of beta b is directly related to the sample correlation between the individual stock's rate of return and the Market rate of return:

$$be\hat{t}a_{stock_i} = b_{stock_i} = r_{stock_i,Market} \times \frac{s_{stock_i}}{s_{Market}}$$

The estimate of beta is a direct function of the sample correlation between an individual stock's rate of return and the Market rate of return, as well as Market sample variance.

Stocks with rates of return that are more strongly correlated with the Market rate of return and those with larger standard deviations have larger betas.

Notice in Figure 8.2 that Wal-Mart returns have a smaller variance than Apple returns. Wal-Mart is a less risky investment. Notice also that both stocks tend to move with the Market, though Apple tends to more closely follow the Market moves. And when both stocks move with the Market, Apple moves more, and Wal-Mart moves less.

It would not be surprising to find that Apple stock is riskier than Wal-Mart stock, since high end electronics are relatively expensive, luxuries. In boom cycles, companies that sell luxuries do more business. Wal-Mart sells a large number of inexpensive basics and necessities. The demand for these products is affected less by economic swings, making Wal-Mart stock relatively less correlated with Market swings, and, hence, less risky.

Table 8.2 contains sample correlation coefficients, standard deviations, and betas for both of the stocks using five years of monthly data.

Table 8.2 Correlations, standard deviations, covariances and betas for July 2010 to July 2015

	correlation with the Market $r_{stock,Market}$	*SD*	*beta* b_{stock}
SP500 RR		.034	
Wal-Mart RR	.30	.044	.39[b,c]
Apple RR	.45	.071	.94[a]

[a]Significant at .01.
[b]Significant at .05
[b]Significantly less than 1.0 at a 95% confidence level.

Correlations between each of the stocks' returns and the Market are positive, indicating that they do move with the Market.

Apple returns are more strongly correlated with the Market index returns than Wal-Mart returns. Apple returns are also more volatile. Because Apple rates of return are both more strongly correlated with the Market and more volatile than Wal-Mart returns, Apple stock will have a larger beta than Wal-Mart stock.

Betas b_{stocki} are shown in the last column of Table 8.3. A percent increase in the Market produces

- less than one percent expected increase in the Wal-Mart stock price, and
- a one percent expected increase in the Apple stock price,

Beta estimates are shown in Table 8.3 and Figure 8.3.

Table 8.3 Estimates of betas

Wal-Mart						
SUMMARY OUTPUT						
Regression Statistics						
R Square	.091					
Standard Error	.042					
Observations	60					
ANOVA	*Df*	*SS*	*MS*	*F*	*Significance F*	
Regression	1	.010	.010	5.8	.019	
Residual	58	.102	.0018			
Total	59	.113				
	Coefficients	*Standard Error*	*t Stat*	*p value*	*Lower 95%*	*Upper 95%*
Intercept	.004	.006	.6	.526	−.008	.015
S&P RR	.390	.162	2.4	.019	.067	.714

Apple						
SUMMARY OUTPUT						
Regression Statistics						
R Square	.201					
Standard Error	.064					
Observations	60					
ANOVA	*Df*	*SS*	*MS*	*F*	*Significance F*	
Regression	1	.060	.060	14.6	.0003	
Residual	58	.237	.004			
Total	59	.297				
	Coefficients	*Standard Error*	*t Stat*	*p value*	*Lower 95%*	*Upper 95%*
Intercept	.013	.009	1.4	.164	−.005	.030
S&P RR	.937	.245	3.8	.0003	.447	1.428

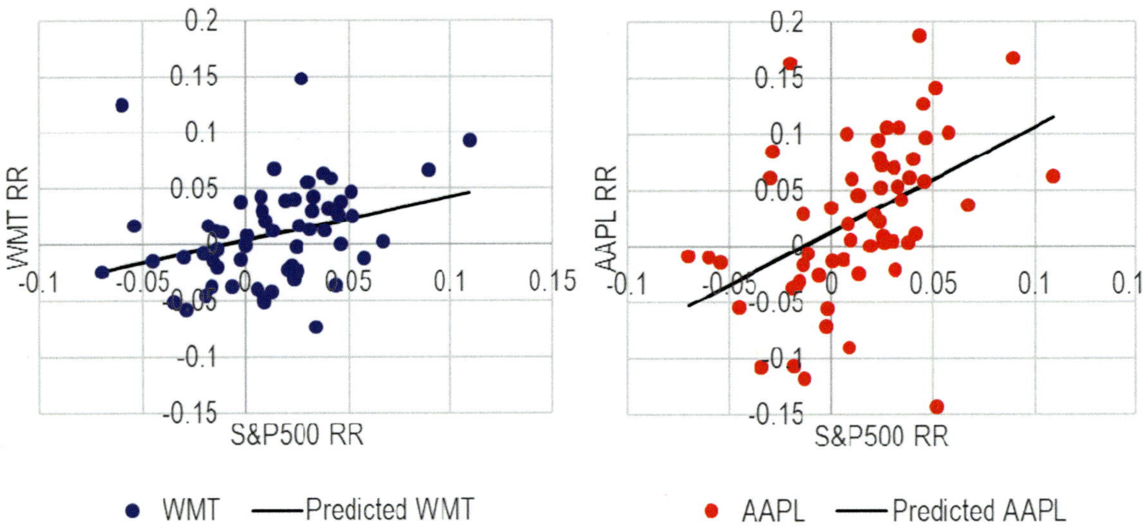

$\hat{R}R_{Wal-Mart_t} = .0037 + .39^b \times S\&P500_t$
RSquare: $.09^b$

$\hat{R}R_{Apple_t} = .\mathbf{013}^a + .\mathbf{94}^a \times S\&P\mathbf{500}_t$
RSquare: $.20^a$

[a] *Significant at* .01
[b] *Signficant at .05*

Figure 8.3 Response of Wal-Mart and Apple stocks to the market

Relative to Wal-Mart, Apple rates of return have both a higher correlation with The Market and a larger standard deviation, components of specific risk, producing the larger beta.

Comparing betas, a potential investor would conclude:

"Wal-Mart stock, with an estimated beta less than one ($b_{Wal\text{-}Mart}$ = .39), is a low risk investment. Wal-Mart returns dampen Market swings. With a percent increase in the Market, we expect to see an average increase of .39% in Wal-Mart's price.

Apple stock, with an estimated beta of one (b_{Apple}= .94) is riskier than Wal-Mart, and mirrors Market movement. With a percent increase in the Market, we expect to see an average increase of about one percent, .94%, in Apple's price."

8.4 A Portfolio Expected Return, Risk and Beta Are Weighted Averages of Individual Stocks

An investor is really interested in the expected return and risk of her portfolio of stocks. These are weighted averages of the expected returns and betas of the individual stocks in a portfolio:

$$E(RR_P) = \sum_i w_i \times E(RR_i)$$

$$b_P = \sum_i w_i \times b_i$$

Where $E(RR_P)$ is the expected portfolio rate of return,
w_i is the percent of investment in the ith stock,
$E(RR_i)$ is the expected rate of return of the ith stock,
b_P is the portfolio beta estimate,
b_i is the beta estimate of the ith stock,

Example 8.2 Three Alternate Portfolios. An Investment Manager has been asked to suggest a portfolio of two stocks from three being considered by a client: Wal-Mart, Google and IBM. The prospective investor wanted to include computer stock in his portfolio and had heard that IBM was a desirable "Blue Chip." She suspected that holding both Google and IBM stocks might be risky, were the computer industry to falter.

To confidently advise her client, the Investment Manager compared three portfolios of two equally weighted stocks from the three requested options. Individual stock weights in each portfolio equal one half. Table 8.4 contains the expected portfolio rates of return and betas for the three possible combinations:

Table 8.4 Expected portfolio returns and beta estimates

Portfolio	Expected Portfolio Return		Portfolio Beta Estimate	
	$\sum E(RR_i)/2$	$E(RR_P)$	$\sum b_i/2$	b_P
Wal-Mart+Google	(.009 + .024)/2	.015	(.39 +1.10)/2	.75
Wal-Mart+IBM	(.009 + .007)/2	.009	(.39 + .70)/2	.54
Google+IBM	(.024 + .007)/2	.008	(1.10 + .70)/2	.90

Alternatively, she could find expected portfolio returns and betas with software, and this would be the practical way to compare more than a few portfolios. Figure 8.4 shows expected (mean) rates of return and regression beta estimates for the three portfolios from Excel:

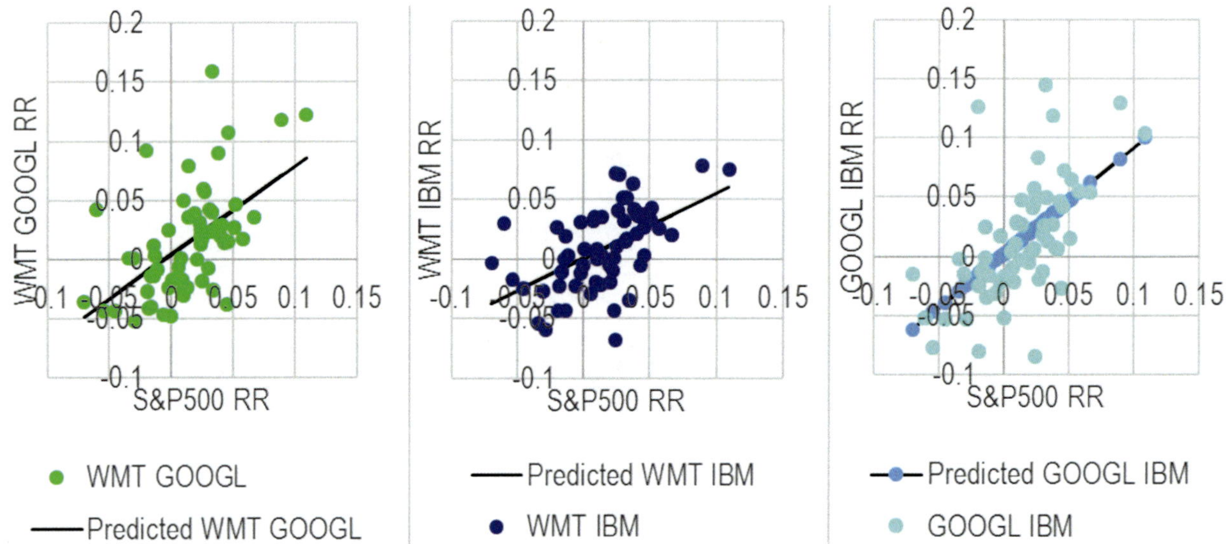

$\hat{R}R_{W+G_t} = .0047 + .75^{a} \times S\&P500_t$ $\qquad$ $\hat{R}R_{W+I_t} = .0006 + .54^{a,b} \times S\&P500_t$ $\qquad$ $\hat{R}R_{G+I_t} = .0016 + .90^{a} \times S\&P500_t$

RSquare: .41[a] *RSquare*: .47[a] *RSquare*: .49[a]

[a]*Significant at* .01
[b]*Significantly less than* 1.

Figure 8.4 Beta estimates of three alternate portfolios

8.5 Better Portfolios Define the Efficient Frontier

In the comparison of alternative portfolios, the Investment Manager wanted to identify alternatives which promised greater expected return without greater risk—or, alternatively, those which reduced risk without reducing return. Better portfolios, which promise the highest return for a given level of risk, define the *Efficient Frontier*. To see the Efficient Frontier, she made a scatterplot of portfolio expected rates of return by portfolio risk. Those relatively efficient portfolios lie in the upper left.

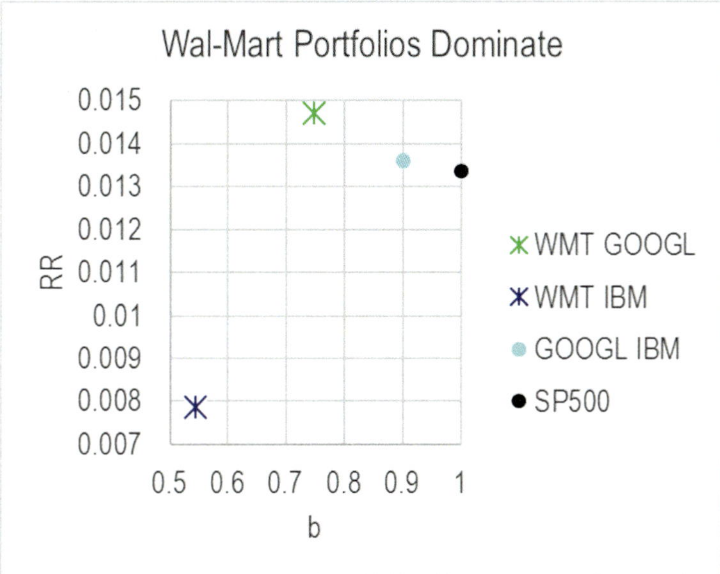

Figure 8.5 Relatively efficient portfolios offer greater expected return and lower risk

Comparing portfolios in Figure 8.5, the Investment Manager found that the two IBM portfolios outperform the Market, offering higher expected rates of return at lower levels of risk, though the two Wal-Mart portfolios outperform the Google IBM combination.

$$RR_{WMT+IBM} = .008 > RR_{IBM+GOOG} = .014 < RR_{WMT+GOOG} = .015$$

$$b_{WMT+IBM} = .54 < b_{WMT+GOOG} = .75 < b_{IBM+GOOG} = .90$$

The better choice would depend on the prospective investor's risk preference. The Wal-Mart IBM combination is less risky than that Wal-Mart Google combination, but offers about half as large a return. The combination of both computer stocks, IBM and Google, is the least diversified and the riskiest.

The Investment Manager presented results of her analysis with recommendations in this memo to her client:

MEMO

Re: Recommended Portfolio is Diversified
To: Ms. Rich N. Vest
From: Christine Kasper, Investment Advisor, Stellar Investments
Date: July 2012

The portfolios combining Wal-Mart with IBM or Google stocks are expected to outperform the less diversified combination of IBM with Google.

Alternate portfolios were compared

Portfolios containing two from the candidate set of three stocks, Wal-Mart, IBM, and Google have been compared to assess their expected returns and risk levels. Assessments were based on five years of monthly prices, July 2010 through July 2015, and movement relative to the S&P500 Market Index during this period.

Wal-Mart Combinations Dominate

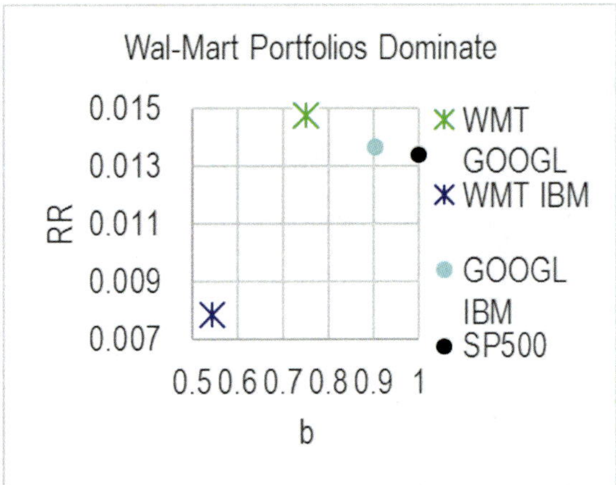

Expected monthly rates of return range from .8 to 1.5%. While both of the Google portfolios outperform the Market, the Wal-Mart combinations offer higher expected rates of return for a given level of risk. The Wal-Mart andGoogle combination yields the highest expected return, 1.5%, while the Wal-Mart and IBM combination is lowest risk, with beta .54.

In response to a 1% change in the S&P500, the Wal-Mart + IBM combination is expected to move less, .54%, dampening Market movement. This is a conservative choice. The Wal-Mart+Google combination mirrors The Market and is expected to move more .75% in response to a 1% change in The Market.

Choose Wal-Mart+IBM for lower risk, Wal-Mart+Google for higher expected return

The choice of Wal-Mart with IBM promises a positive expected return, as well as the lowest risk. Wal-Mart with Google promise the highest expected return, but with added risk. Combining the two tech stocks either reduce expected return or increase risk.

A larger number of stocks are suggested

You may wish to consider a portfolio with a larger number of stocks to increase your diversification and reduce your risk.

8.6 Portfolio Risk Depends on Correlations with the Market and Stock Variability

Both the expected rate of return of a portfolio and its risk, measured by its beta, depend on the expected rates of return and betas of the individual stocks in the portfolio. Individual stock betas are direct functions of

- the correlation between a stock's rate of return and the Market index rate of return, and
- the standard deviation of a stock's rate of return

Beta for a stock or a portfolio is estimated by regressing the stock or portfolio monthly rates of return against monthly Market rates of return. The resulting simple linear regression slopes are estimates of the stock or portfolio beta.

Excel 8.1 Estimate Portfolio Expected Rate of Return and Risk

Three Portfolios with Wal-Mart, IBM and Google. Monthly rates of return for each of the three stocks and the S&P500 index of the Market are in **Excel 8.1 Three Portfolios.**.

Monthly portfolio returns formula. Insert new columns , C, D and E, for each of the threes portfolio containing equally weighted pairs of the three stocks, which will be the average of rates of return of each of pair of stocks in a portfolio. In the second row of each new column enter a formula for the average of two stocks. Select the three new cells and down fill the three new columns.

In F1,
WMT GOOGL
In G1,
WMT IBM
In H1,
GOOGL IBM
In F2,
=average(c2,d2)
In G2,
=average(c2,e2)
In H2,
=average(d2,e2)
In F2,
Cntl+shift+right
Double click the lower right corner to fill down columns

Expected monthly rates of return. Find the expected monthly return for the three portfolios in the first row following the data, row 62.

In E62,
E(RR)
In F62,
Alt MUA
In F62,
Shift+right right
Cntl+R

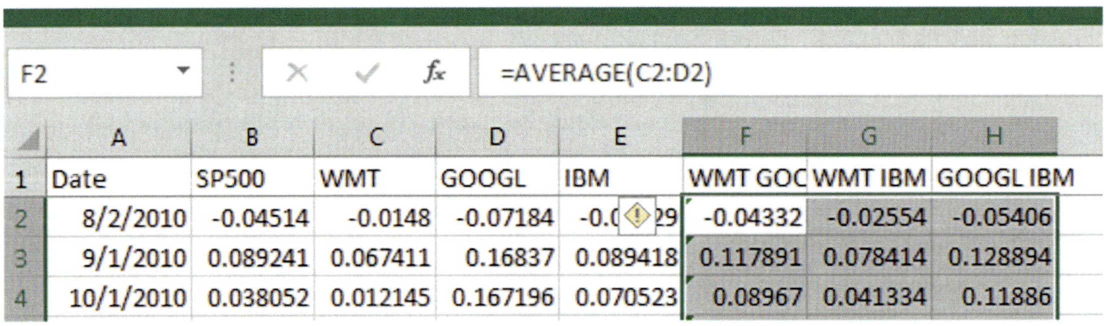

F2 =AVERAGE(C2:D2)

	A	B	C	D	E	F	G	H
1	Date	SP500	WMT	GOOGL	IBM	WMT GOC	WMT IBM	GOOGL IBM
2	8/2/2010	-0.04514	-0.0148	-0.07184	-0.(29	-0.04332	-0.02554	-0.05406
3	9/1/2010	0.089241	0.067411	0.16837	0.089418	0.117891	0.078414	0.128894
4	10/1/2010	0.038052	0.012145	0.167196	0.070523	0.08967	0.041334	0.11886

F62 =AVERAGE(F2:F61)

	E	F	G	H	I	J
1	IBM	WMT GOC	WMT IBM	GOOGL IBM		
60	-0.0412	-0.02733	-0.04309	-0.02544		
61	0.002521	0.159092	0.016557	0.145055		
62	E(RR)	0.014721	0.00785	0.013614		

Estimate betas from simple regression. To find the Market specific risk, *beta,* find the simple regression slope of each portfolio rate of return with *S&P500.*

For the first portfolio, *WMT+GOOGL,* run regression with *WMT+GOOGL* in the **Input Y Range,** and *S&P500* in the **Input X Range:**

	A	B	C	D	E	F	G
1	SUMMARY OUTPUT						
2							
3	*Regression Statistics*						
4	Multiple F	0.549309					
5	R Square	0.30174					
6	Adjusted I	0.289701					
7	Standard E	0.038932					
8	Observati	60					
9							
10	ANOVA						
11		*df*	*SS*	*MS*	*F*	*gnificance F*	
12	Regressio	1	0.037989	0.037989	25.06365	5.49E-06	
13	Residual	58	0.087911	0.001516			
14	Total	59	0.125899				
15							
16		*Coefficients*	*andard Err*	*t Stat*	*P-value*	*Lower 95%*	*Upper 95%*
17	Intercept	0.004727	0.005408	0.874096	0.385671	-0.0061	0.015552
18	SP500	0.747389	0.149288	5.00636	5.49E-06	0.448557	1.046221

Excel 8.2 Plot Return by Risk to Identify Dominant Portfolios and the Efficient Frontier

To compare the expected rates of return and estimated risk of the three portfolios, create a summary of the portfolio betas and expected returns below the data, and then plot the portfolio rates of return against their betas (one portfolio at a time) to identify the Efficient Frontier.

Copy row 1 containing portfolio labels and paste into row 62. Below the expected rates of return, use the Excel slope function to produce the portfolio betas.

In row 1,
Shift+spacebar
Cntl+C
Cntl+down
Shift+spacebar
Alt HIE
In E64,
B
In F64,
=slope(f2:f61,b2 f4 f4 f4 :b61 f4 f4 f4 f4)
In F64,
Shift+right right
Cntl+R

(Pressing **f4** three times locks the row reference, so that the slope function will use the S&P500 returns in column B to find betas for portfolios in columns F, G and H.)

F64 =SLOPE(F2:F61,$B2:$B61)

	A	B	C	D	E	F	G	H
1	Date	SP500	WMT	GOOGL	IBM	WMT GOC	WMT IBM	GOOGL IBM
60	6/1/2015	-0.01936	-0.04497	-0.00968	-0.0412	-0.02733	-0.04309	-0.02544
61	7/1/2015	0.032365	0.030594	0.28759	0.002521	0.159092	0.016557	0.145055
62	Date	SP500	WMT	GOOGL	IBM	WMT GOC	WMT IBM	GOOGL IBM
63					E(RR)	0.014721	0.00785	0.013614
64					b	0.747389	0.543786	0.90086
65								

Move the S&P500 returns to column E, next to the portfolios in order to include the S&P500 in your plot.

In column B,
Cntl+spacebar
Cntl+X
In column E,
Cntl+spacebar
Alt IIIE

Use the Excel average function to add the expected rate of return for the S&P500 and enter 1 for the S&P500 beta.

In E63,
=average(e2:e61)
In E64,
1

	D	E	F	G	H
1	IBM	SP500	WMT GOC	WMT IBM	GOOGL IBM
60	-0.0412	-0.01936	-0.02733	-0.04309	-0.02544
61	0.002521	0.032365	0.159092	0.016557	0.145055
62	IBM	SP500	WMT GOC	WMT IBM	GOOGL IBM
63	E(RR)	0.013371	0.014721	0.00785	0.013614
64	b	1	0.747389	0.543786	0.90086

Select the S&P500 label, expected rate of return and beta in column E and request a scatterplot of the point.

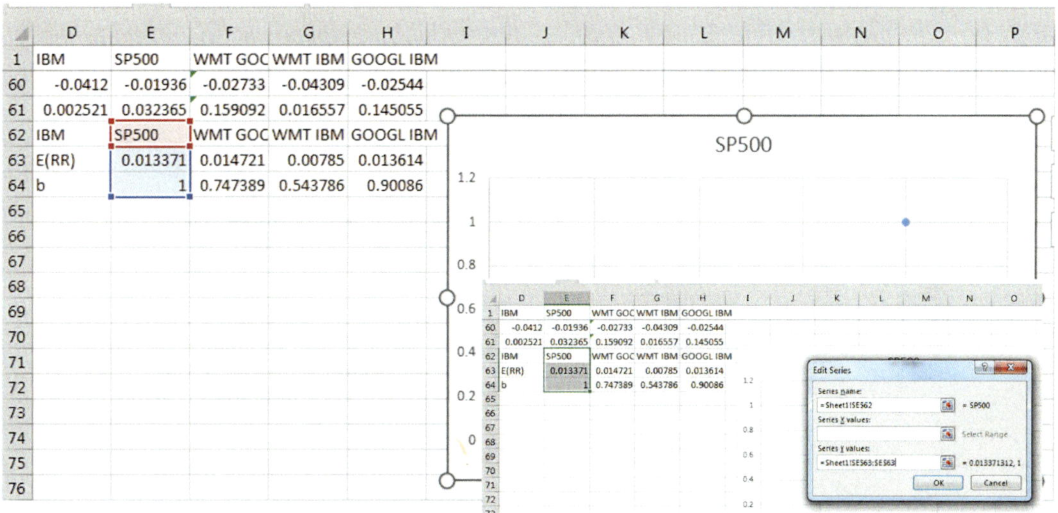

When we are plotting a single point, Excel will read both *b* and the *E(RR)* as a single series, plotting two points. To correct this, edit the series,

Alt JC E Tab Tab Tab Enter Tab Tab
Change 64 to 65
Tab Enter
OK

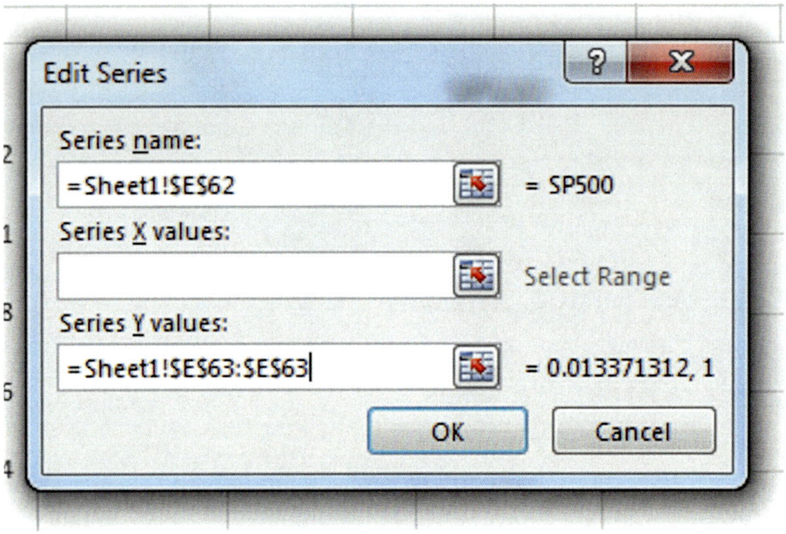

Add each of the three portfolios as separate series, inputting the portfolio label in row 62 for name, beta in row 63 for X and the expected rate of return in row 64 for Y.

Alt JC E Tab Tab Tab A

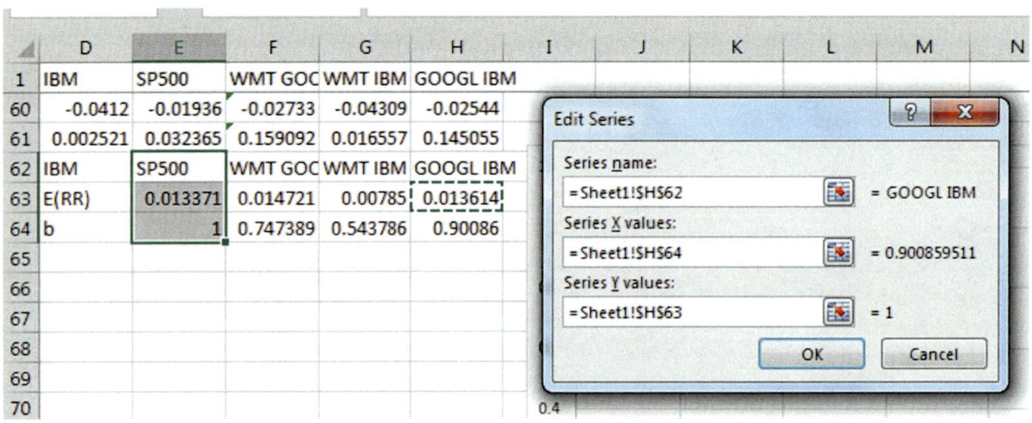

Add chart and axes titles, adjust axes and set fontsize. Add the legend.

Alt JCAL

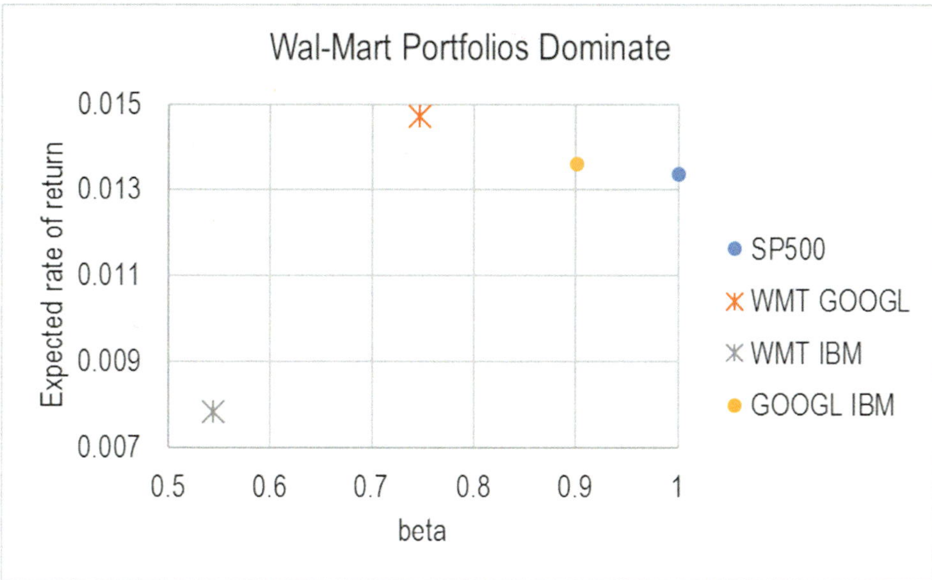
Wal-Mart Portfolios Dominate
Expected rate of return
0.015
0.013
0.011
0.009
0.007
0.5
0.6
0.7
0.8
0.9
1
beta
SP500
WMT GOOGL
WMT IBM
GOOGL IBM

Lab 8 Portfolio Risk and Return

8 stocks contains a five year times series of monthly prices for 36 stocks and the S&P500 Market index for months July 2010 through July 2015. Stocks include ten "blue chips," as well as several others that have been in recent news.

Find the *rate of return* for each of the stocks.

Find the *beta* for each of the stocks.

Which stocks performed worse than the Market, with lower expected rates of return and a higher expected *beta?*

____

Use logic to choose two stocks to combine in an equally weighted portfolio, and then add a column with portfolio rates of return.

The expected rate of return of my portfolio: ____________

Use regression to find the 95% confidence interval for beta of your portfolio

The Market specific risk, beta, of my portfolio: _____ to _____

My portfolio ___dampens, ___mirrors, ___exaggerates
Market swings.

Create an alternative portfolio with two other stocks.

Find the expected rate of return expected *beta*.

Plot *E(RR)* by *b* for the S&P500 and your portfolios to see the *Efficient Frontier*, then compare the two portfolios:

Portfolio______ dominates Portfolio______ OR ___neither dominates the other

Assignment 8 Portfolio Risk and Return

An investor is considering investment in both Costco stock. She would like to invest 50% in Costco and 50% in a second stock. She is considering American Airlines, Avis, Kroger and Wal-Mart. She has asked for your recommendation. **COST stocks** contains five years of monthly stock prices and rates of return.

1. Which of the stocks dampen Market swings?

___ AAL ___CAR ___COST ___ KR ___WMT

2. Which of the stocks exaggerate Market swings?

___ AAL ___CAR ___COST ___ KR ___WMT

3. Which are better choices for the proposed two stock portfolio?

___ AAL ___CAR ___ KR ___WMT

4. Plot the expected rates of return by betas for the four two-stock portfolios, to illustrate your recommendations, and embed here:

Chapter 9
Association Between Two Categorical Variables: Contingency Analysis with Chi Square

Categorical variables, including nominal and ordinal variables, are described by tabulating their frequencies or probability. If two categorical variables are associated, the frequencies of values of one will depend on the frequencies of values of the other. Chi square tests the hypothesized association between two categorical variables and contingency analysis quantifies their association.

9.1 When Conditional Probabilities Differ from Joint Probabilities, There Is Evidence of Association

Contingency analysis begins with the crosstabulation of frequencies of two categorical variables. Figure 9.1 shows a crosstabulation of sandwich spreads and topping combinations chosen by forty students:

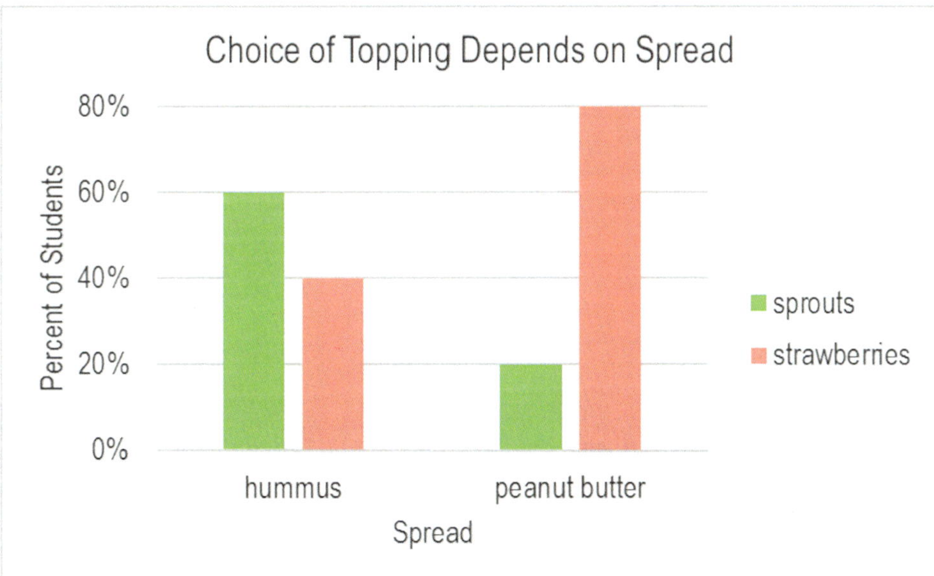

Counts	*sprouts*	*strawberries*	*total*
hummus	12	8	20
peanut butter	4	16	20
total	20	20	40

%Row	*sprouts*	*strawberries*	*total*
Hummus	60	40	100
peanut butter	20	80	100
Total	50	50	100

Figure 9.1 Crosstabulation: sandwich topping depends on spread

C. Fraser, *Business Statistics for Competitive Advantage with Excel 2016*,
DOI 10.1007/978-3-319-32185-1_9

If the unconditional probabilities of category levels, such as sprouts versus strawberries topping, differ from the probabilities, conditional on levels of another category, such as hummus or peanut butter spread, we have evidence of association. In this sandwich example, sprouts were chosen by half the students, making its unconditional probability .5. If a student chose hummus spread, the conditional probability of sprouts topping was higher (.60). If a student chose peanut butter spread, sprouts was the less likely topping choice (.40).

Example 9.1 Recruiting Stars. Human Resource managers are hoping to improve the odds of hiring outstanding performers and to reduce the odds of hiring poor performers by targeting recruiting efforts. Management believes that recruiting at the schools closer to firm headquarters may improve the odds of hiring stars. Students familiar with local customs may feel more confident at the firm. Removing schools far from headquarters may reduce the odds of hiring poor performers. Management's hypotheses are:

> H_0: Job performance is not associated with undergraduate program location.
> H_1: Job performance is associated with undergraduate program location.

To test these hypotheses, department supervisors throughout the firm sorted a sample of forty recent hires into three categories based on job performance: poor, average, and outstanding. The sample employees were also categorized by the proximity to headquarters: Home State, Same Region, and Outside Region. These cross-tabulations are shown in the PivotChart and PivotTable in Figure 9.2.

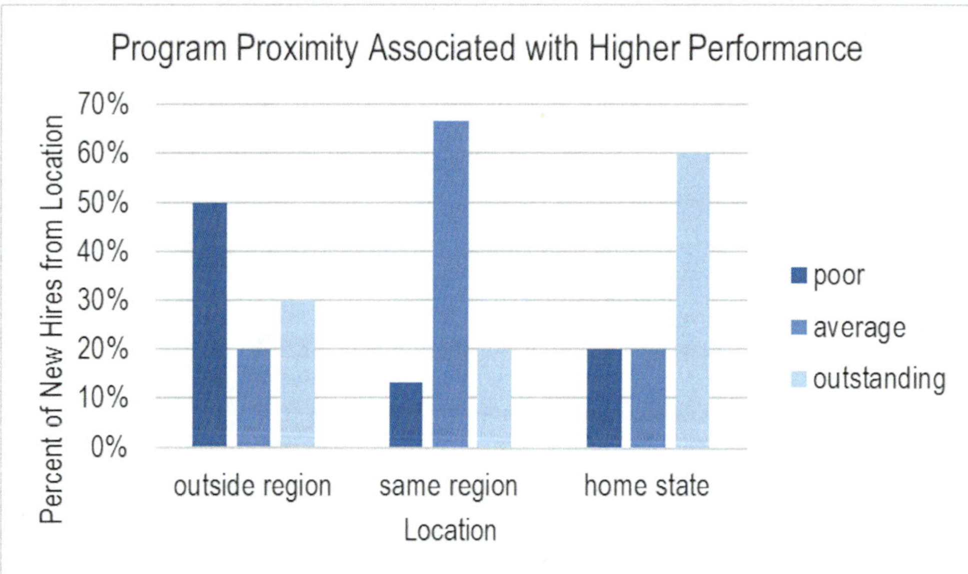

Count	*Performance*			
Location	*Poor*	*Average*	*Outstanding*	*Total*
Outside Region	5	2	3	10
Same Region	2	10	3	15
Home State	3	3	9	15
Total	10	15	15	40
% of Row	*Performance*			
Location	*Poor*	*Average*	*Outstanding*	*Total*
Outside Region	50%	20%	30%	100%
Same Region	13%	67%	20%	100%
Home State	20%	20%	60%	100%
Total	25%	38%	38%	100%

χ^2_4	12.5	*p value*	.02

Figure 9.2 Job performance depends on program location

The crosstabs indicate that a quarter of the firm's new employees are *Poor* performers, about forty percent are *Average* performers, and about forty percent are *Outstanding* performers. From the PivotChart we see that more than a quarter of employees from programs *Outside Region* are *Poor* performers, and more than forty percent of employees from *Home State* programs are *Outstanding* performers. Were program location and performance *not* associated, a quarter of the recruits from each location would be *Poor* performers. We would, for example, expect a quarter of ten employees recruited from *Outside Region* to be Poor performers, or 2.5 (=.25(10)). Instead, there are actually five (*Outside Region, Poor*) employees. There is a greater chance, 50%, of *Poor* performance, given *Outside Region*, relative to *Same Region* or *Home State*. Ignoring program location, the probability of poor performance is .25; acknowledging program location, this probability of poor performance varies from .13 (*Same Region*) to .50 (*Outside Region*). These differences in row percentages suggest an association between program rank and performance.

9.2 Chi Square Tests Association Between Two Categorical Variables

The chi square (χ^2) statistic tests the significance of the association between performance and program location, by comparing expected cell counts with actual cell counts, squaring the differences, and weighting each cell by the inverse of expected cell frequency.

$$\chi^2_{(R-1),(C-1)} = \sum_{ij}^{RC} (e_{ij} - n_{ij})^2 / e_{ij},$$

Where R is the number of row categories,

C is the number of column categories,

n is the number in the ith row and jth column,

e is the number expected in the ith row and jth column.

χ^2 gives more weight to the least likely cells. χ^2 distributions are skewed and with means equal to the number of degrees of freedom. Several χ^2 distributions with a range of degrees of freedom are shown in Figure 9.3.

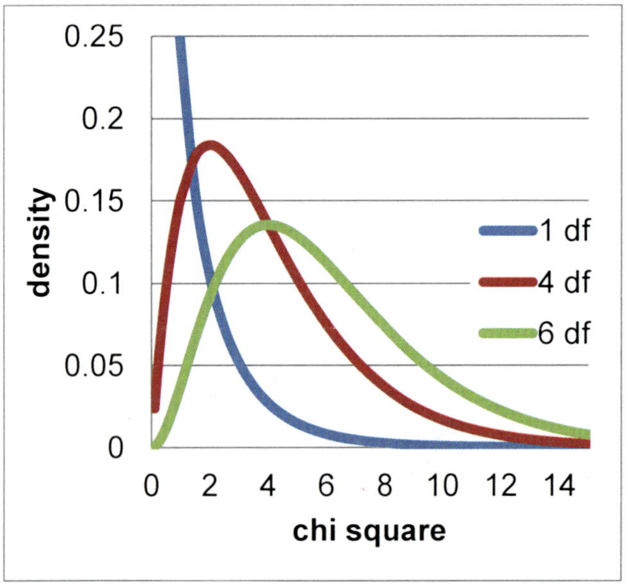

Figure 9.3 Chi square distributions for a range of degrees of freedom

In the **Recruiting Stars** example, **Figure 9.2**, chi square, χ_4^2, is 12.5, which can be verified using the formula:

$$\begin{aligned}\chi^2 = \; & (2.5 - 5)^2/2.5 + (3.8 - 2)^2/3.8 + (3.8 - 3)^2/3.8 \\ & + (3.8 - 2)^2/3.8 + (5.6 - 10)^2/5.6 + (5.6 - 3)^2/5.6 \\ & + (3.8 - 3)^2/3.8 + (5.6 - 3)^2/5.6 + (5.6 - 9)^2/5.6 \\ = \; & 2.5 + .9 + .2 \\ + \; & .9 + 3.5 + 1.2 \\ + \; & .2 + 1.2 + 2.0 = 12.5\end{aligned}$$

From a table of χ_4^2 distributions, we find that for a crosstabulation of this size, with three rows and three columns, (df=(Rows−1) × (Columns − 1)=2 × 2 = 4), $\chi_4^2 = 12.5$ indicates that the p-value is .02. Two percent of the distribution lies right of 12.5. There is little chance that of observing the sample data were performance and program tier not associated. The null hypothesis of lack of association is rejected.

Those cells which contribute more to chi square indicate the nature of association. In this example, we see in Table 9.1 that these are the (*Outside Region*, *Poor*), (*Same Region*, *Average*), and (*Home State*, *Outstanding*) cells:

Table 9.1 Contribution to chi square by cell

$$\chi^2 = \mathbf{2.5} + .9 + .2 + .9 + \mathbf{3.5} + 1.2 + .2 + 1.2 + \mathbf{2.0} = 12.5$$

	Poor	*Average*	*Outstanding*
Outside Region	**2.5**	.9	.2
Same Region	.9	**3.5**	1.2
Home State	.2	1.2	**2.0**

Poor performance is more likely if a new employee came from a program *Outside Region*, *Average* performance is more likely if a new employee came from a program in the *Same Region*, and *Outstanding* performance is more likely if a new employee came from a *Home State* program. Job performance is associated with program location.

9.3 Chi Square Is Unreliable If Cell Counts Are Sparse

There are two possible reasons why the chi square statistic is large and apparently significant. The first reason is the likely actual association between program location and performance. The second reason is that there are few (less than five) expected employees in five of the nine cells, shown in Table 9.2.

Table 9.2 Expected counts by cell

	Poor	*Average*	*Outstanding*
Outside Region	2.5	3.8	3.8
Same Region	3.8	5.6	5.6
Home State	3.8	5.6	5.6

Since the chi square components include expected cell counts in the denominator, *sparse* (with expected counts less than five) cells inflate chi square. When sparse cells exist, we must either combine categories or collect more data.

In the **Recruiting Stars** example, management was most interested in increasing the chances of hiring *Outstanding* performers. Since some believed that *Outstanding* performers were recruited from programs in the *Home State*, these categories were preserved. *Same Region* and *Outside Region* program locations were combined. *Poor* and *Average* performance categories were combined. We are left with a 2 × 2 contingency analysis, Figure 9.4.

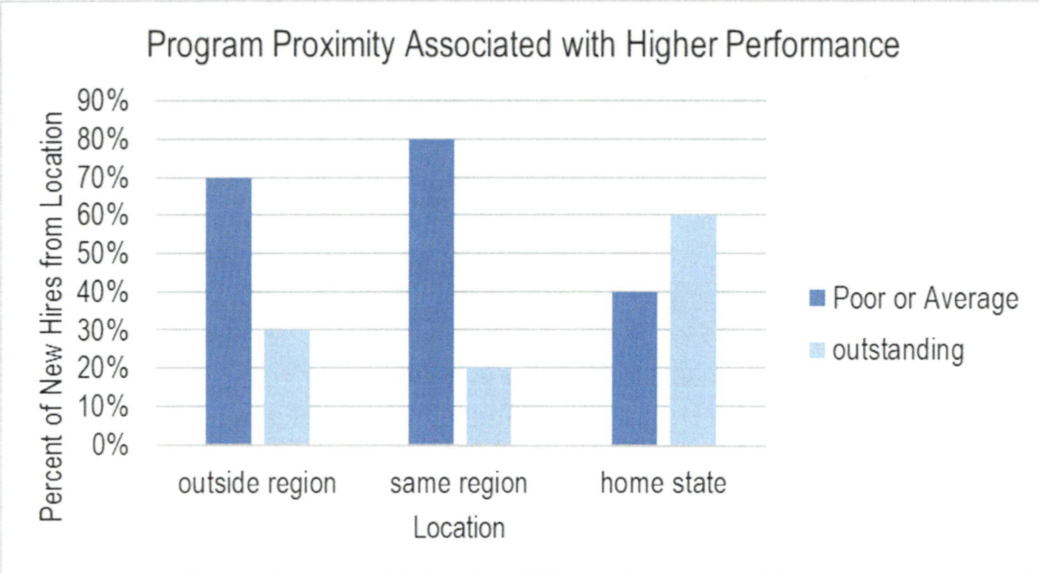

Count	*performance*			*% Row*	*performance*		
Location	*poor/ average*	*outstanding*	*total*	*Location*	*poor/ average*	*outstanding*	*Total*
Same or Outside Region	19	6	25	*Same or Outside Region*	76%	24%	100%
Home State	6	9	15	*Home State*	40%	60%	100%
Total	25	15	40	*Total*	63%	38%	100%

Chi Square	5.2
df	1
p value	.02

Figure 9.4 PivotChart of performance by program location with fewer categories

With fewer categories, all expected cell counts are now greater than five, providing a reliable $\chi_1^2 = 5.2$, which remains significant (*p value*=.02). The PivotChart continues to suggest that the incidence of *Outstanding* performance is greater among employees recruited from *Home State* programs. The impact of program location on *Poor* performance is unknown, since *Poor* and *Average* categories were combined. Also unknown is the difference between employees from *Same* and *Outside Regions* programs, since these categories were likewise combined.

Recruiters would conclude:

"Job performance of newly hired employees is associated with undergraduate program location. Twenty-four percent of our new employees recruited from Same or Outside Region undergraduate programs have been identified as Outstanding performers. Within the group recruited from Home State undergraduate programs, more than twice this percentage, 60%, are

Outstanding performers, a significant difference. Results suggest that in order to achieve a larger percent of Outstanding performers, recruiting should be focused on Home State programs."

9.4 Simpson's Paradox Can Mislead

Using contingency analysis to study the association between two variables can be potentially misleading, since all other related variables are ignored. If a third variable is related to the two being analyzed, contingency analysis may indicate that they are associated, when they may not actually be. Two variables may appear to be associated because they are both related to a third, ignored variable.

Example 9.2 American Cars. The CEO of American Car Company was concerned that the oldest segments of car buyers were avoiding cars that his firm assembles in Mexico. Production and labor costs are much cheaper in Mexico, and his long term plan is to shift production of all models to Mexico. If older, more educated and more experienced buyers avoid cars produced in Mexico, American Car could lose a major market segment unless production remained in The States.

The CEO's hypotheses were:

H_0: Choice between cars assembled in the U.S. and cars assembled in Mexico is not associated with age category.
H_1: Choice between cars assembled in the U.S. and cars assembled in Mexico is associated with age category.

He asked Travis Henderson, Director of Quantitative Analysis, to analyze the association between age category and choice of U.S. made versus Mexican made cars. The research staff drew a random sample of 263 recent car buyers, identified by age category. After preliminary analysis, age categories were combined to insure that all expected cell counts in an [Age Category × Origin Choice] crosstabulation were each at least five. Contingency analysis is shown in the PivotChart and Pivot Tables in Figure 9.5.

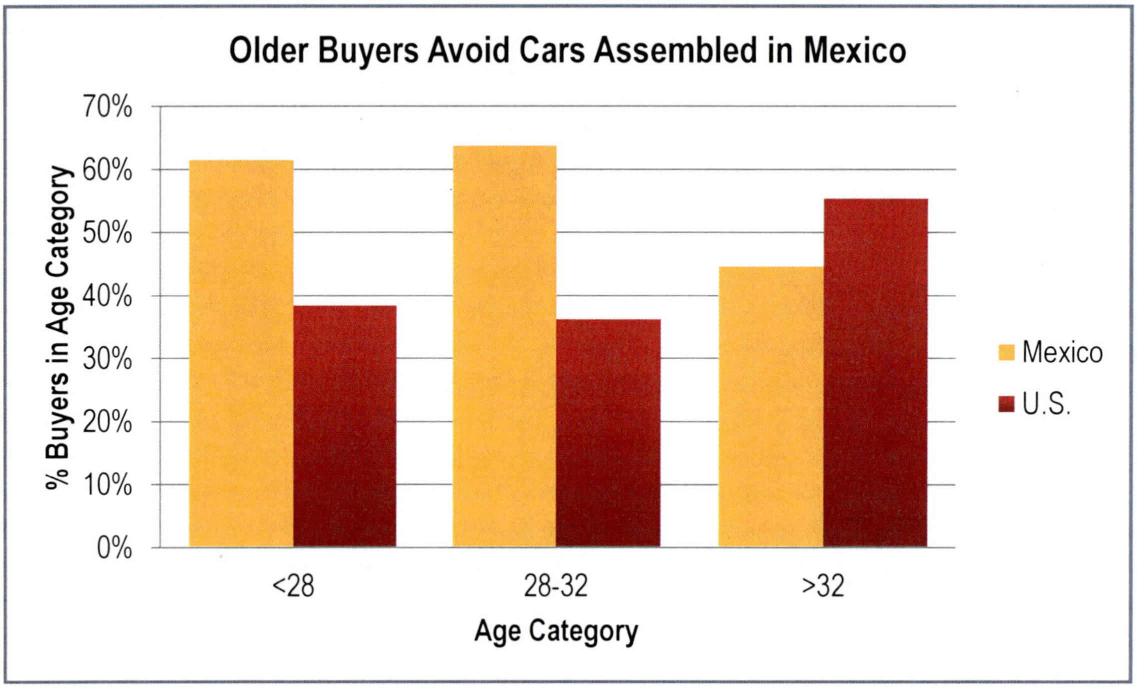

Count	*Assembled in*				*% Rows*		
Age	*Mexico*	*U.S.*	*Total*	*Age*	*Mexico*	*U.S.*	*Total*
Under 28	56	35	91	*Under 28*	62%	38%	100%
28 to 32	51	29	80	*28 to 32*	64%	36%	100%
33 Plus	41	51	92	*33 Plus*	45%	55%	100%
Total	148	115	263	*Total*	56%	44%	100%

Chi Square	8.0
df	2
p value	.02

Figure 9.5 Contingency analysis of U.S. vs. Mexican made car choices by age

A glimpse of the PivotChart confirmed suspicions that older buyers did seem to be rejecting cars assembled in Mexico. The *p value* for chi square was .02, indicating that the null hypothesis, lack of association, ought to be rejected. Choice between U.S. and Mexican made cars seemed to be associated with age category. Fifty six percent of the entire sample across all ages chose cars assembled in Mexico. Within the oldest segment, however, the Mexican assembled car share was lower: 45%. While nearly two thirds of the younger segments chose cars assembled in Mexico, less than half of the oldest buyers chose Mexican made cars.

The CEO was alarmed with these results. His company could lose the business of older, more experienced buyers if production were shifted South of the Border. Brand managers were about to begin planning "Made in the U.S.A." promotional campaigns targeted at the oldest car buyers. Emily Ernst, the Director of Strategy and Planning, suggested that age was probably not the correct basis for segmentation. She explained that the older buyers shop for a particular *type* of car—a family sedan or station wagon—and few family sedans or wagons were being assembled in Mexico. Models assembled at home in the U.S. tended to be large sedans and station wagons—styles sought by older buyers. She proposed that it was *style* that influenced the U.S. versus Mexican assembled choice, and not age, and that it was *style* that was dependent on age. Her hypotheses were:

H_0: Choice of car style is not associated with age category.
H_1: Choice of car style is associated with age category.

To explore this alternate hypothesis, the research team ran contingency analysis of style choice (SUV, Sedan/Wagon and Coupe) by age category, Figure 9.6.

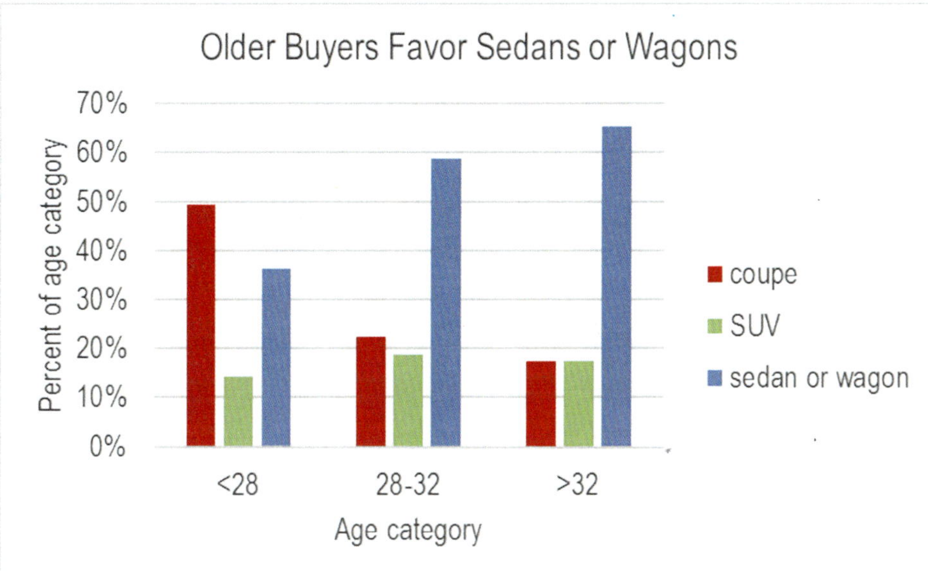

Count	*Style*			
Age	*sedan/ wagon*	*coupe*	*SUV*	*Total*
< 28	33	45	13	91
28 to 32	47	18	15	80
33+	60	16	16	92
Total	140	79	44	263

Row%	*Style*			
Age	*sedan/ wagon*	*coupe*	*SUV*	*Total*
< 28	36%	49%	14%	100%
28 to 32	59%	23%	19%	100%
33+	65%	17%	17%	100%
Total	53%	30%	17%	100%

χ^2_4	**26.2**	***p value***	**.0000**

Figure 9.6 Contingency analysis of car style choice by age category

Contingency analysis of this sample indicates that choice of style is associated with age category. More than half (53%) of the car buyers chose a sedan or wagon, though only about a third (36%) of the younger buyers chose a sedan or wagon, and nearly twice as many (65%) older buyers chose a sedan or wagon. Thirty percent of the sample bought a coupe, and just nearly half (49%) of the younger buyers chose a coupe. Only 17% of the oldest buyers bought a coupe. These are significant differences supporting the conclusion that style of car chosen is associated with age category.

This is the news that the CEO was looking for. If older car buyers are choosing U.S.-made cars because they desire family styles, sedans and wagons, which tend to be assembled in the U.S., then perhaps these older buyers aren't shunning Mexican-made cars. His hypotheses were:

H_0: Given choice of a sedan or wagon, choice of U.S. versus Mexican assembled is not associated with age category.

H_1: Given choice of a sedan or wagon, choice of U.S. versus Mexican assembled is associated with age category.

To test these hypotheses, the analysis team conducted three contingency analyses of origin choice (U.S. versus Mexican assembled) by age category, looking at each style separately in Figure 9.7 and Table 9.3.

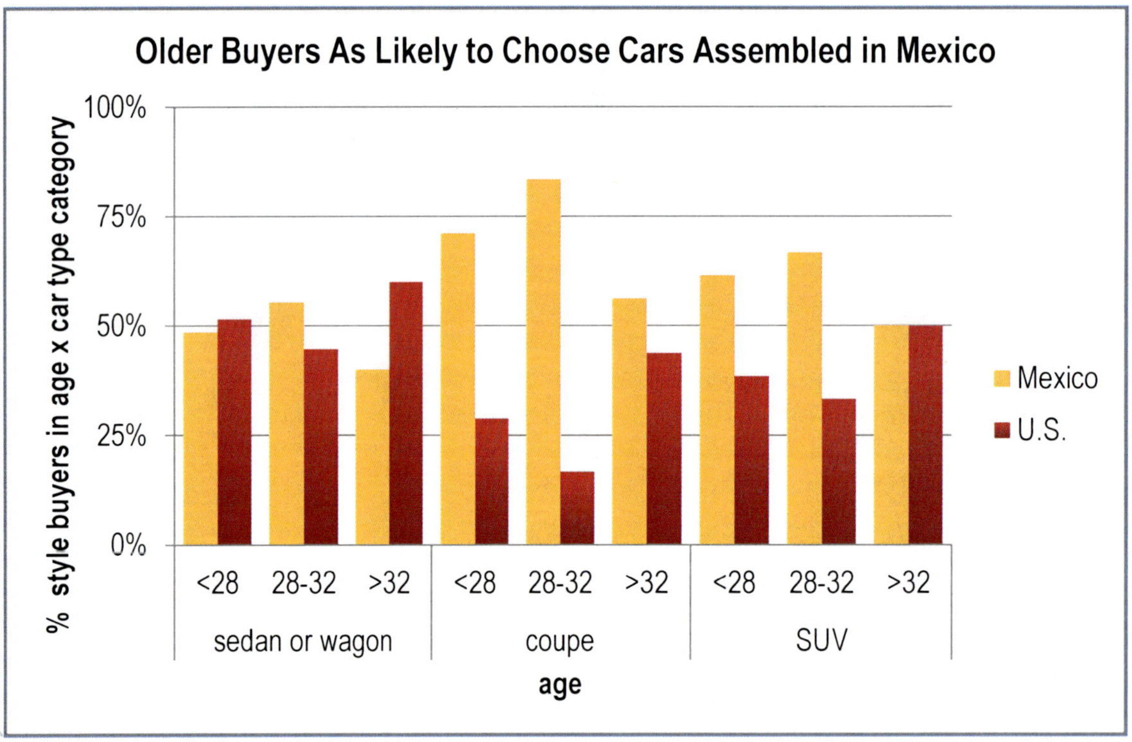

Figure 9.7 Contingency analysis: origin choice given age by style

Table 9.3 Contingency analysis: origin choice by age given style

		% Age by Style *Assembled In*					
					χ^2	*df*	*p value*
Style	*Age*	*Mexico*	*U.S.*	*Total*			
sedan or wagon	*under 28*	48%	52%	100%			
	28 to 32	55%	45%	100%			
	33 plus	40%	60%	100%			
total		47%	53%	100%	2.5	2	.29
coupe	*under 28*	71%	29%	100%			
	28 to 32	83%	17%	100%			
	33 plus	56%	44%	100%			
total		71%	29%	100%	3.0	2	.22
SUV	*under 28*	62%	38%	100%			
	28 to 32	67%	33%	100%			
	33 plus	50%	50%	100%			
total		59%	41%	100%	.9	2	.63
Grand Total		56%	44%	100%			

Controlling for style of car by looking at each style separately reveals lack of association between origin preference for U.S. versus Mexican made cars and age category. Across all three car styles, *p values* are greater than .05. There is not sufficient evidence in this sample to reject the null hypothesis. We conclude from this sample that the U.S. versus Mexican assembled choice is not associated with age category. The domestic automobile manufacturer should therefore not alter plans to move production South.

Simpson's Paradox describes the situation where two variables appear to be associated only because of their mutual association with a third variable. If the third variable is ignored, results are misleading. Because contingency analysis focuses upon just two variables at a time, analysts should be aware that apparent associations may come from confounding variables, as the **American Cars** example illustrates.

The Research Team summarized these results in this memo:

MEMO

Re.: Country of Assembly Does Not Affect Older Buyers' Choices
To: CEO, American Car Company
Emily Ernst, Director of Planning and Strategy
Brand Management
From: Travis Hendershott, Director of Quantitative Analysis

Analysis of a sample of new car buyers reveals that styles of car drive the choices of distinct age segments. Choices of all ages of buyers are independent of country of manufacture.

Contingency Analysis. Choices of 263 new car buyers were analyzed to assess the dependence of choice on country of manufacture, U.S. or Mexico, and age category.

Choice between Mexican and U.S. cars does not depend on buyer age.

Car choices between Mexican and U.S. cars do not depend on buyer age, though choices between *styles* are age dependent.

Younger buyers are more likely to choose a sporty coupe. Older buyers are more likely to buy a sedan or wagon.

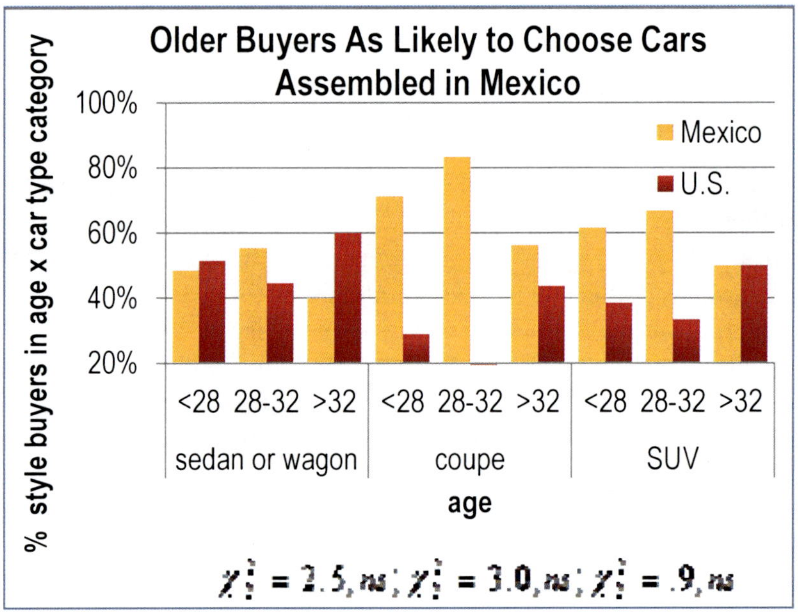

Production in Mexico will not influence car choices

Production in Mexico is not expected to affect car buyer choices, providing the opportunity to shift assembly South to take advantage of cheaper labor.

A larger, more diverse sample may provide additional information

A larger sample would enable examination of more representative age categories, and specifically, a broader middle segment and older oldest segment.

9.5 Contingency Analysis Is Demanding

Contingency analysis requires a large and balanced dataset to insure a stable chi square. Even large samples may contain small proportions of particular categories, forcing combinations that aren't ideal. In the **American Cars** example, a broad category was used for the oldest age segment, combining fairly different ages, 33 through 60, and a narrow category was defined for the middle age segment, ages 28 through 32. The sample, though large, was not balanced and contained a large proportion of car buyers ages 30 through 39. This group was split and combined with sparse younger and older age categories to allow expected cell counts greater than five. With smaller samples, just two categories for a variable may remain, which may limit hypothesis testing. In the **Recruiting Stars** example, final results could not be used to assess the association between recruiting and poor employee performance after Poor and Average performing employees were combined.

9.6 Contingency Analysis Is Quick, Easy, and Readily Understood

Despite the fairly demanding data requirements, contingency analysis is appealing because it is simple, and results are easily understood. For very large samples, sparse cells are not a problem and many categories may be used, increasing the specificity of results and allowing a range of hypothesis tests.

For smaller samples, other alternatives, such as logit analysis, exist for analyzing categorical variable associations. These carry fewer data demands and allow incorporation of multiple variables. Multivariate analysis helps us avoid drawing incorrect conclusions in cases where Simpson's Paradox might mislead.

Excel 9.1 Construct Crosstabulations and Assess Association Between Categorical Variables with PivotTables and PivotCharts

American Cars. In order to explore the possible association between choice of U.S.-assembled and Mexican-assembled cars by age, begin by making a PivotTable to see the crosstabulation.

Open **Excel 9.1 American Cars.**

Select filled cells in the *Age* and *Made In* columns and then insert a PivotTable.

Drag *Age* to **ROW,** *Made In* to **COLUMN,** and *Made In* to ∑ **Values.**

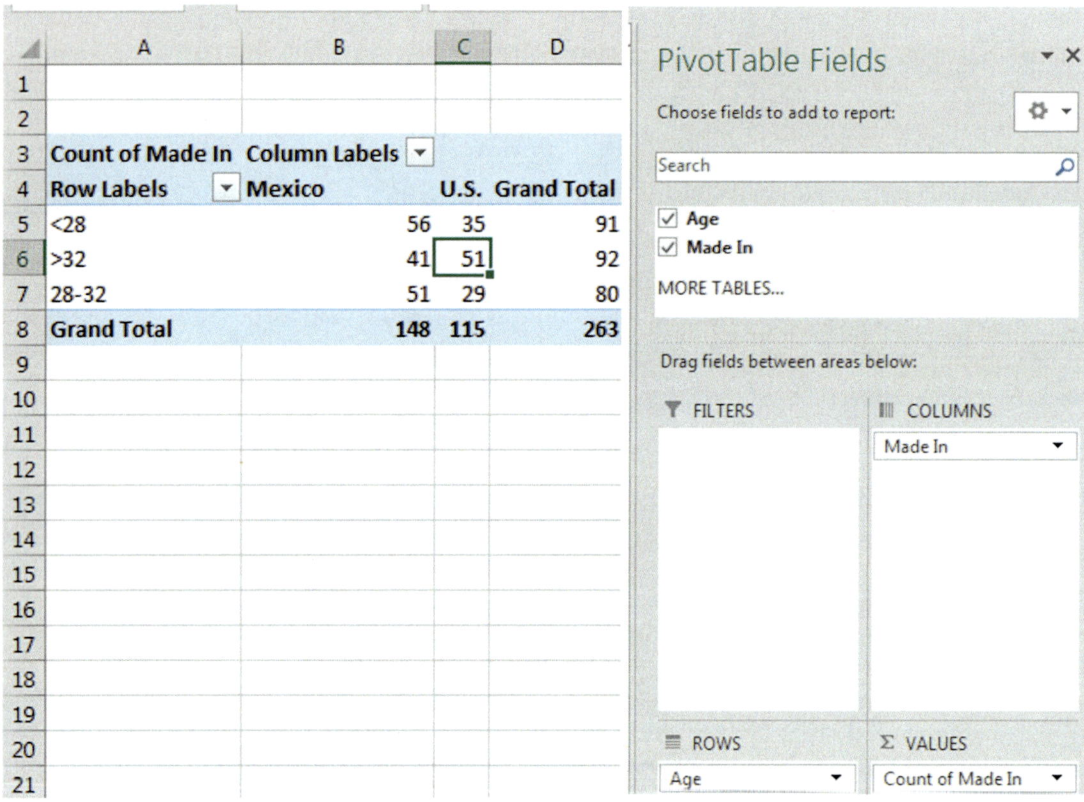

To see the conditional probabilities of choice of cars *Made In* the U.S. and Mexico given *age* category, convert cell counts to **% of row**.

From a data cell in the table:

Alt JT G Tab > Tab down to **% of Row Total, Enter**

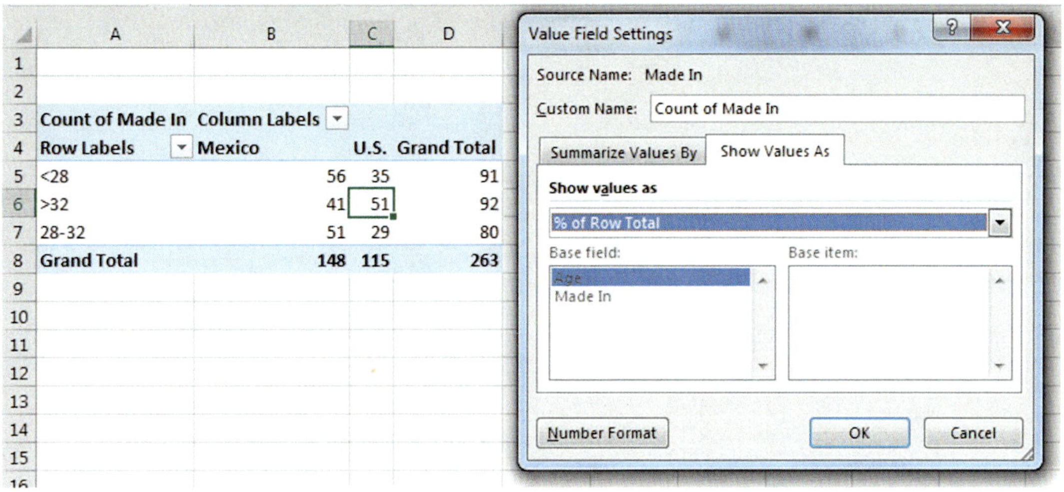

To put the age categories in order, select drag the *>32* cell to the end.

Count of Made In	Column Labels		
Row Labels	Mexico	U.S.	Grand Total
<28	61.54%	38.46%	100.00%
28-32	63.75%	36.25%	100.00%
>32	44.57%	55.43%	100.00%
Grand Total	**56.27%**	**43.73%**	**100.00%**

Make a PivotChart of *Made In* by *Age*: | **Alt JT C**

Choose a design and style, and add a chart title that reflects your conclusion.

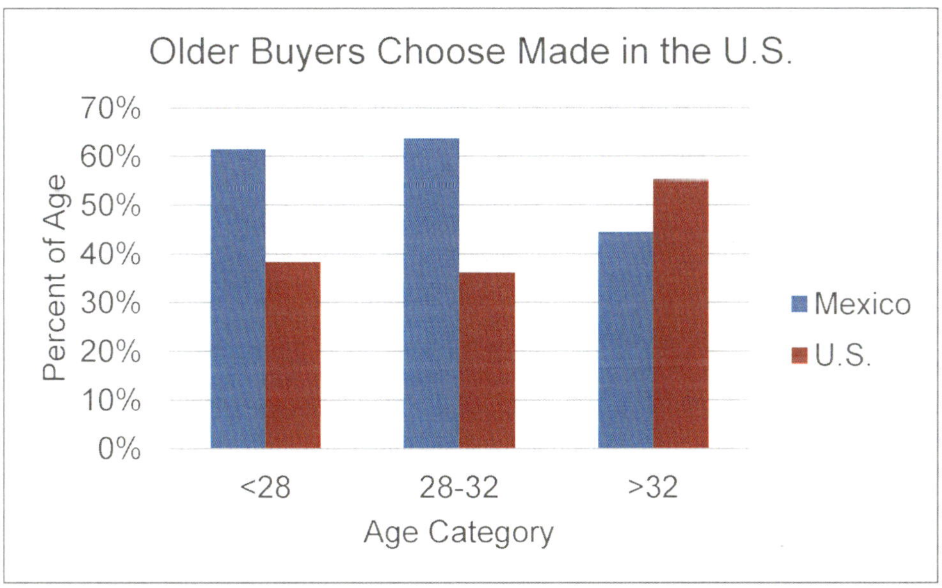

Excel 9.2 Use Chi Square to Test Association

To find the chi square statistic, change the PivotTable cells back to counts.

From a cell in the PivotTable,

Alt JT G Tab > Tab up to **No Calculation, Enter**

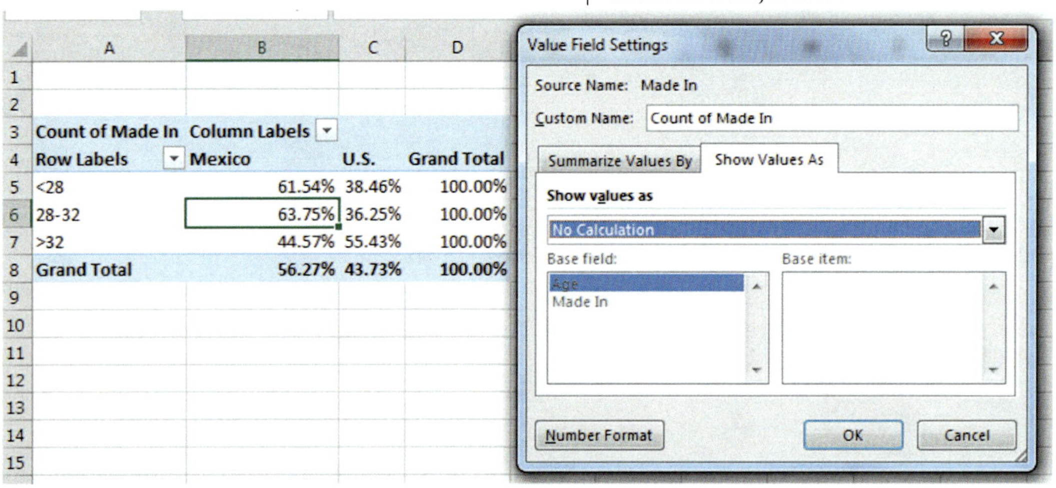

For chi square, make a table of *expected* cell counts and a table of cell contributions to chi square.

Select B3:C7, copy, and paste right of the PivotTable with values and formats, but not formulas,

Alt H V S U

Repeat to paste in a second copy.

	A	B	C	D	E	F	G	H
1								
2								
3	Count of Made In	Column Labels			Column Labels		Column Labels	
4	Row Labels	Mexico	U.S.	Grand Total	Mexico	U.S.	Mexico	U.S.
5	<28	56	35	91	56	35	56	35
6	28-32	51	29	80	51	29	51	29
7	>32	41	51	92	41	51	41	51

Find expected counts from the row, column and grand totals.

In E5, enter the formula for the expected count, multiplying cells containing the Grand Total of buyers < 28, D5, and the Grand Total of cars assembled in Mexico, B8, and then dividing by the Grand Total, D8:Mexico, B8, and then dividing by the Grand Total, D8:

=D5 fn4 fn4 fn4*B8 fn4 fn4/D8 fn4.

Pressing **fn 4** three times locks the column, pressing **fn 4** twice locks the row, and pressing **fn 4** once locks both, so that we can fill in the remaining cells in the table with this formula.

Fill in the column, and then fill in the adjacent row.

	A	B	C	D	E	F	
1							
2							
3	**Count of Made In**	**Column Labels**			Column Labels		C
4	**Row Labels**	**Mexico**	**U.S.**	**Grand Total**	Mexico	U.S.	N
5	<28	56	35	91	51.20913	39.79087	
6	28-32	51	29	80	45.01901	34.98099	
7	>32	41	51	92	51.77186	40.22814	
8	**Grand Total**	**148**	**115**	**263**			

In the third table, find each cell's contribution to chi square, the squared difference between expected counts, in the second table, and actual counts, in the first table, divided by expected counts in the second table.

In the first cell, G5, of the third contributions to chi square table, **e**nter:

=**(E5-B5)^2/E5.**

Fill in the column and the rows:

G5 | f_x =(E5-B5)^2/E5

	A	B	C	D	E	F	G	H
1								
2								
3	**Count of Made In**	**Column Labels**			Column Labels		Column Labels	
4	**Row Labels**	**Mexico**	**U.S.**	**Grand Total**	Mexico	U.S.	Mexico	U.S.
5	<28	56	35	91	51.20913	39.79087	0.448211	0.576828
6	28-32	51	29	80	45.01901	34.98099	0.794603	1.022619
7	>32	41	51	92	51.77186	40.22814	2.241237	2.884375
8	**Grand Total**	**148**	**115**	**263**				

Use the Excel function **SUM(***array1,array2)* to find contributions to chi square from each of the age categories.

In I5,
=sum(G5:H5)

Add the cell contributions in the Mexico column, I:

In the Grand Total row, find the Mexico sum. In G8, | **Alt MUS**

Right fill to add contributions to chi square from U.S. cars and chi square:

G8 =SUM(G5:G7)

	A	B	C	D	E	F	G	H	I
1									
2									
3	Count of Made In	Column Labels			Column Labels		Column Labels		
4	Row Labels	Mexico	U.S.	Grand Total	Mexico	U.S.	Mexico	U.S.	
5	<28	56	35	91	51.20913	39.79087	0.448211	0.576828	1.025038
6	28-32	51	29	80	45.01901	34.98099	0.794603	1.022619	1.817222
7	>32	41	51	92	51.77186	40.22814	2.241237	2.884375	5.125612
8	Grand Total	148	115	263			3.484051	4.483822	7.967873
9									

Find the *p value* for this chi square using the Excel function

CHISQ.DIST.RT(*chisquare,df***)** with degrees of freedom *df* of 2

I9 =CHISQ.DIST.RT(I8,2)

	G	H	I	J	K	L
4	Mexico	U.S.				
5	0.448211	0.576828	1.025038			
6	0.794603	1.022619	1.817222			
7	2.241237	2.884375	5.125612			
8	3.484051	4.483822	7.967873	chisquare		
9			0.018612	pvalue		
10						

Excel 9.3 Conduct Contingency Analysis with Summary Data

Sometimes data are in summary form. That is, we know the sample size, and we know the percent of the sample in each category.

Marketing Cereal to Children. Kooldogg expects that many Saturday morning cartoon viewers would be attracted to their sugared cereals. A heavy advertising budget for sugared cereals is allocated to Saturday morning television. We will use contingency analysis to analyze the association between Saturday morning cartoon viewing and frequent consumption of

Kooldogg cereal with sugar added. From a survey of 300 households, researchers know whether or not children ages 2 through 5 *Watch Saturday Morning Cartoons* on a regular basis (at least twice a month) and whether or not those children *Eat Kooldogg Cereal with Added Sugar* (at least once a week).

Open **Excel 9 Kooldogg Kids Ads.**

Select the summary data and make a PivotTable, with *Watches Saturday Morning Cartoons* in **Rows,** *Eats Kooldogg Sugary Cereal* in **Columns,** and *Number of Children* in **Σ Values:**

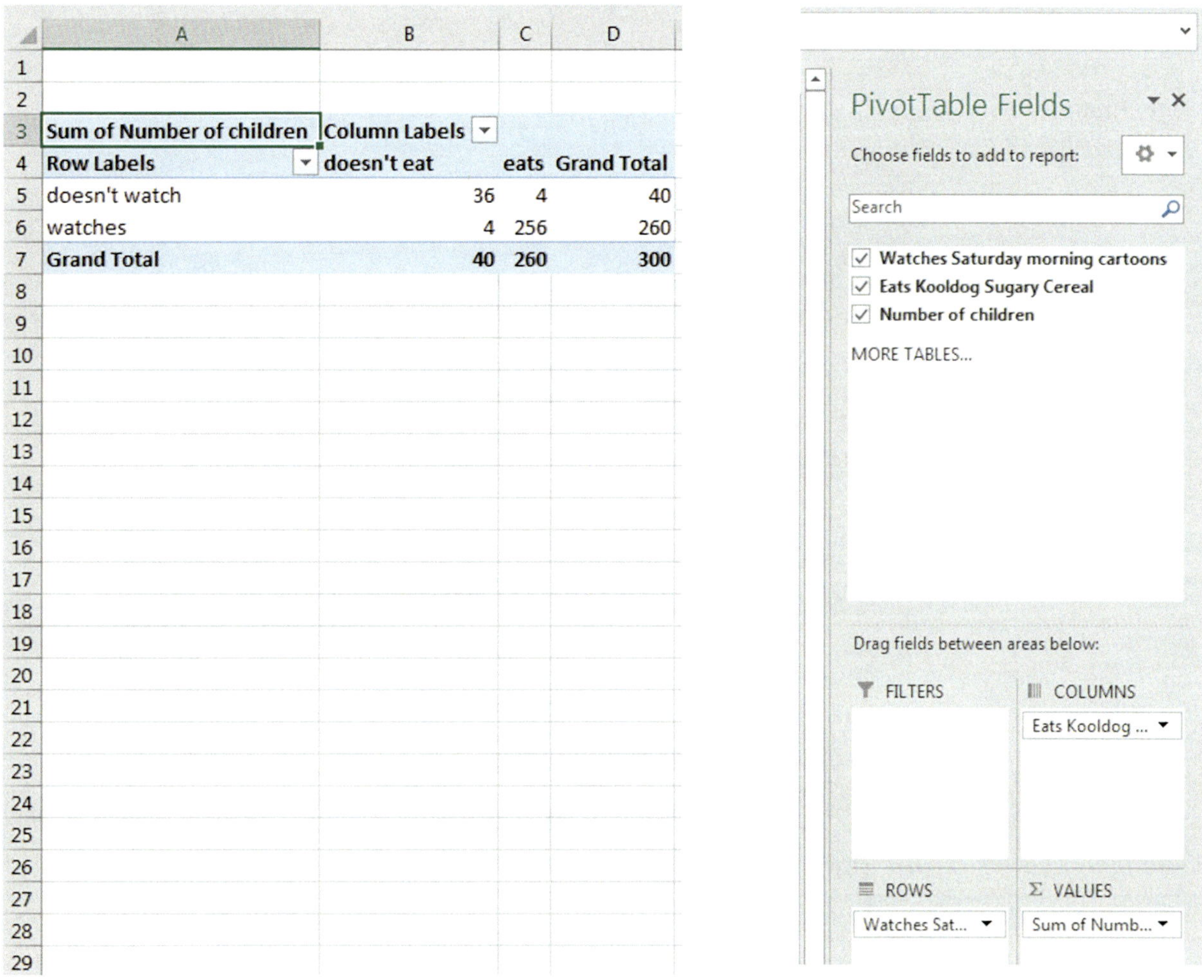

Copy B3:C6 and paste twice with formats and values, **Alt HVSU:**

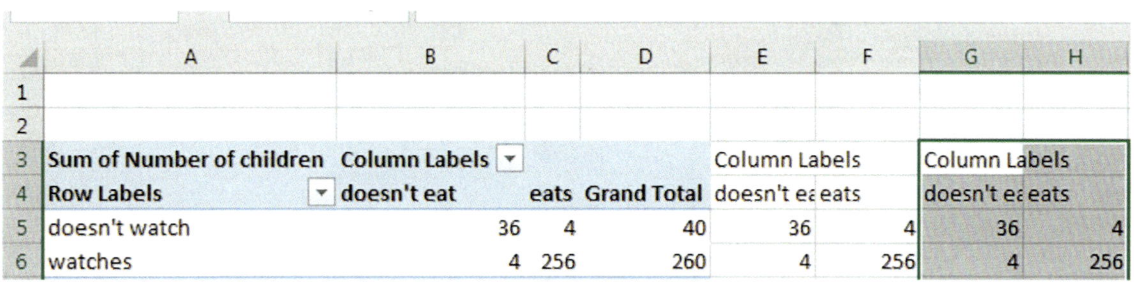

In the second table, find the expected cell counts under the assumption that Kooldog cereal consumption is independent of Saturday morning TV viewing.

E5 =$D5*B$7/$D$7

	A	B	C	D	E	F
1						
2						
3	Sum of Number of children	Column Labels			Column Labels	
4	Row Labels	doesn't eat	eats	Grand Total	doesn't ea	eats
5	doesn't watch	36	4	40	5.333333	34.66667
6	watches	4	256	260	34.66667	225.3333
7	Grand Total	40	260	300		

In the third table, find cell contributions to chi square with squared differences between expected cell counts and actual cell counts, divided by expected cell counts.

G5 =(E5-B5)^2/E5

	A	B	C	D	E	F	G	H
1								
2								
3	Sum of Number of children	Column Labels			Column Labels		Column Labels	
4	Row Labels	doesn't eat	eats	Grand Total	doesn't ea	eats	doesn't ea	eats
5	doesn't watch	36	4	40	5.333333	34.66667	176.3333	27.12821
6	watches	4	256	260	34.66667	225.3333	27.12821	4.17357

Sum the cell contributions to chi square in to find chisquare.

G7 =SUM(G5:G6)

	G	H	I	J	K
4	doesn't ea	eats			
5	176.3333	27.12821	203.4615		
6	27.12821	4.17357	31.30178		
7	203.4615	31.30178	234.7633		

Use **CHISQ.DIST.RT()** to find the *p value* of chi square:

I8 =CHISQ.DIST.RT(I7,1)

	G	H	I	J	K	L
4	doesn't ea	eats				
5	176.3333	27.12821	203.4615385			
6	27.12821	4.17357	31.30177515			
7	203.4615	31.30178	234.7633136	chisquare		
8			5.45241E-53	p value		

Based on sample evidence, the null hypothesis of independence is rejected. Eating cereal with added sugar is associated with Saturday morning cartoon viewing.

To see the association, change the cell counts to percents of row.

From a cell in the PivotTable,
Alt JT G Tab > Tab dn to **% Row Total**

	A	B	C	D
4	**Row Labels**	**doesn't eat**	**eats**	**Grand Total**
5	doesn't watch	90.00%	10.00%	100.00%
6	watches	1.54%	98.46%	100.00%
7	**Grand Total**	**13.33%**	**86.67%**	**100.00%**

Make a **PivotChart** with shortcuts **Alt JT C** to see the association.

Choose a design and style, add axes titles and a chart title which summarizes your conclusion:

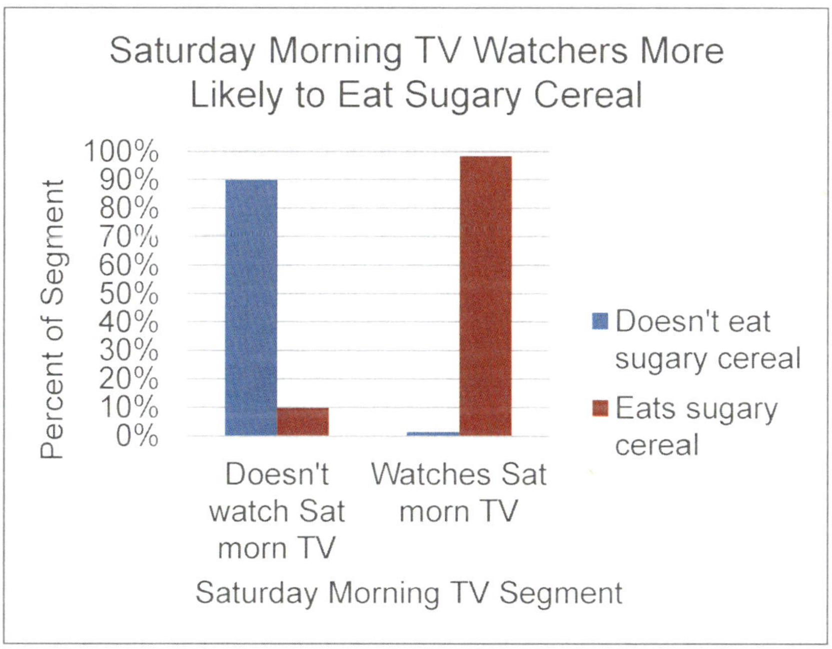

Lab 9 Skype Appeal

Following the launch of Google's Android phone, rumors surfaced that Google was considering a joint venture with Skype. Skype boasts more than 330 million users worldwide, with as many as 80 million being connected at one time. Google management believes that Skype appeals most to younger consumers.

Google conducted a survey of 133 randomly chosen consumers, from ages 14 through 65. Consumers were asked which they relied on more: (i) cell phone, or (ii) Skype.

Responses are in **Lab 9 Skype Appeal.**

I. Skype Use by Age Category

1. What are Google's hypotheses?

   Make a PivotTable of the cross tabs of callers by age and choice.

2. Make a table with the ***expected*** number of callers by age and type of call:

expected count in row *i* column *j* = *number in row i* × *percent in column j*

$$e_{ij} = n_i p_j$$

OR *expected users in age segment i of call type j* = *number in age segment i* × *percent who use calltype j*

How many of the 26 18 to 21 year olds would you expect to rely on Skype ***if*** call choice is ***not*** associated with age? _________

3. Mark (X) cells which are *sparse:*

Age	Cell phone	Skype
14–17		
18–21		
22–29		
30–39		
40–49		
50–59		
60–65		

   Group (circle) age segments so that no cells are sparse.
   Update your *expected* cell counts.

4. Make a table of cell contributions to chi square: $(e_{ij} - n_{ij})^2/e_{ij}$

Sum the cell contributions to chi square to find chi square: ___________

5. Find the p value for your chi square with __ (=(rows−1)×(columns−1)) df: _________

6. Is choice of call type dependent on age? Y or N

7. To see the conditional probabilities of choice by age category, change cell counts to percents of row.

   What percent of callers surveyed rely on more Skype? ____________%

   What percent of 18 to 21 year olds rely more on Skype? _________%

8. Illustrate your results using a column chart.

II. Skype Use by Age Category and Distance from Family

Management believes that Skype choice may differ between consumers with and without distant family. Repeat your analysis, using only consumers with distant family.

9. For consumers with distant family, is choice of call type dependent on age? Y or N

10. *Choice* most dependent on age: ______________

11. *Choices* depend most on *age segment*: __________

12. Make a PivotChart to illustrate your results.

13. For those with distant family, which *age segment*(s) are more likely than average to rely on

    Skype? ____________

Assignment 9.1 Wine Preferences by Global Region

Concha y Toro would like to know if preferences for South American, Old World (European) and New World (U.S.) wines are associated with a consumer's global region of residence. 95 consumers from South America, California, Virginia, other parts of the U.S., France, Spain, Italy, and other European countries were asked to indicate their most preferred wine. They are interested in learning the probabilities that New World and Old World consumers will prefer South American wine. These data are in **Concha y Toro wine prefs**.

1. Is country of origin of wine preferred associated with residence? ___Y ___N
   Cite your evidence, including the statistic and pvalue: ________________
2. Present a column chart illustrating your conclusion:
3. What is the probability that a consumer will prefer South American wine? _____
4. What is the probability that a U.S. consumer will prefer South American wine? _____

Assignment 9.2 Fit Matters

Procter & Gamble management would like to know whether intent to try their new preemie diaper concept is associated with the importance of fit. If Likely Triers value fit more than Unlikely Triers, fit could be emphasized in advertisements.

Assignment 9.2 Fit Matters contains data from a concept test of 97 mothers of preemie diapers, including trial *Intention* and *Fit Importance*, measured on a 9-point scale.

You may decide to combine categories.

- Use contingency analysis to test the hypothesis that *intent* to try is associated with the *importance of fit*.
- If the association is significant, explain the nature of association.
- Include a PivotChart and explain what it illustrates.

Assignment 9.3 Netbooks in Color

Dell managers want to know whether college students' preferences for light weight netbooks and wide color selection are associated with major. Dell's netbook is lighter than many competing netbooks and comes in more colors than any other netbook. Managers believe that light weight and wide choice of colors may appeal to Arts & Sciences and Commerce students more than to Engineering students, which would give Dell an advantage to be promoted in those segments.

A sample of netbook and iPad owners was drawn from each of three schools on the UVA campus, Commerce, Arts & Sciences, and Engineering. Netbook or iPad brands owned by students were recorded. Those data, with number of *colors* available and *weight* are in **netbooks.**

Determine whether preference for light weight and variety of colors are associated with college major.

1. State the hypotheses that you are testing.
2. What are your conclusions? (Include the statistical tests that you used to form your conclusions.)
3. What is the probability that:
   a. A netbook or iPad owner will own a light weight brand?
   b. An Arts & Science student will own a light weight brand?
   c. A Commerce student will own a light weight brand?
   d. An Engineering student will own a light weight brand?
4. What is the probability that:
   a. A netbook or iPad buyer will own a brand available in at least 6 colors?
   b. An Arts & Science student will own a brand available in at least 6 colors?
   c. A Commerce student will own a brand available in at least 6 colors?
   d. An Engineering student will own a brand available in at least 6 colors?

A Dell Intern believes that conclusions may differ if only netbooks owners are considered, excluding the unique segment of iPad owners.

5. Repeat your analyses using excluding iPad owners. Summarize your conclusions, including the statistics that you used.
6. Illustrate your netbook results (excluding iPad owners) with PivotCharts.

Case 9.1 Hybrids for American Car

Environmental concerns have led some customers to switch to hybrid cars.

American Car (AC) offers two hybrids, AC Sapphire and AC Durado, an SUV and a pickup. AC offers no hybrid automobiles. Major competitors, Ford, Toyota and Honda, offer hybrid automobiles. AC executives believe that with their hybrid SUV and pickup, they will be able to attract loyal AC customers who desire a hybrid. Shawn Green, AC Division Head, is worried that customers who were driving conventional sedans, coupes or wagons may not want a truck or an SUV. They might switch from AC to Ford, Toyota or Honda in order to purchase a hybrid car.

To investigate further, Mr. Green commissioned a survey of car buyers. The new car purchases of a representative random sample of 4000 buyers were sorted into eight groups, based on the type of conventional car they had owned and *Traded* (Prestige Sport, Compact SUV, Large, and Full-size SUV) and whether or not they bought *Hybrid* or Conventional. These data are in **Case 9.1 Hybrid**. The number of *Buyers* indicates popularity of each *Traded*, *Hybrid* combination.

Conduct contingency analysis with this data to determine whether *choice of hybrid vehicles* depends on *type of vehicle owned previously*.

Specifically,

1. Is there an association between the *type of car owned and Traded* and *choice of a Hybrid* instead of a Conventional car?
2. What is the probability that a new car buyer will choose a *hybrid*?
3. How likely is each of the segments to switch to hybrids?
4. Illustrate your results with a PivotChart. Include a bottom-line title.
5. What are the implications of results for American Car Division?
   What is your advice to Mr. Green?

Case 9.2 Tony's GREAT Advertising

Kellogg spends a hefty proportion of its advertising budget to expose children to ads for sweetened cereal on Saturday mornings. Kellogg brand ads feature cartoon hero characters similar to the cartoon hero characters that children watch on Saturday morning shows. This following press release is an example:

Advertising Age
Kellogg pounces on toddlers; Tiger Power to wrest tot monopoly away from General Mills' $500M Cheerios brand. (News) *Stephanie Thompson.*

Byline: STEPHANIE THOMPSON

In the first serious challenge to General Mills' $500 million Cheerios juggernaut, Kellogg is launching a toddler cereal dubbed Tiger Power.

The cereal, to arrive on shelves in January, will be endorsed by none other than Frosted Flakes icon Tony the Tiger and will be "one of our biggest launches next year," according to Kellogg spokeswoman Jenny Enochson. Kellogg will position the cereal-high in calcium, fiber and protein-as "food to grow" for the 2-to-5 set in a mom-targeted roughly $20 million TV and print campaign that begins in March from Publicis Groupe's Leo Burnett, Chicago.

Cereal category leader Kellogg is banking on Tiger Power's nutritional profile as well as the friendly face of its tiger icon, a new shape and a supposed "great taste with or without milk" to make a big showing in take-along treats for tots.

Tony Grate, the brand manager for Frosted Flakes would like to know whether there is an association between Saturday morning cartoon viewing and consumption of his brand.

The Saturday morning TV viewing behaviors, *Saturday Morning Cartoons*, and consumption of Frosted Flakes, *Frosted Flake Eater*, are contained in **Case 9.2 Frosted Flakes.** A random sample of 300 children ages 2 through 5 were sorted into four groups based on whether or not each watches at least 3 hours of television on Saturday morning at least twice a month and whether or not each consumes Frosted Flakes at least twice times a week. The number of *Children* indicates popularity of each *Saturday Morning Cartoons*, *Frosted Flake Eater* combination.[1]

1. Is there an association between watching *Saturday morning cartoons* and consumption of Frosted Flakes?
2. What is the probability that a *cartoon watcher* consumes Frosted Flakes?
3. How likely is each segment to consume Frosted Flakes?
4. Illustrate your results with a properly labeled PivotChart. Include a bottom-line title.
5. What are the implications of results for Tony Grate?

Case 9.3 Hybrid Motivations

American car executives have asked you to analyze data collected from a stratified sample of 301 car owners. The goal is to develop a profile of hybrid owners, which distinguishes them from conventional car owners. If differences are identified, those differences will be used to promote American's hybrid models.

[1] These data are fictitious, though designed to reflect a realistic scenario.

One third of the owners surveyed own Priuses, one third own other hybrids, and one third own a conventional car. Car owners were asked to indicate the primary motivation which led to the last car choice. Possible responses included three *functional* benefits (fuel economy, lower emissions, tax incentives), *aesthetics/style* and *vanity* ("makes a statement about me"). Data are in **9 hybrid motivations.**

It is thought that choice of a hybrid is associated with vanity… the desire to make a statement.

1. What is the probability that a car buyer was motived by vanity?
2. What is the probability that a buyer motivated by vanity chose a Prius?
3. What is the probability that a buyer motivated by vanity chose another hybrid model?
4. What is the probability that buyer motivated by vanity chose a conventional car?
5. Is choice of car type associated with motivation? Y or N
6. Statistic and p value you used to reach your conclusion in 5:
7. Embed a column chart which illustrates the association between motivation and car choice. Include axes labels with units and a title which describes what the conclusions which the audience should see:

Some American executives believe that Prius owners are unique.

8. Excluding Prius owners, is choice of car type associated with motivation? Y or N
9. Statistic and p value you used to reach your conclusion in 8:

Chapter 10
Building Multiple Regression Models

Explanatory multiple regression models are used to accomplish *two* complementary goals: *identification of key drivers of performance* and *prediction of performance under alternative scenarios*. The variables selected affect both the explanatory accuracy and power of models, as well as forecasting precision. In this chapter, the focus is on variable selection, the first step in the process used to build powerful and accurate multiple regression models.

Multiple regression offers a major advantage over simple regression. Multiple regression enables us to account for the joint impact of multiple drivers. Accounting for the influence of multiple drivers provides a truer estimate of the impact of each one individually. In real world situations, multiple drivers together influence performance. Looking at just one driver, as we do with simple regression, we are very likely to conclude that its impact is much greater than it actually is. A single driver takes the credit for the joint influence of multiple drivers working together. For this reason, multiple regression provides a clearer picture of influence.

We use logic to choose variables initially. Some of the variables which logically belong in a model may be insignificant, either because they truly have no impact, or because their influence is part of the joint influence of a correlated set of predictors which together drive performance. *Multicollinear* predictors create the illusion that important variables are insignificant.

If an insignificant predictor adds little explanatory power, it is removed from the model. It is either not a performance driver, or it is a driver, but it is a redundant driver, because other variables reflect the same driving dimension. Simple regression of the potential driver in question help to distinguish whether or not it is multicollinearity that is producing insignificance for a variable.

10.1 Explanatory Multiple Regression Models Identify Drivers and Forecast

Explanatory multiple regression models are used to achieve two complementary goals: identification of key *drivers* of performance and prediction of performance under alternative scenarios. This prediction can be either what would have happened had an alternate course of action been taken, or what can be expected to happen under alternative scenarios in the future.

Decision makers want to know, given uncontrollable external influences, which controllable variables make a difference in performance. We also want to know the nature and extent of each of the influences when considered together with the full set of important influences. A multiple regression model will provide this information.

Once key drivers of performance have been identified, a model can be used to compare performance predictions under alternative scenarios. This *sensitivity analysis* allows managers to compare expected performance levels and to make better decisions.

C. Fraser, *Business Statistics for Competitive Advantage with Excel 2016*,
DOI 10.1007/978-3-319-32185-1_10

10.2 Use Your Logic to Choose Model Components

The first step in model building happens before looking at data or using software. Using logic, personal experience, and others' experiences, we first decide which of the potential influences ought to be included in a model. From the set of variables with available data, which could reasonably be expected to influence performance? In most cases, a reason is needed for including each independent variable in a model. Independent variables tend to be related to each other in our correlated world, and models are unnecessarily complicated if variables are included which don't logically affect the dependent performance variable. This complication from correlated predictors, *multicollinearity,* is explored later in the chapter.

Example 10.1 Sakura Motors Quest for Cleaner Cars. The new product development group at Sakura Motors is in the midst of designing a new line of cars which will offer reduced greenhouse gas emissions for sale to drivers in global markets where air pollution is a major concern. They expect to develop a car that will emit only 5 tons of greenhouse gases per year.

What car characteristics drive emissions? The management team believes that smaller, lighter cars with smaller, more fuel efficient engines will be cleaner. The U.S. Government publishes data on the fuel economy of car models sold in the U.S. (fueleconomy.gov), which includes *manufacturer, model*, engine size (*cylinders*), and gas mileage (*MPG*) for each category of car. This data source also includes *emissions* of tons of greenhouse gases per year. A second database, consumerreports.org, provides data on acceleration in *seconds* to go from 0 to 60 miles per hour, which reflects car model sluggishness, and two measures of size, *passengers* and curb *weight*. Management believes that responsiveness and size may have to be sacrificed to build a cleaner car.

The multiple linear regression model of *emissions* will include these car characteristics, *miles per gallon (MPG), seconds* to accelerate from 0 to 60, *horsepower, liters, cylinders, passenger* capacity, and weight in *pounds(K),* each thought to drive *emissions*:

$$\hat{emissions}\left(\frac{tons}{yr}\right)_i = b_0\left(\frac{tons}{yr}\right) + b_1\frac{tons/yr}{mi/gal} \times MPG_i + b_2\frac{tons/yr}{secs} \times seconds_i$$

$$+ b_3\frac{tons/yr}{hp} \times horsepower_i + b_4\frac{tons/yr}{liter} \times liters_i$$

$$+ b_5\frac{tons/yr}{cyl} \times cylinders_i + b_6\frac{tons/yr}{pass} \times passengers_i$$

$$+ b_7\frac{tons/yr}{lbs} \times pounds_i$$

where $\widehat{emissions}_i$ is the expected tons of annual emissions of the *i*th car model,

b_0 is the intercept indicating expected emissions if *MPG, seconds, pounds(K), passengers, horsepower, cylinders* and *liters* were zero,

$b_1, b_2, b_3, b_4, b_5, b_6, b_7$ are the regression coefficient estimates indicating the expected marginal impact on emissions of a unit change in each car characteristic when other characteristics are at average levels, and

MPG_i, $seconds_i$, $horsepower_i$, $cylinders_i$, $liters_i$, $passengers_i$, $pounds(K)_i$ are characteristics of the *i*th car model.

When more than one independent variable in included in a linear regression, the coefficient estimates, or parameters estimates, are *marginal*. They estimate the marginal impact of each predictor on performance, given average levels of each of the other predictors.

The new product development team asked the model builder to choose a sample of car models which represents extremes of emissions, worst and best. Thirty five car models were included in the sample. These included imported and domestic cars, subcompacts, compacts, intermediates, full size sedans, wagons, SUVs, and pickups. Within this set there are considerable differences in all of the car characteristics, shown in Table 10.1 and Figure 10.1.

Table 10.1 Car characteristics in the Sakura Motors sample

Car Characteristic	*Minimum*	*Median*	*Maximum*
Emissions (tons)	5.2	8.7	12.5
MPG	15	22	34
Seconds (0 to 60)	7	9	12
Passengers	4	5	9
Pounds(K)	2.5	4.0	5.9
Horsepower	108	224	300
Cylinders	4	6	8
Liters	1.5	3.3	6.0

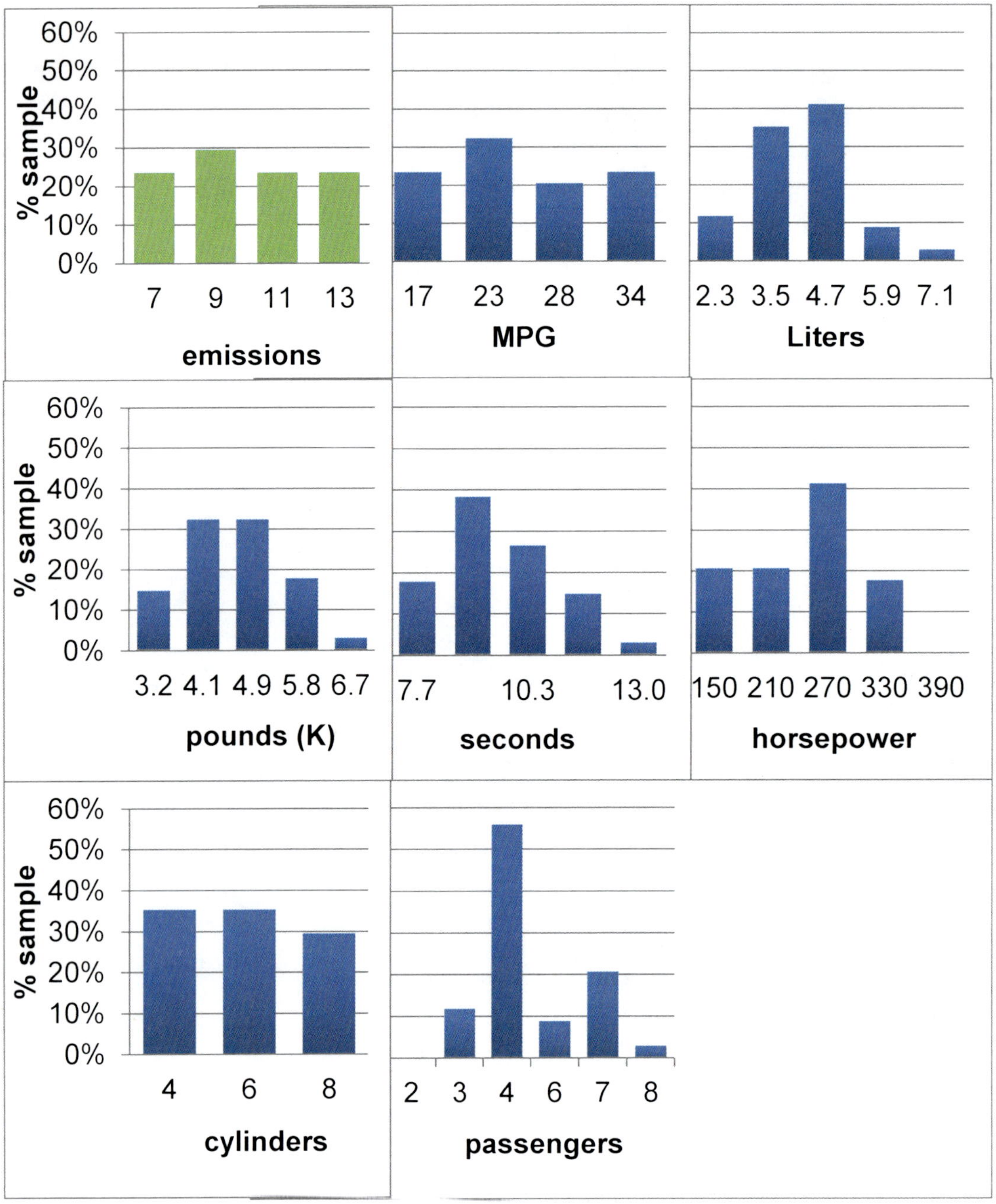

Figure 10.1 Car characteristics in the Sakura Motors sample

10.3 Multicollinear Variables Are Likely When Few Variable Combinations Are Popular in a Sample

Since these data come directly from the set of cars actually available in the market, many characteristic combinations do not exist. For example, there is no car with a 1.5 liter engine that weighs 4000 pounds. The seven car characteristics tend to be related to each other and come in particular combinations in existing cars. We are knowingly introducing correlated independent variables, also called *multicollinear* independent variables, into our model, because the characteristic combinations which are not represented do not exist.

Results from Excel are shown in Table 10.2.

Table 10.2 Multiple linear regression of emissions with seven car characteristics

SUMMARY OUTPUT					
Regression Statistics					
R Square		.928			
Standard Error		.644			
Observations		34			
ANOVA	*df*	*SS*	*MS*	*F*	*Significance F*
Regression	7	138	19.8	47.7	.0001
Residual	26	11	.4		
Total	33	149			

RSquare is .928, or 93%, indicating that, *together*, variation in the seven car characteristics accounts for 93% of the variation in emissions. The *standard error* is .64, which indicates that forecasts of emissions would be within about 1.3 tons of average actual emissions for a particular car configuration.

10.4 *F* Tests the Joint Significance of the Set of Independent Variables

F tests the null hypothesis that *RSquare* is 0%, or, equivalently, that all of the coefficients are zero:

$$\mathbf{H_0}\text{: } Rsquare = 0$$

Versus

$$\mathbf{H_1}\text{: } Rsquare > 0$$

OR

$$\mathbf{H_0}\text{: All of the coefficients are equal to zero, } \beta_i = 0$$

Versus

$$\mathbf{H_1}\text{: At least one of the coefficients is not equal to zero.}$$

The F test compares explained to unexplained variation, which would be zero, under the null hypothesis. Sample evidence, shown in Figure 10.2, will enable rejection of the null hypothesis if the explained slice is large enough.

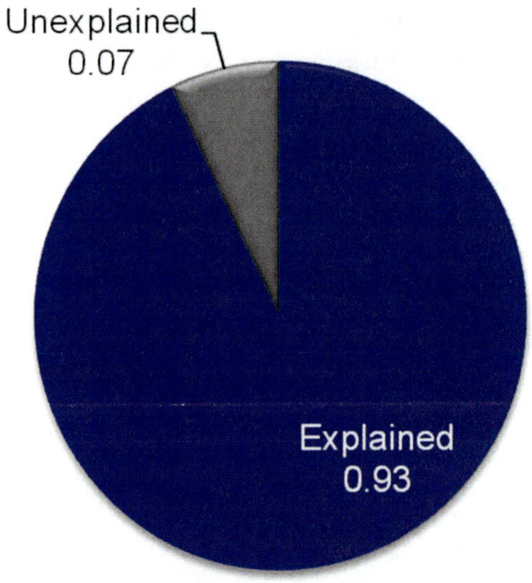

Figure 10.2 Explained and unexplained variation

More drivers increase potential variation explained. The *F* test accounts for the number of drivers, as well as the sample size, with the comparison of explained variation per predictor (the *regression degrees of freedom*) with unexplained variation for a given sample and model size, the residual degrees of freedom.comparison for **Sakura** is shown in Figure 10.3.

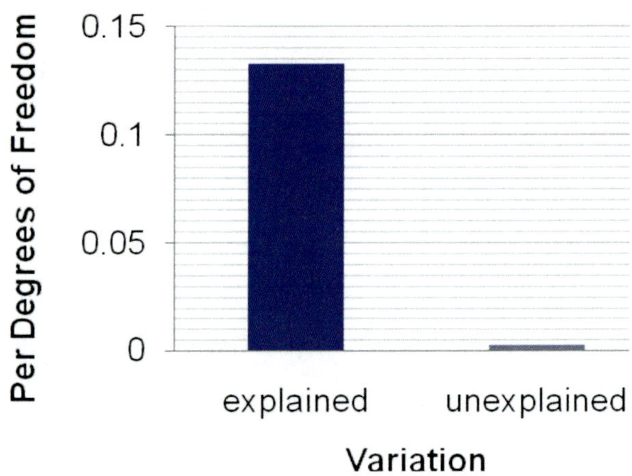

Figure 10.3 Explained and unexplained variation

The ratio of these produces an F statistic with the regression and residual degrees of freedom:

$$F_{regression\ df, residual\ df} = \frac{RSquare/regression\ df}{(1 - RSquare)/residual\ df}$$

For **Sakura**, the *F* statistic is 48, with 7 and 26 degrees of freedom:

$$F_{7,26} = \frac{.928/7}{(1 - .928)/26} = \frac{.133}{.0028} = 48$$

The *F* statistic is compared to the *F* distribution with the same degrees of freedom. Figure 10.4 illustrates *F* distributions for 1, 2, 4, and 7 predictors with a sample of 30.

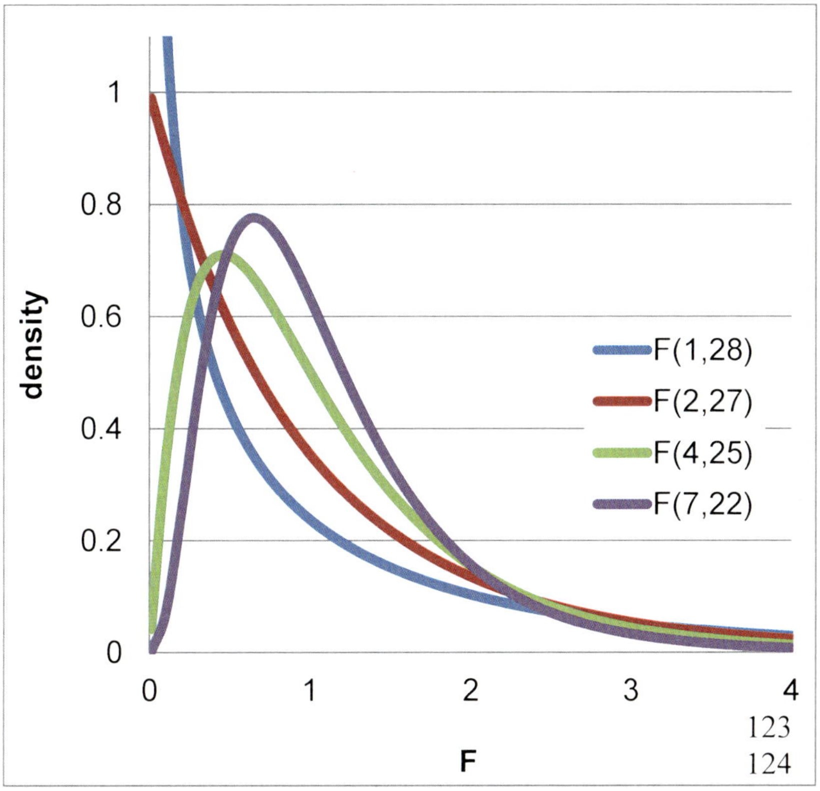

Figure 10.4 A family of F distributions for a regression with sample of 30

The **Sakura** model *F* is 48, which lies to the extreme right of the $F_{7,26}$ distribution. The *p value,* labeled *Significance F* in Excel, is .0001, indicating that it is unlikely that we would observe these data patterns, were none of the seven car characteristics driving emissions. It may be that just one of the seven characteristics drives emissions, or it may be that all seven are significant influences. With this set of seven predictors, some of the variation in *emissions* has been explained.

10.5 Insignificant Parameter Estimates Signal Multicollinearity

To determine which of the seven car characteristics are significant drivers of emissions, we initially look at the significance of *t tests* of the individual regression parameter estimates. A *t* statistic in multiple regression is used to test the hypothesis that a marginal coefficient is zero.

When we have no information about the direction of influence, a two tail test of each marginal slope is used:

$$\mathbf{H_0:}\ \beta_i = 0$$

Versus

$$\mathbf{H_1:}\ \beta_i \neq 0$$

In the more likely case that, when, from theory or experience, we know the likely direction of influence, a one tail test is used. When the suspected direction of influence is positive, the null and alternate hypotheses are:

$$\mathbf{H_0}\text{:}\ \beta_i \leq 0$$

Versus

$$\mathbf{H_1}\text{:}\ \beta_i > 0.$$

Conversely, when the expected direction of influence is negative the hypotheses are:

$$\mathbf{H_0}\text{:}\ \beta_i \geq 0,$$

Versus

$$\mathbf{H_1}\text{:}\ \beta_i < 0.$$

Excel provides a two tail *t* statistic for each marginal slope by making calculating the number of standard errors each marginal slope is from zero:

$$t_{residual\ df,i} = b_i / s_{b_i}$$

Notice that a *t* statistic of a marginal slope in multiple regression is compared with the *t* distribution for the residual degrees of freedom. For each predictor in the model, we lose one degree of freedom. Excel provides the corresponding *p value* for the two tail *t* test of each marginal slope. In the case that we want to use a one tail test, the *p value* is divided by two. The *t* distribution used in the *emissions* model, with 26 degrees of freedom, is shown in Figure 10.5.

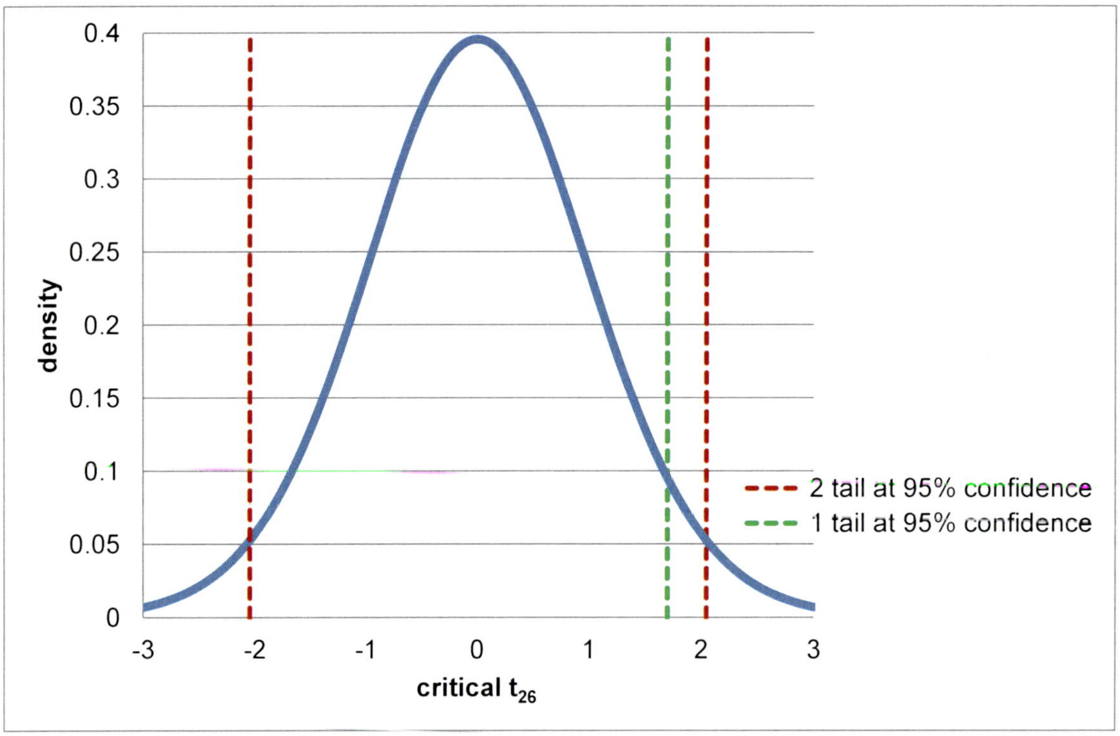

Figure 10.5 *t* distribution with 26 degrees of freedom

In the emissions model, Sakura analysts were confident that the impact of *MPG* on emissions ought to be negative, and that each of the influences of *horsepower, cylinders, liters, weight,* and *passengers* on *emissions* ought to be positive. For these six potential drivers, one tail *t* tests could be used. Sakura managers were not sure of the direction of influence of acceleration on *emissions*, and so a two tail test would be used for the *seconds* slope.

Table 10.3 Marginal slopes and their *t* tests

	Coefficients	*Standard Error*	*t Stat*	*p value*	*1 tail p value*[a]
Intercept	9.2	1.90	4.8	<.0001	
seconds	.23	.099	2.3	.03	
mpg	−.23	.037	−6.2	.0001	<.0001
liters	.41	.29	1.4	.17	.08
cylinders	−.035	.19	−.2	.85	.43
horsepower	−.00052	.0037	−.1	.89	.44
pounds (K)	.54	.30	1.9	.08	.04
passengers	−.086	.12	−.7	.48	.24

[a]*p values* corresponding to one tail tests are not provided by Excel and have been added here.

Excel *t* tests of the marginal slopes, shown in Table 10.3, suggest that only *seconds* to accelerate 0 to 60, *MPG,* and *pounds* drive differences in emissions. Neither engine size characteristics, *horsepower, cylinders* and *liters,* nor car size characteristic, *passengers,* appears to influence *emissions*. Coefficient estimates for *cylinders*, *horsepower,* and *passengers* have the "wrong signs." Cars with more cylinders, larger, more powerful engines and more passenger capacity are expected to emit more pollutants. These are surprising and nonintuitive results.

When predictors which ought to be significant drivers appear to be insignificant, or when parameter estimates are of the wrong sign, we suspect *multicollinearity*. Multicollinearity, the correlation between predictors, thwarts driver identification. When the independent variables are themselves related, they jointly influence performance. It is difficult to tell which individual variables are more important drivers, since they vary together. Because of their correlation, the standard errors s_{b_i} of the marginal slope coefficient estimates, b_i, are inflated. We are not very certain of each true influence in the population is since their influence is joint. The confidence intervals of the true partial slopes are large, since these are multiples of the standard errors of the partial slope estimates. Individual predictors seem to be insignificant though they may be truly significant. In some cases, coefficient signs may be "wrong."

10.6 Combine or Eliminate Collinear Predictors

We have two remedies for multicollinearity cloudiness:

- We can combine correlated variables, or
- we can eliminate variables that are contributing redundant information.

Correlations between the predictors are shown in Table 10.4.

Table 10.4 Pairwise correlations between predictors

	MPG	*seconds*	*liters*	*horsepower*	*cylinders*	*pounds (K)*	*passengers*
MPG	1						
seconds	−.05	1					
liters	−.81	−.17	1				
horsepower	−.53	−.36	.76	1			
cylinders	−.74	−.19	.92	.77	1		
pounds (K)	−.77	−.01	.84	.72	.81	1	
passengers	−.53	−.05	.59	.55	.60	.70	1

Some of correlated predictors can be eliminated, assuming that several reflect a common dimension. If *liters, horsepower,* and *cylinders* each reflect engine size, two are possibly redundant and may be represented by the third. The alternative is to combine correlated predictors, either by constructing an index from a weighted average of the correlated predictors, or by forming ratios of pairs of correlated predictors.

An index of engine size could be made from a weighted average of *liters, cylinders,* and *horsepower*. *Factor analysis* is a statistical procedure that would provide the weights to form such an index. The challenge associated with use of an index is in its interpretation. Sakura managers need to know how much difference particular car characteristics make, and they may not be satisfied knowing that an *engine size index* influences *emissions.* Factor analysis is beyond the scope of this text, but does enable construction of indices from correlated predictors.

Ratios of correlated predictors are used when they make intuitive sense. For example, economic models sometimes use the ratio of *GDP* and *population* to make *GDP per capita,* an intuitively appealing measure of personal wealth.

We will eliminate the seemingly redundant predictors to build a model for Sakura, though combining correlated predictors would be an acceptable alternative. This will not eliminate multicollinearity, but it will reduce multicollinearity by removing correlated predictors.

We will remove one driver at a time, since the removal of any one reduces multicollinearity and changes coefficient estimates and significance levels. We'll choose the driver with the coefficient estimate that is furthest from significance, the one with the largest p value, which is *horsepower*. The coefficient for *horsepower* has an incorrect sign, the opposite of what we expect from logic and experience, and *horsepower* is highly correlated with *liters, cylinders* and *pounds.* Its removal may increase significance of the coefficients for those drivers.

Regression statistics for the first, full model with *horsepower,* and the second model without *horsepower,* are shown in Table 10.5. Removing *horsepower* improved the model. The standard error, our measure of precision, not smaller. The reduction in RSquare is negligible, .01%. *Horsepower* appears to have been redundant, since *liters* and *cylinders* continue to represent the impact of engine power in the model.

Table 10.5. Regression statistics of the original, full model and the model without horsepower

Regression Statistics			
	full model	*Horsepower out*	change
R Square	0.92782	0.92776	0.0001
Standard Error	0.644	0.632	0.012

Coefficient estimates and significance levels are shown in Table 10.6. Though coefficient standard errors are slightly smaller, t statistics are slightly larger, and p values are slightly smaller, three drivers remain insignificant and two have incorrect signs. The driver with coefficient furthest from significance, *cynlinders,* will be removed next. *Cylinders* is highly correlated with *mpg, liters* and *pounds*, and has an incorrect sign. *Liters,* which is highly correlated with both *horsepower* and *cylinders,* will represent all three, and take the credit for the impact of engine size on emissions.

Table 10.6 Coefficients after horsepower has been removed

	Coefficients	*Standard Error*	*t Stat*	*p value*	*1 tail p value*[a]
Intercept	9.2	1.90	4.9	<.0001	
seconds	.23	.090	2.6	.02	
mpg	−.23	.035	−6.5	.0001	<.0001
liters	.41	.28	1.4	.16	.08
cylinders	−.040	.18	−.2	.84	.42
pounds (K)	.53	.28	1.9	.06	.03
passengers	−.086	.12	−.7	.47	.24

Regression statistics for the third model, without *horsepower* and *cylinders,* are shown in Table 10.7. The standard error has improved, and the loss of explanatory power, measured with RSquare, is minimal, .01%.

Table 10.7 Regression without horsepower and cylinders

Regression Statistics			
	horsepower out	*horsepower & cylinders out*	change
R Square	.9278	.9276	.0001
Standard Error	.632	.621	.011

Table 10.8 presents coefficient estimates and p values for the third regression. Standard errors a generally slightly smaller, t statistics are larger, and p values are smaller. *Cylinders,* like *horsepower*, was redundant. *Liters,* reflects engine size, and represents *horsepower* and *cylinders.* However, the coefficient for *passengers* has an incorrect sign and remains insignificant. *Passenger* capacity is highly correlated with weight. Larger, more spacious cars weigh more. *Passengers* will be removed from the model, expecting that it is a redundant measure of car size. If explanatory power is not sacrificed, *pounds(K)* will reflect car size.

Pounds (K) is chosen to represent car size, since its coefficient sign is as expected and significant, while the sign for the *passengers* coefficient is "wrong."

Table 10.8 Coefficients after horsepower and cylinders have been removed

	Coefficients	*Standard Error*	*t Stat*	*p value*	*1 tail p value*[a]
Intercept	9.1	1.82	5.0	<.0001	
seconds	.24	.088	2.7	.01	
mpg	−.23	.034	−6.6	.0001	<.0001
liters	.36	.20	1.8	.08	.04
pounds (K)	.52	.27	2.0	.06	.03
passengers	−.089	.11	−.8	.44	.22

The revised *partial* model becomes:

$$emi\hat{s}sions_i = b_0 + b_1 \times MPG_i + b_2 \times seconds_i + b_3 \times liters_i + b_4 \times pounds_i$$

Regression results using this *partial* model are shown in Table 10.9.

Table 10.9 Regression of emissions with four car characteristics

SUMMARY OUTPUT						
Regression Statistics						
R Square		.926				
Standard Error		.617				
Observations		34				
ANOVA	*df*	*SS*	*MS*	*F*	*Significance F*	
Regression	4	138	34.5	90.8	.0000	
Residual	29	11	.4			
Total	33	149				
	Coefficients	*Standard Error*		*t Stat*	*p value*	*1 tail p value*
Intercept	9.0	1.8		5.0	<.0001	
seconds	.24	.087		2.8	.01	
mpg	−.23	.034		−6.7	<.0001	<.0001
liters	.36	.20		1.8	.08	.04
pounds (K)	.43	.24		1.8	.08	.04

The *partial* model *RSquare*, .926, is less than one percentage point lower than the original *full* model *RSquare*, .929. With just four of the seven car characteristics, we can account for 93% of the variation in emissions. Little explanatory power has been lost, and the standard error has dropped from .644 to .617, reducing the margin of error in forecasts by 5% (= (1.32 − 1.26)/1.32). Model *F* is significant, suggesting that one or more of the four predictors influences emissions. All four of the predictors are significant drivers. All coefficient estimates have correct signs. As was the case in the full model, *emissions* are lower for smaller, responsive cars with higher fuel economy. By reducing multicollinearity, it can now also be concluded that *emissions* are lower for smaller cars with smaller engines.

The final multiple linear regression model of emissions is:

$$\widehat{emissions}\left(\frac{tons}{yr}\right)_i = 9.0^a\left(\frac{tons}{yr}\right) + .24^b\left(\frac{tons/yr}{second}\right) \times seconds_i$$

$$-.23^a\left(\frac{tons/yr}{mi/gal}\right) \times MPG_i + .36^b\left(\frac{tons/yr}{liters}\right) \times liters_i$$

$$+.43^b\left(\frac{tons/yr}{lbs}\right) \times pounds_i$$

$RSquare^a = .93$
[a]*Significant at a .01 level or better.*
[b]*Significant at a .05 level or better.*

To determine whether or not our model satisfies the assumptions of linear regression, the distribution of residuals is examined, just as with a simple regression model. In Figure 10.6, the residuals, with skewness of .46, are approximately *Normal*.

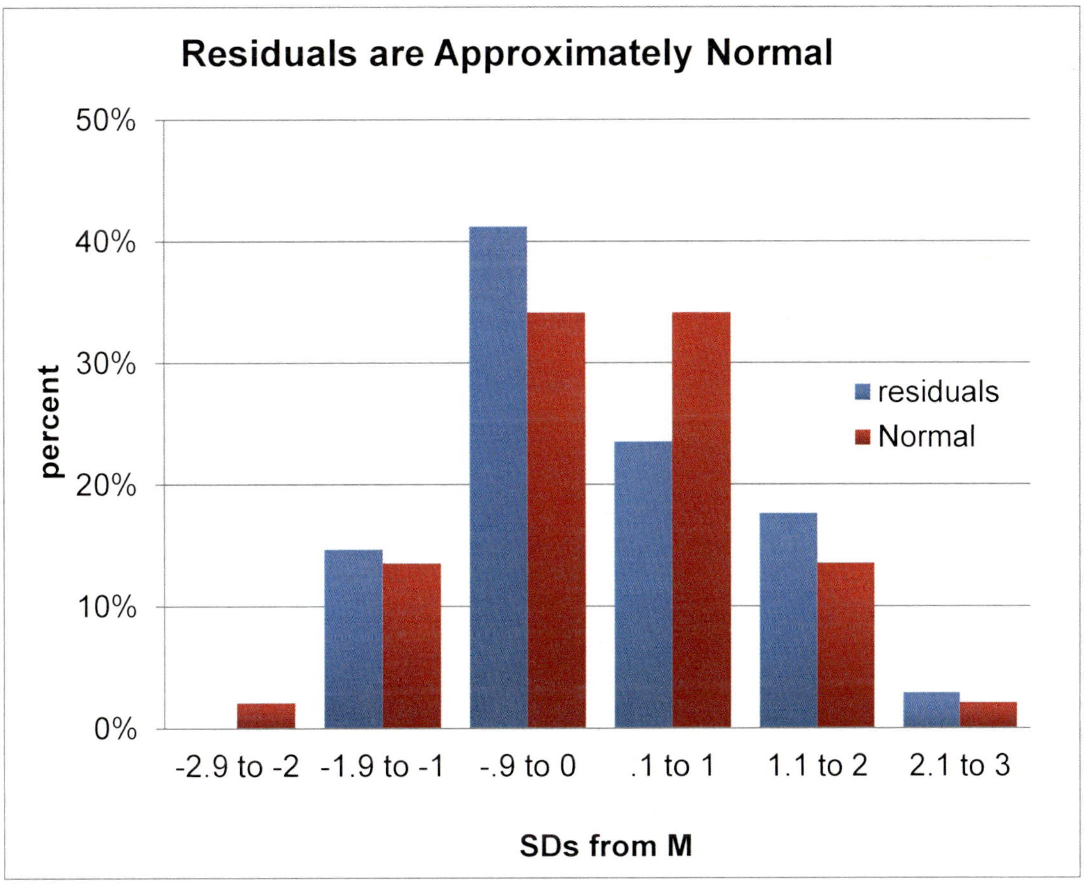

Figure 10.6 Distribution of residuals

10.7 Decide Whether Insignificant Drivers Matter

It is typical to find that potential drivers, which logically ought to matter, are insignificant and fail to improve a model. The model is improved by removing those insignificant drivers. However, those apparently insignificant drivers should not be forgotten. They may be performance drivers that do matter, but their correlation with other drivers makes them redundant. Management will want to know whether potential drivers that were removed from a model are drivers that do matter, or whether they simply do not matter.

Potential drivers which do matter, but are insignificant and appear not to matter, can be identified with correlations and simple regression. If multicollinearity has diminished the significance of a driver,

- that driver will be correlated with drivers remaining in the model, and
- that driver will be significant in a simple regression

Using simple regression to assess significance of a variable that appears not to matter gives that variable an "unfair" chance to claim credit for the joint impact of the set of correlated drivers. Simple regression will overstate a driver's importance, and we would not use the simple regression model in sensitivity analysis or forecasts. However, simple regression can provide important evidence that a potential driver matters, but is redundant.

In **Sakura,** three variables were removed from the multiple regression model: *horsepower, cylinders*, and *passengers*. All three were correlated with drivers that remained in the model. It seemed likely that all three were drivers of emissions that mattered, but simply could not be included in the multiple regression model, because of redundancy. The modeling team ran three simple regressions to decide whether or not this was the case.

Results of the three simple regressions, shown in Tables 10.10, 10.11, and 10.12 provide evidence that all three of the car characteristics drive emissions: All three simple regression models are significant, and all three slopes are in the expected positive direction. Sakura cannot ignore horsepower, cylinders, nor passengers in their car designs.

Table 10.10 Simple regression with horsepower alone

SUMMARY OUTPUT	
Regression Statistics	
Multiple R	.553
R Square	.306
Standard Error	1.798
Observations	34

ANOVA					
	df	*SS*	*MS*	*F*	*Significance F*
Regression	1	45.7	45.7	14.1	.001
Residual	32	103.5	3.2		
Total	33	149.2			

	Coefficients	Standard Error	t Stat	p value	Lower 95%	Upper 95%
Intercept	4.53	1.22	3.7	.0008	2.04	7.02
horsepower	.021	.006	3.8	.0007	.010	.032

Table 10.11 Simple regression with cylinders alone

SUMMARY OUTPUT	
Regression statistics	
Multiple R	.76
R Square	.58
Standard error	1.39
Observations	34

ANOVA

	df	SS	MS	F	Significance F
Regression	1	87.1	87.1	44.9	.0000
Residual	32	62.1	1.9		
Total	33	149.2			

	Coefficients	Standard error	t Stat	p value	Lower 95%	Upper 95%
Intercept	3.16	.90	3.5	.0013	1.33	5.00
Cylinders	.99	.15	6.7	.0000	.69	1.30

Table 10.12 Simple regression with passengers alone

SUMMARY OUTPUT	
Regression Statistics	
Multiple R	.540
R Square	.291
Standard Error	1.818
Observations	34

ANOVA

	df	SS	MS	F	Significance F
Regression	1	43.5	43.5	13.2	.001
Residual	32	105.7	3.3		
Total	33	149.2			

	Coefficients	Standard Error	t Stat	p value	Lower 95%	Upper 95%
Intercept	4.12	1.38	3.0	.005	1.31	6.92
passengers	.87	.24	3.6	.001	.38	1.35

10.8 Sensitivity Analysis Quantifies the Marginal Impact of Drivers

We want to compare influences of the significant drivers to identify those which make the greatest difference. The coefficient estimates, indicating the impacts of unit differences in the drivers, are not directly comparable, since the drivers are measured in different units, seconds, liters, pounds, and miles per gallon. However, the *part worths*, the products of coefficient estimates and driver values, are directly comparable, since the part worths are in common units, tons of emissions per year. We will forecast emissions at average levels of each of the car characteristics. Predicted emissions are equal to the sum of the intercept and the part worths, which in the case of Sakura, is the sum of the intercept and the impacts of *miles per gallon, seconds, liters* and *pounds.* Below is predicted emissions for a car with median *MPG, seconds, liters* and *pounds,* from the sum of the intercept and the four part worths:

Intercept	b_{MPG}	*Med MPG*	b_{secs}	*Med seconds*	b_{liters}	*Med liters*	b_{lbs}	*Med pounds (K)*	*Predicted emissions*
8.99 +	−0.23 ×	22 +	0.24 ×	8.8 +	0.36 ×	3.5 +	0.43 ×	4.11 =	9.09
8.99 +	−5.02	+	2.11	+	1.26	+	1.75	=	9.09

To compare the driver impacts, we will compare the improvements in predicted emissions which would result from improvement of each of the drivers to the most desirable level in the sample, which is the minimum *seconds, liters, pounds* and the maximum *mpg.* These five scenarios are compared in Table 10.13

Table 10.13 Part worths and predited emissions for five scenarios

Scenario	*Inter-cept*		*MPG pw (tons /yr)*		*secs pw (tons /yr)*		*liters pw (tons /yr)*		*pounds pw (tons /yr)*		*predicted emissions*	*improvement from baseline*
baseline	8.99	+	−5.02	+	2.11	+	1.26	+	1.7539	=	9.09	
highest MPG	8.99	+	**−7.65**	+	2.11	+	1.26	+	1.755	=	6.46	2.63
quickest acceleration	8.99	+	−5.02	+	**1.60**	+	1.26	+	1.755	=	8.59	0.50
smallest engine	8.99	+	−5.02	+	2.11	+	**0.54**	+	1.755	=	8.37	0.72
lightest weight	8.99	+	−5.02	+	2.11	+	1.26	+	**1.0611**	=	8.40	0.69

MPG. Improving a car with median *mpg* to the sample maximum *mpg* reduces emissions by 2.63 tons per year to 6.46 tons per year. Within a representative range of values for each of the car characteristics, fuel economy makes the largest difference in emissions, shown in Figure 10.7. (Solid lines show response to improvements from the sample median to the sample maximum for *mpg,* and to the sample minimums for *seconds, liters* and *pounds.* Dotted lines show response to additional improvement which technological advances might achieve.)

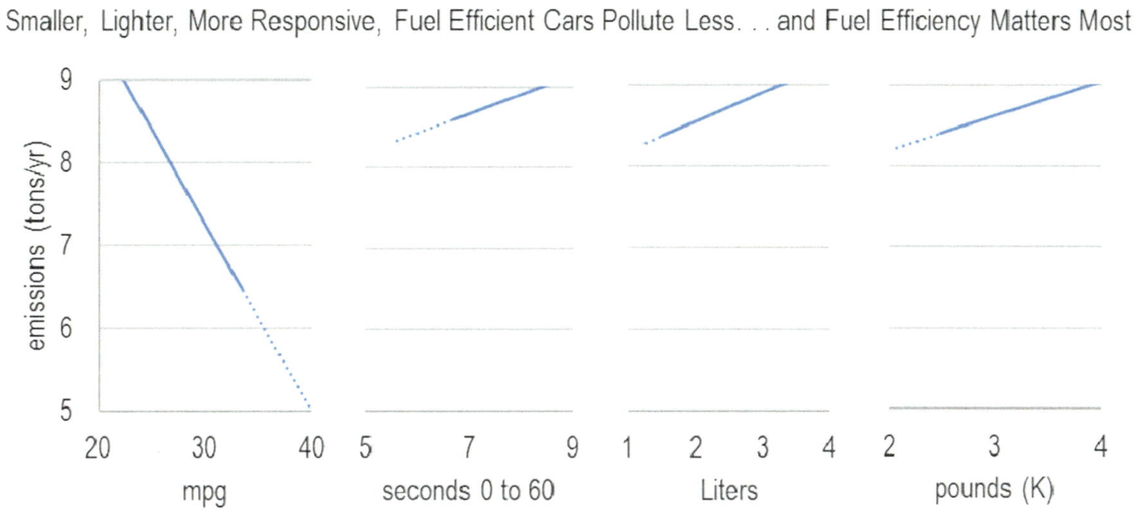

Figure 10.7 Predicted emissions by car characteristic

This is a large improvement, though not enough alone to meet the 5.0 tons per year goal. Fuel economy improvements will need to be made in conjunction with improvements in one or more of the other car characteristics, or technological improvements will need to improve *mpg* more than that of existing cars. What improvement in *mpg* would it take to achieve the 5.0 tons per year goal (holding all other car characteristics constant at the median)? To reduce emissions from the predicted value of 6.46 to 5.00 tons per year, at gains of .23 tons per year for each additional *mpg* requires an additional improvement of 6.4 *mpg* over the sample maximum of 33.5, of 39.9 *mpg*:

Improvement in emissions desired:

$$\Delta \widehat{emissions}\left(\frac{tons}{yr}\right) = 5\left(\frac{tons}{yr}\right) - 6.46\left(\frac{tons}{yr}\right) = -.54\left(\frac{tons}{yr}\right)$$

Improvement in *mpg* required to achieve goal:

$$\frac{\Delta \widehat{emissions}}{b_{mpg}} = \frac{-.54(\frac{tons}{yr})}{-.23\left(\frac{\frac{tons}{yr}}{mpg}\right)} = 6.4\ \Delta mpg$$

The impacts of drivers can also be compared by identifying the improvement in each that could produce a given level of improvement in the performance variable. In the Sakura case, how much improvement in *mpg, seconds, liters* or *pounds* would be needed to reduce *emissions* by 1 ton per year? Dividing the given level of improvement by a driver's coefficient estimate provides the level of driver improvement required. Improvement in fuel economy of 4.4 *mpg* would deliver an expected reduction in emissions of 1 ton per year:

$$\Delta mpg = \frac{\Delta emis\hat{s}ions(\frac{tons}{yr})}{b_{mpg}\left(\frac{tons/yr}{mpg}\right)} = \frac{-1.00(\frac{tons}{yr})}{-.23\left(\frac{tons/yr}{mpg}\right)} = 4.4\ mpg$$

Pounds(K) and Liters. Reducing car weight or reducing engine size to the sample minimum improves expected emissions by about .7 ton per year. Even the combination of a lighter car with a smaller engine is probably not enough to reach the emissions goal of 5 tons per year. A reduction in engine size of 2.8 liters would improve expected emissions by 1 ton per year. A reduction in car size by 2.3K pounds would produce a comparable improvement in emissions. In combination with fuel economy improvements, either car weight or engine size improvements could make the goal attainable.

Seconds. Improving car responsiveness by reducing the time to accelerate from 0 to 60 to the sample minimum could improve expected emissions about half a ton. To achieve a reduction in expected emissions of 1 ton per year, seconds to accelerate from 0 to 60 would need to quicken by 4.2 seconds. Combined with any of the other car characteristics, responsiveness could help Sakura achieve their emissions goal, though acceleration alone makes the least difference in emissions.

The model provides clear indications for the new product development team. To improve emissions, they will need to design more responsive, lighter-weight cars with smaller engines and superior fuel economy. Changing just one car characteristic will not be enough to meet the goal of 5 tons per year.

The Quantitative Analysis Director summarized model results in the following memo to Sakura Management:

MEMO

Re: Light, responsive, fuel efficient cars with smaller engines are cleanest
To: Sakura Product Development Director
From: Will Little, Quantitative Analysis Director
Date: July 2015

Improvements in gas mileage and responsiveness, with reductions in weight and engine size will allow Sakura to achieve the emissions target of 5 tons per year.

A regression model of emissions was built from a representative sample of 34 diverse car models, considering fuel economy, acceleration, engine size and car size.

Fuel economy matters most.

Differences in mpg, weight, engine size, and acceleration account for 93% of the variation in emissions. Emissions forecasts are expected to be no further than 1.3 ton per year from actual average emissions.

Mpg is the most powerful driver of emissions. Increasing mpg by four is expected to reduce emissions by 1 ton per year. Car and engine size and responsiveness matter, but make less difference. A 2K pound reduction in weight is expected to reduce emissions by about 1 ton per year. Reducing engine size by three liters reduces expected emissions by about 1 ton per year. Passenger capacity, horsepower, and cylinders matter, but were not included in the model, since these are similar to weight and engine size in liters. Reduction in passenger capacity, horsepower or cylinders is expected to reduce emissions. Responsiveness makes the smallest difference in emissions. Reducing acceleration from 0 to 60 by 4 seconds would improve emissions about 1 ton per year.

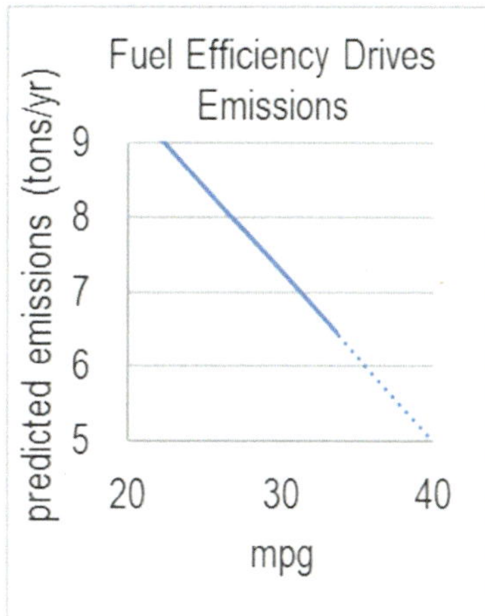

$$\begin{aligned} \widehat{emissions(tons/yr)}_i = \; & 9.0^a\,(tons/yr) \\ & +.24^b\left(\frac{tons/yr}{\text{sec}}\right)\times sec_i \\ & -.23^a\left(\frac{tons/yr}{mpg}\right)\times MPG_i \\ & +.36^b\left(\frac{tons/yr}{litres}\right)\times litres_i \\ & +.43^b\left(\frac{tons/yr}{lbs}\right)\times lbs(K)_i \end{aligned}$$

The 5 ton per year goal is achievable

To achieve emissions of 5 tons per year within existing characteristic ranges, more than one car characteristic must be changed. Improvements in mpg and responsiveness, with reductions in car or engine size will enable Sakura to meet the target.

New technology may help to reduce emissions

Model results assume existing engine technology. With development of cleaner, more fuel efficient, responsive technologies, even lower emissions could possibly be achieved.

10.9 Model Building Begins with Logic and Considers Multicollinearity

Novice model builders sometimes mistakenly think that the computer can choose those variables which belong in a model. Computers have no experience making decisions and can never replace decision makers' logic. (Have you ever tried holding a conversation with a computer?) The first step in superior model building is to use your head. Use logic and experience to identify independent variables which ought to influence the performance variable which you are interested in explaining and forecasting. Both your height and GDP increased over the past ten years. Given data on your annual height and annual GDP, the computer could churn out a significant parameter estimate relating variation in your height to variation in GDP (or variation in GDP to variation in your height). Decision makers must use their logic and experience to select model variables. Software will quantify and calibrate the influences that we know, from theory or experience, ought to exist.

It is a multicollinear world. Sets of variables together jointly influence performance. Using ratios of collinear predictors reduces multicollinearity. Removing redundant predictors allows us to more accurately explain performance and forecast. Correlations and simple regressions are used to determine whether insignificant variables matter, but simply look as though they don't because of multicollinearity—or whether they simply do not matter.

From the logically sound set of variables, pruned to eliminate redundancies and reduce multicollinearity, we have a solid base for superior model building. To this we will consider adding variables to account for seasonality or cyclicality in time series in Chapter 11 and the use of indicators to build in influences of segment differences, structural shifts and shocks in Chapter 12. In Chapters 13 and 14, alternative nonlinear models are considered, for situations where response is not constant.

Excel 10.1 Build and Fit a Multiple Linear Regression Model

Sakura Motors Quest for a Clean Car.

Open the **Sakura** dataset and run multiple regression with the dependent variable *emissions* and the independent variables, *MPG, seconds, cylinders, liters, horsepower, passengers,* and *pounds*.

Alt AY*n*R down down C1:c35 tab d1:j35 tab LR

	A	B	C	D	E	F	G
1	SUMMARY OUTPUT						
2							
3	*Regression Statistics*						
4	Multiple F	0.963233					
5	R Square	0.927817					
6	Adjusted I	0.908383					
7	Standard E	0.643631					
8	Observati	34					
9							
10	ANOVA						
11		*df*	*SS*	*MS*	*F*	*gnificance F*	
12	Regressio	7	138.4448	19.77783	47.74239	3.05E-13	
13	Residual	26	10.7708	0.414261			
14	Total	33	149.2156				
15							
16		*Coefficients*	*andard Err*	*t Stat*	*P-value*	*Lower 95%*	*Upper 95%*
17	Intercept	9.164994	1.903614	4.814523	5.48E-05	5.252059	13.07793
18	MPG	-0.22612	0.036559	-6.18514	1.53E-06	-0.30127	-0.15097
19	seconds	0.229146	0.09864	2.323068	0.028265	0.02639	0.431903
20	liters	0.414272	0.293367	1.412128	0.16977	-0.18875	1.017296
21	pounds (K	0.544133	0.294646	1.846736	0.076198	-0.06152	1.149786
22	passenger	-0.08552	0.119738	-0.7142	0.481466	-0.33164	0.160607
23	cylinders	-0.03513	0.188284	-0.18656	0.853456	-0.42215	0.351897
24	horsepow	-0.00052	0.003688	-0.14111	0.888875	-0.0081	0.00706

Add a column for one tail *t* tests of *MPG, liters, horsepower, cylinders, pound (K)* and *passengers*, and the fill in *p values* by dividing Excel's two tail *p values* by 2.	In J16, One tail p value In J17, **=e17/2** In J17, **Shift+down to J24** **Cntl+D**

J17 =E17/2

	G	H	I	J
16	*Upper 95%*	*ower 95.0%*	*pper 95.0%*	*one tail p value*
17	13.07793	5.252059	13.07793	2.74167E-05
18	-0.15097	-0.30127	-0.15097	7.64946E-07
19	0.431903	0.02639	0.431903	0.014132587
20	1.017296	-0.18875	1.017296	0.084884914
21	1.149786	-0.06152	1.149786	0.038099027
22	0.160607	-0.33164	0.160607	0.240733231
23	0.351897	-0.42215	0.351897	0.426728152
24	0.00706	-0.0081	0.00706	0.444437286

Multicollinearity symptoms. While the model is significant (*Significance F* <.0001), *only three of the car characteristics are significant* (*p value* <.05). We are not certain that *liters, cylinders, passengers,* and *horsepower* are influential, since their *p values* ≥.05. *Horsepower, cylinders* and *passengers* have "incorrect" negative signs. Cars with greater horsepower, more cylinders, and more passenger space ought to be bigger polluters. Together, the lack of significance of seemingly important predictors and the three sign reversals signal multicollinearity.

Look at correlations to confirm suspicions that *liters, horsepower* and *cylinders* are correlated (and together reflect car power) and that *pounds(K)* and *passengers* are correlated (and together reflect car size).

Move back to the data sheet and run correlations between the car characteristics.	**Cntl+Page Down** **Alt AY*n*C** **D1:j35 tab L**

	A	B	C	D	E	F	G	H
1		*MPG*	*seconds*	*liters*	*ɔounds (K)*	*ɔassengers*	*cylinders*	*ɪorsepowe*
2	MPG	1						
3	seconds	-0.04901	1					
4	liters	-0.81008	-0.17065	1				
5	pounds (K	-0.76895	-0.01287	0.835189	1			
6	passengeı	-0.52581	-0.04871	0.592526	0.702565	1		
7	cylinders	-0.74357	-0.18889	0.924148	0.807687	0.602062	1	
8	horsepow	-0.52917	-0.36347	0.762617	0.71753	0.545275	0.770781	1
9								

Horsepower is furthest from significance, with the highest p value in the regression, its coefficient has an incorrect negative sign, and it is highly correlated with *liters, pounds* and *cylinders.* Move back to the data sheet and run regression without *horsepower.*

Cntl+page down
Alt AY*n*R down down
Tab d1:i35

	A	B	C	D	E	F	G
1	SUMMARY OUTPUT						
2							
3	*Regression Statistics*						
4	Multiple F	0.963204					
5	R Square	0.927762					
6	Adjusted I	0.911709					
7	Standard E	0.631842					
8	Observati	34					
9							
10	ANOVA						
11		*df*	*SS*	*MS*	*F*	*gnificance F*	
12	Regressio	6	138.4365	23.07276	57.79404	3.83E-14	
13	Residual	27	10.77904	0.399224			
14	Total	33	149.2156				
15							
16		*Coefficients*	*andard Err*	*t Stat*	*P-value*	*Lower 95%*	*Upper 95%*
17	Intercept	9.145025	1.863573	4.907252	3.91E-05	5.321288	12.96876
18	MPG	-0.22728	0.034968	-6.49968	5.73E-07	-0.29903	-0.15553
19	seconds	0.234394	0.089688	2.613437	0.014475	0.050369	0.418418
20	liters	0.406937	0.283436	1.435727	0.162565	-0.17463	0.988499
21	pounds (K	0.531272	0.275062	1.931463	0.063985	-0.03311	1.095653
22	passengeı	-0.08625	0.117435	-0.73442	0.469025	-0.3272	0.15471
23	cylinders	-0.04028	0.181323	-0.22214	0.825874	-0.41232	0.331765

Add one tail *p values*. Move to the first regression sheet, select and copy the one tail p values in column J, move back to the second regression sheet, and then paste into column J.

Cntl+page up up
In J16,
Cntl+Shift+down
Cntl+C
Cntl+page down down
In J16,
Cntl+V

	J
16	*one tail p value*
17	1.95E-05
18	2.87E-07
19	0.007238
20	0.081283
21	0.031993
22	0.234513
23	0.412937

Three driver coefficients are insignificant, and two have incorrect signs.

Move back to the data sheet and run regression without *cylinders,* the driver furthest from significance, which is highly correlated with *mpg, liters,* and *pounds.* From the third regression, move to the second regression, select and copy one tail p values in column J, and then paste into the third regression in column J.

	A	B	C	D	E	F	G	H	I	J
1	SUMMARY OUTPUT									
2										
3	*Regression Statistics*									
4	Multiple F	0.963135								
5	R Square	0.92763								
6	Adjusted I	0.914707								
7	Standard E	0.621023								
8	Observati	34								
9										
10	ANOVA									
11		*df*	*SS*	*MS*	*F*	*gnificance F*				
12	Regressio	5	138.4168	27.68337	71.78004	4.36E-15				
13	Residual	28	10.79874	0.385669						
14	Total	33	149.2156							
15										
16		*Coefficients*	*andard Err*	*t Stat*	*P-value*	*Lower 95%*	*Upper 95%*	*ower 95.0%*	*pper 95.0%*	*one tail p value*
17	Intercept	9.098958	1.820288	4.998636	2.78E-05	5.370266	12.82765	5.370266	12.82765	1.39E-05
18	MPG	-0.22759	0.034343	-6.62686	3.45E-07	-0.29793	-0.15724	-0.29793	-0.15724	1.72E-07
19	seconds	0.236568	0.087626	2.699764	0.011635	0.057075	0.416061	0.057075	0.416061	0.005818
20	liters	0.362828	0.198798	1.825106	0.078671	-0.04439	0.770047	-0.04439	0.770047	0.039336
21	pounds (K	0.523137	0.267946	1.9524	0.060955	-0.02572	1.071999	-0.02572	1.071999	0.030477
22	passenger	-0.08897	0.114793	-0.77505	0.444806	-0.32411	0.146172	-0.32411	0.146172	0.222403

Passengers remains insignificant, and with the incorrect sign. *Passengers* is highly correlated with *pounds*.

Move back to the data sheet and run regression without *passengers*, and then add one tail p values in column J.

	A	B	C	D	E	F	G	H	I	J
1	SUMMARY OUTPUT									
2										
3	*Regression Statistics*									
4	Multiple F	0.962329								
5	R Square	0.926077								
6	Adjusted I	0.915881								
7	Standard E	0.616733								
8	Observati	34								
9										
10	ANOVA									
11		*df*	*SS*	*MS*	*F*	*gnificance F*				
12	Regressio	4	138.1852	34.54629	90.82543	5.71E-16				
13	Residual	29	11.03042	0.380359						
14	Total	33	149.2156							
15										
16		*Coefficients*	*andard Err*	*t Stat*	*P-value*	*Lower 95%*	*Upper 95%*	*ower 95.0%*	*pper 95.0%*	*one tail p value*
17	Intercept	8.98999	1.802313	4.988029	2.62E-05	5.303846	12.67613	5.303846	12.67613	1.31E-05
18	MPG	-0.2284	0.03409	-6.69989	2.38E-07	-0.29812	-0.15868	-0.29812	-0.15868	1.19E-07
19	seconds	0.239516	0.086938	2.755019	0.010033	0.061708	0.417325	0.061708	0.417325	0.005017
20	liters	0.360546	0.197403	1.826446	0.078094	-0.04319	0.764281	-0.04319	0.764281	0.039047
21	pounds (K	0.426995	0.235862	1.810361	0.080614	-0.0554	0.909388	-0.0554	0.909388	0.040307

Look at residuals to check model assumptions. Find the skewness of the residuals to assess their symmetry, evidence of a Normal distribution.
In C62,
=skew(c28:c61)

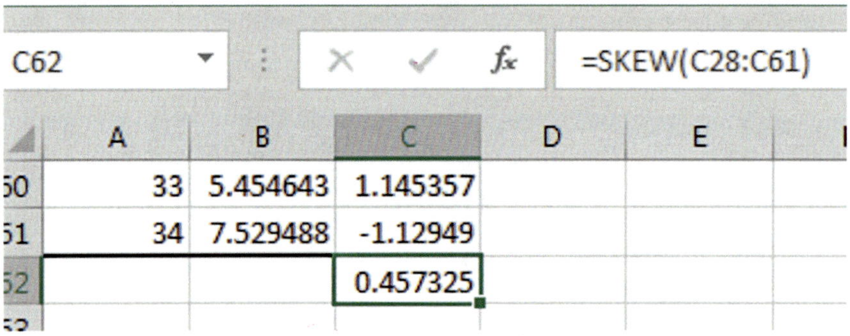

Excel 10.2 Use Sensitivity Analysis to Compare the Marginal Impacts of Drivers

For sensitivity analysis, to determine the impact of improvements in each of the four car characteristics, begin with a hypothetical baseline car, such as a car with median MPG, acceleration, engine size, and weight. For that hypothetical car, find predicted emissions from the intercept and the four part worths.

Move to the data sheet, select and copy the car characteristics data in columns D through G, move back to the fourth (final) regression sheet, and past next to residuals in D27.

Cntl+page down
In D1,
Shift+right right right
Cntl+shift+down
Cntl+C
Cntl+page up
In D27,
Cntl+V

	C	D	E	F	G
25					
26					
27	*Residuals*	*MPG*	*seconds*	*liters*	*pounds (K)*
28	0.107058	20	8.2	3.5	4.555
29	0.225012	24.5	6.7	3.2	3.565
30	-0.22103	24	7.4	3	3.65
31	0.209116	15.5	8.3	6	4.66
32	-0.20328	15.5	7.6	5.7	5.335

Copy the intercept into column H, find the part worths in columns I through L, and sum the intercept and part worths to find predicted emissions in column M.

In H27,
Intercept
In H28,
=b17 f4
In I27,
Mpg pw
In I28,
=b18 f4 *d28
In J27,
Seconds pw
In J28,
=b19 f4 *e28
In K27,
Liters pw
In K28,
=b20 f4 *f28
In L27,
Pounds pw
In L28,
=b21 f4 *g28
In M27,
Predicted emissions
In M28,
=sum(h28:l28)
In H27,
Cntl+shift+right
Double click lower right corner to down fill

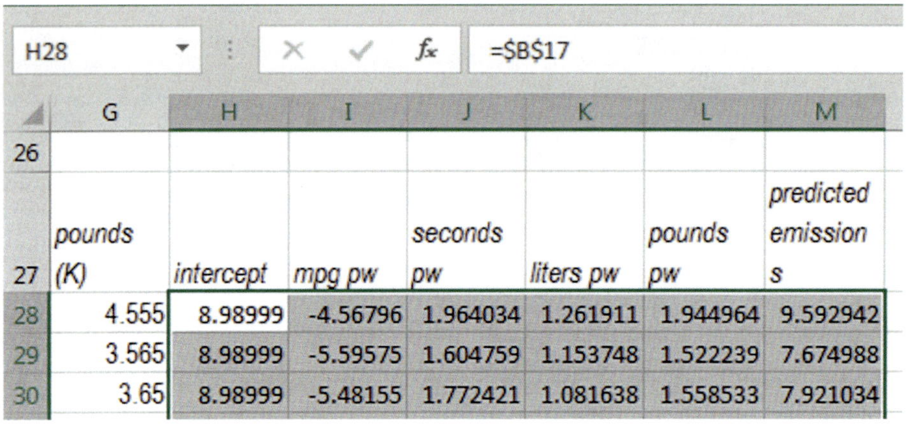

H28 =$B$17

	G	H	I	J	K	L	M
26							
27	pounds (K)	intercept	mpg pw	seconds pw	liters pw	pounds pw	predicted emissions
28	4.555	8.98999	-4.56796	1.964034	1.261911	1.944964	9.592942
29	3.565	8.98999	-5.59575	1.604759	1.153748	1.522239	7.674988
30	3.65	8.98999	-5.48155	1.772421	1.081638	1.558533	7.921034

Find predicted emissions for a hypothetical baseline car. Insert cells in column D where hypothetical scenario labels can be entered, and thn find medians in row 62, pressing function 4 twice for the first and last rows of the array to lock the row references.

In D27,
Cntl+shift+down
Alt HIII
In D62,
Baseline
In E62,
=median(e28 f4 f4 :e61 f4 f4)
Shift+right right right
Cntl+R

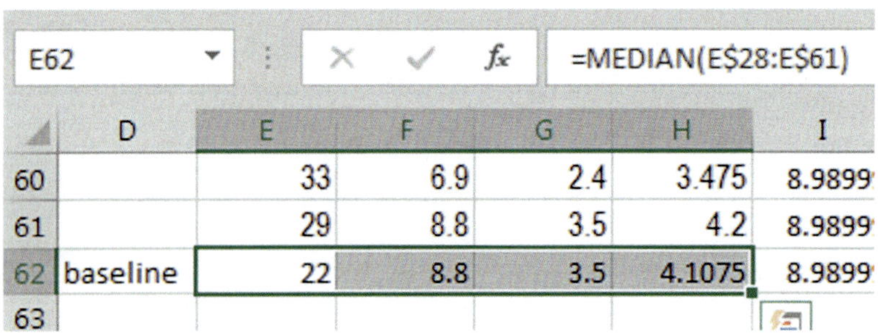

E62 =MEDIAN(E$28:E$61)

	D	E	F	G	H	I
60		33	6.9	2.4	3.475	8.9899
61		29	8.8	3.5	4.2	8.9899
62	baseline	22	8.8	3.5	4.1075	8.9899
63						

For comparison, add four hypothetical cars that are “best” along one of each of the four car characteristics with the baseline, but equivalent along three out of four characteristics. First, make four new hypotheticals that are identical to the baseline (which you’ll change in the next step).

In D63,
Highest mpg
In D64,
Lowest seconds
In D65,
Smallest engine
In D66,
Lightest weight
In E62,
Shift+right right right
Clock lower right corner to down fill

E62 =MEDIAN(E$28:E$61)

	D	E	F	G	H
60		33	6.9	2.4	3.475
61		29	8.8	3.5	4.2
62	baseline	22	8.8	3.5	4.1075
63	highest mpg	22	8.8	3.5	4.1075
64	least seconds	22	8.8	3.5	4.1075
65	smallest engine	22	8.8	3.5	4.1075
66	lightest weight	22	8.8	3.5	4.1075
67					

To complete the four hypotheticals, change *mpg* to the maximum in row 63, change *seconds* to the minimum in row 64, change *liters* to the minimum in row 65, and change *pounds* to the minimum in row 66.

In E63,
=max(e28:e61)
In F64,
=min(f28:f61)
In G65,
=min(g28:g61)
In H66,
=min(h28:h61)

H66 =MIN(H28:H61)

	D	E	F	G	H
60		33	6.9	2.4	3.475
61		29	8.8	3.5	4.2
62	baseline	22	8.8	3.5	4.1075
63	highest mpg	33.5	8.8	3.5	4.1075
64	least seconds	22	6.7	3.5	4.1075
65	smallest engine	22	8.8	1.5	4.1075
66	lightest weight	22	8.8	3.5	2.485

Down fill the part worths and predicted values in columns I through N of rows 62 through 66.

In I61,
Cntl+shift+right
Shift+down down down down down
Cntl+D

I61 | f_x =$B$17

	D	E	F	G	H	I	J	K	L	M	N
60		33	6.9	2.4	3.475	8.98999	-7.53713	1.652662	0.865311	1.483809	5.454643
61		29	8.8	3.5	4.2	8.98999	-6.62354	2.107743	1.261911	1.793381	7.529488
62	baseline	22	8.8	3.5	4.1075	8.98999	-5.02475	2.107743	1.261911	1.753884	9.088776
63	highest mpg	33.5	8.8	3.5	4.1075	8.98999	-7.65133	2.107743	1.261911	1.753884	6.4622
64	least seconds	22	6.7	3.5	4.1075	8.98999	-5.02475	1.604759	1.261911	1.753884	8.585791
65	smallest engine	22	8.8	1.5	4.1075	8.98999	-5.02475	2.107743	0.540819	1.753884	8.367683
66	lightest weight	22	8.8	3.5	2.485	8.98999	-5.02475	2.107743	1.261911	1.061084	8.395976
67											

Find the difference between *predicted emissions* of each scenario and the baseline provides an estimate of the expected difference that an improvement in each characteristic could make.

In O62,
Expected improvement
In O63,
=N63-N62 f4
In O63,
Shift+down down down
Cntl+D

O63 | f_x =N63-$N$62

	D	E	F	G	H	I	J	K	L	M	N	O	P
60		33	6.9	2.4	3.48	8.99	-7.5	1.65	0.87	1.48	5.45		
61		29	8.8	3.5	4.2	8.99	-6.6	2.11	1.26	1.79	7.53		
62	baseline	22	8.8	3.5	4.11	8.99	-5	2.11	1.26	1.75	9.09	expected improvement	
63	highest mpg	33.5	8.8	3.5	4.11	8.99	-7.7	2.11	1.26	1.75	6.46	-2.62658	
64	least seconds	22	6.7	3.5	4.11	8.99	-5	1.6	1.26	1.75	8.59	-0.50298	
65	smallest engine	22	8.8	1.5	4.11	8.99	-5	2.11	0.54	1.75	8.37	-0.72109	
66	lightest weight	22	8.8	3.5	2.49	8.99	-5	2.11	1.26	1.06	8.4	-0.6928	
67													

Scatterplots of marginal response. To see the impact of each driver, plot predicted *emissions* of hypotheticals.

First, focus on *MPG*. Move predicted emissions in column N to column F, adjacent to *mpg* in column E. Select the baseline row and the highest mpg row and request a scatterplot. Excel sometimes reverses our intended axes, plotting x on the y axis, and vice versa. Switch the axes, so that predicted emissions are on the vertical axis.

In N27,
Cntl+shift+down
Cntl+X
In F27,
Alt HIE
In E62,
Shift+down
Shift+right
Alt ND
Alt JCW

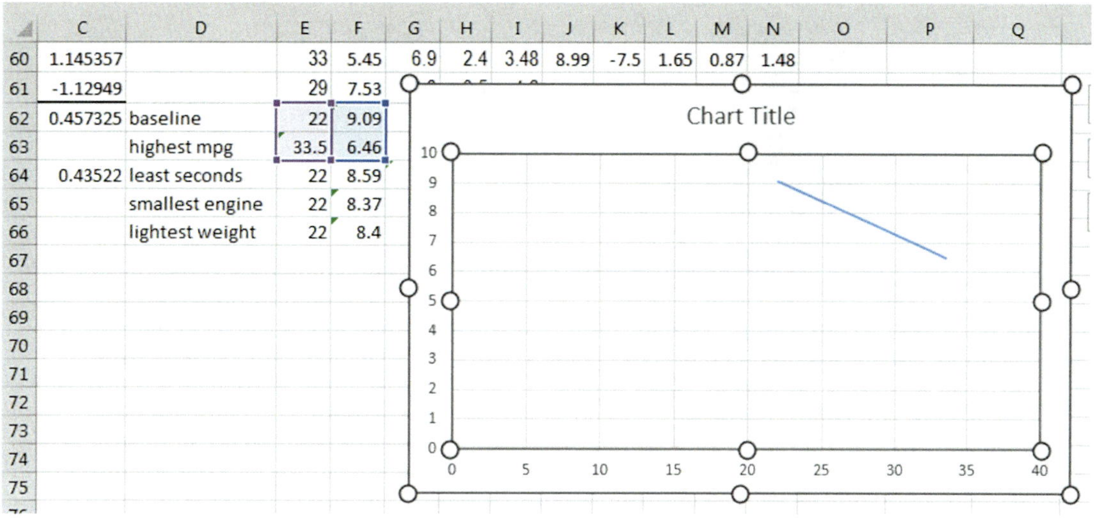

Add axis and chart titles, adjust the horizontal axis, and set fontsize to 12. (Later, you will adjust the vertical axis to match the other three scatterplots.)

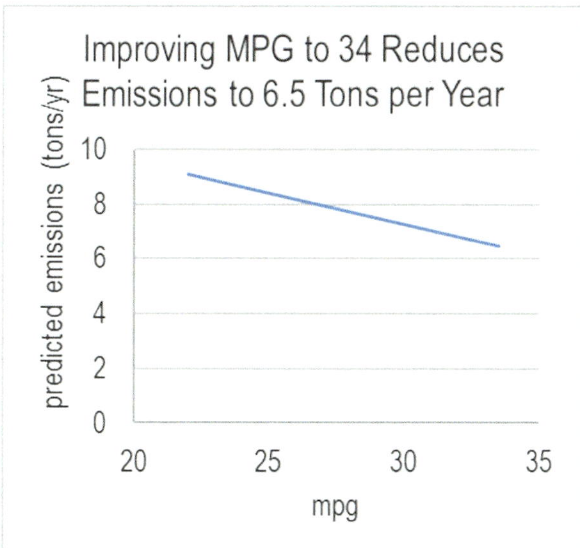

Repeat this process see the impacts of acceleration, engine size, and car size. In each case, move predicted emissions, now in column F, to the column right of the driver that you want to graph. Also in each case, copy the baseline row 62 and paste above the driver scenario that you want to graph. For *seconds,* Excel commands are below.

In F27,
Cntl+shift+down
Cntl+X
In H27,
Alt HIE
In row 62,
Shift+spacebar
Cntl+C
In row 64,
Alt HIE
In F64,
Shift+down
Shift+right
Alt ND
Alt JCW

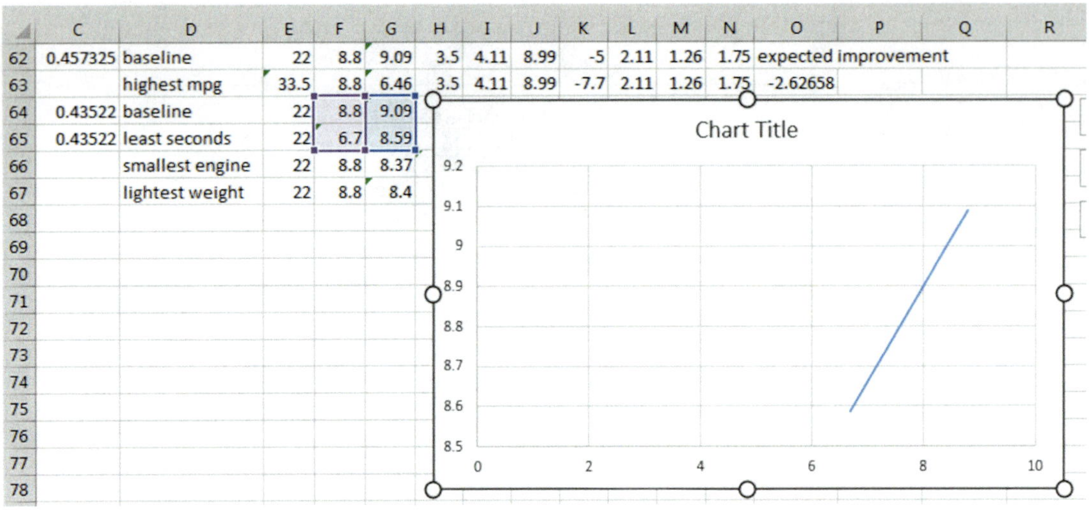

Add a horizontal axis title and chart title, adjust the horizontal axis, and set fontsize to 12.

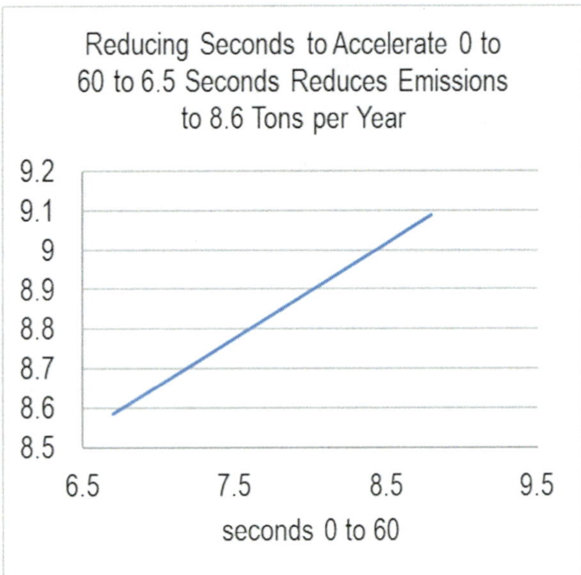

Choose the same minimum and maximum axes values for *predicted emissions* to make comparisons across car characteristics easiest. Delete the vertical axis on three of the four graphs and display side by side.

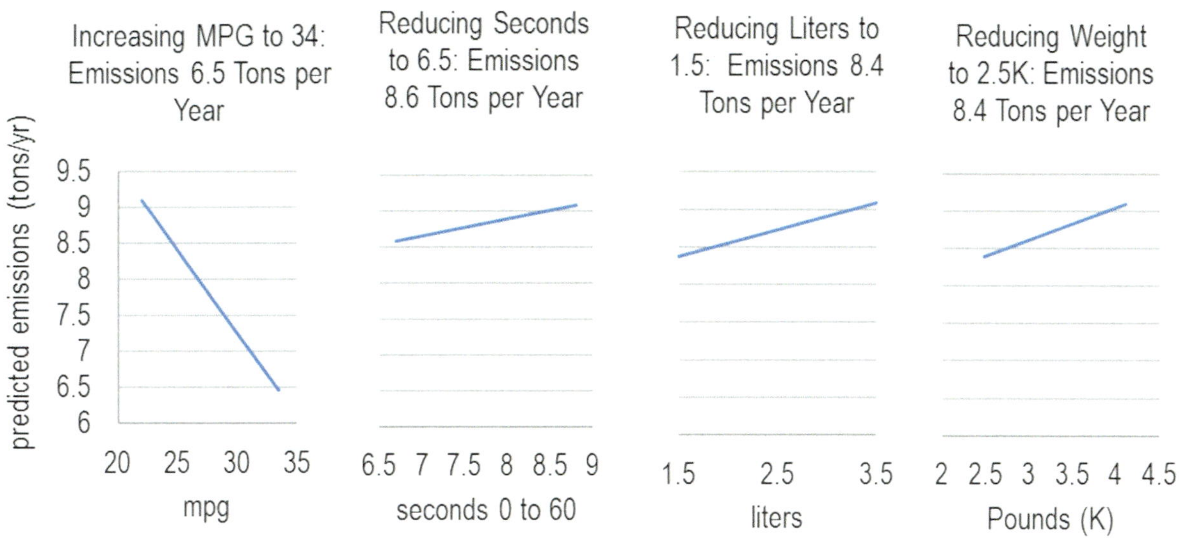
Increasing MPG to 34: Emissions 6.5 Tons per Year
Reducing Seconds to 6.5: Emissions 8.6 Tons per Year
Reducing Liters to 1.5: Emissions 8.4 Tons per Year
Reducing Weight to 2.5K: Emissions 8.4 Tons per Year
predicted emissions (tons/yr)
9.5
9
8.5
8
7.5
7
6.5
6
20 25 30 35
mpg
6.5 7 7.5 8 8.5 9
seconds 0 to 60
1.5 2.5 3.5
liters
2 2.5 3 3.5 4 4.5
Pounds (K)

Lab 10 Model Building with Multiple Regression: Pricing Dell's Navigreat

Dell has experience selling GPS systems built by other firms and plans to introduce a Dell system, the Navigreat. They would like information that will help them set a price.
The Navigreat has

- an innovative, *highly portable* design, *weighing only 5 ounces*, with a *state-of the art display*
- a *3.5" screen*, neither large, nor small, relative to competitors.
- innovative technology which guarantees precise *routing time* estimates,

Dell executives believe that these features, *portability, weight, display* quality, *screen size,* and *routing time* precision, drive the price that customers are willing to pay for a GPS system.

Recent ratings by *Consumer Reports* provide data on the retail *price* of 18 competing brands, as well as

- *portability* (1 to 5 scale), *weight* (ounces), and *display quality* (1 to 5 scale),
- *screen size* (inches)
- *routing time precision* (1 to 5 scale),

These data are in **Lab 10 Dell Navigreat**. Also in the file, in row 21, are the attributes and expected ratings of the Navigreat.

Build a multiple regression model of GPS system *price*, including the characteristics thought by management to be drivers of *price*.

Regression results. Is the model *RSquare* significantly greater than 0? Y N

Evidence: *Significance F*=__________

Which of the potential drivers have slopes significantly different from 0?

	portability	*weight*	*display*	*Screen size*	*Routing time*
Slope different from zero	Y or N	Y or N	Y or N	Y or N	Y or N
Evidence (*p value*)					

Which of the drivers have slopes of unexpected sign?

	portability	*weight*	*display*	*Screen size*	*Routing time*
Slope sign unexpected	Y or N	Y or N	Y or N	Y or N	Y or N

Confirm suspected multicollinearity. The GPS system physical design determines its *screen size, display quality, weight* and *portability*. Run correlations to see if these characteristics are highly correlated.

	Highly correlated ($r_{x1,x2}$>.5)
Portability, weight	Y or N
Portability, display	Y or N
Portability, screen size	Y or N
Weight, display	Y or N
Weight, screen size	Y or N
Display, screen size	Y or N

Remove one redundant characteristic and re-run the regression.

How much did RSquare decline? ____

Determine the improvement in predictive accuracy:

	Full model (1)	Reduced model (2)	Improvement in *margin of error* (3)=(2)-(1)
Standard error	$	$	
Approximate margin of error in 95% predictions	$	$	$

Which of the potential drivers in this partial model have slopes significantly different from 0? (Cross out characteristics that you excluded in this model.)

	portability	*weight*	*display*	*Screen size*	*Routing time*
Slope different from zero	Y or N	Y or N	Y or N	Y or N	Y or N
Evidence (*p value*)					

Which of the drivers have slopes of unexpected sign? (Cross out characteristics that you excluded in this model.)

	portability	*weight*	*display*	*Screen size*	*Routing time*
Slope sign unexpected	Y or N	Y or N	Y or N	Y or N	Y or N

Remove one redundant characteristic and re-run the regression.

How much did RSquare decline? ____

Determine the improvement in predictive accuracy:

	Second model (1)	Third model (2)	Improvement in *margin of error* (3)=(1)−(2)
Standard error	$	$	
Approximate margin of error in 95% predictions	$	$	$

Which of the potential drivers in this partial model have slopes significantly different from 0? (Cross out characteristics that you excluded in this model.)

	portability	*weight*	*display*	*Screen size*	*Routing time*
Slope different from zero	Y or N	Y or N	Y or N	Y or N	Y or N
Evidence (*p value*)					

Which of the drivers have slopes of unexpected sign? (Cross out characteristics that you excluded in this model.)

	portability	*weight*	*display*	*Screen size*	*Routing time*
Slope sign unexpected	Y or N	Y or N	Y or N	Y or N	Y or N

Remove one redundant characteristic and re-run the regression.

How much did RSquare decline? ____

Determine the improvement in predictive accuracy:

	Third model (1)	Fourth model (2)	Improvement in *margin of error* (3)=(1)−(2)
Standard error	$	$	
Approximate margin of error in 95% predictions	$	$	$

Which of the potential drivers in this partial model have slopes significantly different from 0? (Cross out characteristics that you excluded in this model.)

	portability	*weight*	*display*	*Screen size*	*Routing time*
Slope different from zero	Y or N	Y or N	Y or N	Y or N	Y or N
Evidence (*p value*)					

Which of the drivers have slopes of unexpected sign? (Cross out characteristics that you excluded in this model.)

	portability	*weight*	*display*	*Screen size*	*Routing time*
Slope sign unexpected	Y or N	Y or N	Y or N	Y or N	Y or N

Assess residuals.
Are residuals approximately *Normal*? Y or N

Predict prices. Use the regression equation to find *expected prices* for each of the GPS systems, including the Navigreat from the intercept and two part worths.

Find the *critical t* value for 95% prediction intervals with your model *residual degrees of freedom*.

Find the *margin of error* from the *critical t* and regression *standard error.*
Find the *lower* and *upper 95% prediction intervals* for each model, including the Navigreat.

Will Dell be able to charge a retail price of $650 for the Navigreat? Y or N

Do variables removed from the model matter? Run simple regressions with each of the variables removed to decide whether each is a driver of price.

Driver removed	*Driver?*
Portability	Y or N
Display quality	Y or N
Weight	Y or N

Sensitivity analysis: Identify the most important driver of prices by comparing the differences in *expected prices* between three hypothetical GPS systems.
Add these three hypotheticals, and then extend *expected price* to include these.

Screen size	*Route time rating*	*Expected price*	*Difference due to*
Median (3.5″)	Median (3.5="Moderate/Good")	$	Screen size: $__________
Largest (5″)	Median	$	
Median	Best (5="Excellent")	$	Route time rating: $__________

If Dell wants to charge a retail price of $650 for the Navigreat, what product design modification ought to be made?__________________________________

Assignment 10.1 Sakura Motor's Quest for Fuel Efficiency

The new product development team at Sakura Motors has decided that the new car which they are designing will have superior gas mileage on the highway. Use the data in **Assignment 10 Sakura Motors** to build a model to help the team. Variables in the dataset include:

MPGHwy
manufacturer's suggested retail base price
engine size (*liters*)
engine *cylinders*
engine *horsepower*
curb *weight*
acceleration in *seconds* to go from 0 to 60
percent of owners *satisfied* who would buy the model again

Use your logic to choose car characteristics which ought to influence highway gas mileage. Determine which car characteristics influence *highway gas mileage*.

With sensitivity analysis, find the relative importance of significant influences on *highway fuel economy*. Find the car characteristic levels which could be expected to achieve **40 miles per gallon** in highway driving. (Sakura is not limited to existing designs.)

Write a **one page, single spaced** memo presenting your model, sensitivity analysis and design recommendations.

Present your final model in standard format

What is the margin of error of model forecasts of MPG?

Discuss the relative importance of significant influences, including the expected difference in *fuel economy* that differences in each could be expected to make if other characteristics were held at median values

Conduct a sensitivity analysis comparing expected fuel economy with best and median levels of each predictor in your final model when other characteristics are at median levels. Discuss the relative importance of significant influences, referring to

(i) *a table of Fuel Response to Car Characteristics which you have added to the second page of Attachments*
(ii) *a scatterplot of the impact of each significant driver on fuel economy. This plot shows predicted fuel economy on the vertical axis by values of a driver.*

There is a two page limit:

One single spaced page for your memo text with a single embedded scatterplot (of *predicted MPGHwy* by the most important car characteristic).

A second page of Exhibits showing your sensitivity analysis table and plots.

Please use Times New Roman 12 pt font and round your statistics to two or three significant digits.

Case 10.1 Fast Food Nations

Fast food restaurants have been sprouting in diverse global regions. McDonalds has announced plans to open 700+ new restaurants in China. There is some concern that other global regions may be more fruitful locations.

Executives believe that countries with larger urban populations, and more populations under 15 years are likely the most fruitful locations for potential fast food establishments. (They are not sure whether education and economic productivity favors or discourages the fast food industry.)

Fast Food Nations contains data on number of fast food establishments, demographics, and economic productivity of 47 countries. Build a model linking number of fast food establishments to demographic and economic characteristics of countries to help McDonalds choose the most promising locations for new restaurants.

1. Identify the outlier(s), with respect to fast food establishments and list, below:
2. Which variables drive fast food sales in a country?

___ Population under 15 ___ Urban population ___ Expected years in school

___ GDP per capita

3. Present your model equation linking demographics and economic productivity to fast food establishments in a country. (Be sure to specify units and significance levels.) Format matters.
4. Illustrate the impact of the most influential driver in your model and embed below:
5. Describe the power of your model, referring to the appropriate statistic:
6. Describe the precision of your model, referring to the appropriate statistic:
7. Comparing predicted with actual fast food establishments, which country would be the most fruitful choice for future McDonalds restaurants?
8. How do the "Fast Food Nations" differ with respect to fast food, demographics, or in terms of economics, from the other countries? (For each difference that you identify, include reference to the statistic that you used to conclude that a difference exists.)

Case 10.2 Chasing Chipotle's Success

Yum brands (Taco Bell) management is considering an attempt to replicate Chipotle's success by implementing a process for selecting new locations that mirrors Chipotle's selection process, detailed in their website:

Want a Chipotle in your backyard?

If you have a suggestion for our next location, please review the criteria on this page and submit a request using the form below. We will answer you with many thanks and send along your suggestion to our real estate folks in your area. Keep in mind we are not looking to franchise at this time.

General Location Needs

✓ Urban and suburban with strong residential and daytime population.

✓ Preferred generators include residential, office, retail, university, recreation and hospitals.

Yum! Brands believes that States in the Heartland may be the most promising for new locations, since residents of Heartland States are thought earn more college *degrees per person,* and since Heartland States would offer access to local meat and produce.

Data in **Chipotle Locations** contains these data for each State:

Heartland State or not
Chipotle locations in 2012
colleges
college *degrees* conferred in 2010
2010 population
college *degrees* conferred *per person* in 2010
2010 people per sq mi

Help Yum! Brands strategists by identifying drivers of Chipotle location selection (*locations* in a State).

1. What drives number of *locations* in a State?

   ___ *colleges* ___*college degrees* ___*population* ___*people per sq mi*

   ___*degrees per person*

2. Write the equation for your final model of *locations* in a State. (Note: Format matters.)
3. Embed a graph illustrating the impact on *locations* of the most important driver. (Note: Format matters.)
4. Identify and describe differences between *Heartland* States and other States. Present and interpret the statistics that you used:
5. Based on the Chipotle selection process revealed by your model results, in which two States will Chipotle likely add new locations?
6. Based on your results, explain why Yum! Brands should or should not focus on Heartland States for new locations in a single sentence:

Case 10.3 Costco's Warehouse Location Scheme

Wal-Mart executives have been tracking Costco's success. They would like to identify the drivers of Costco's location choices. Will Costco enter China? Will Costco enter the other BRICs (Brazil, Russia and India)?

From Costco press releases, they have learned that Costco targets locations

- which are surrounded by affluent households,
- where there are a lot of college educated consumers.

A Costco spokesperson has also noted that Costco likes to have a gas station at every warehouse, since gas stations drive business. From Costco's press releases,
Wal-Mart executives believe that Costco locations may be in countries with (i) high GDP per capita, (ii) high percentage of urban population, (iii) a large percentage of college educated consumers, and (iv) a high percentage of cars per person.

Warehouse locations contains GDP per capita, urban population, percent 25 to 34 with college degrees, and vehicles per K population for each of 14 countries.

A. Identify drivers of Costco locations and quantify their importance

1. Which drive the number of Costco locations? ___ GDP per capita

   ___ urban population ___ % 25 to 34 w college degrees ___ vehicles per person

2. Present your equation of Costco locations:
3. Illustrate the impact of differences in the two most important drivers on differences in predicted Costco locations, comparing a hypothetical baseline country at median driver levels with two hypothetical countries at the maximum (97.5%) level of one of the two most important drivers.
4. Based on your analysis, in which two countries will Costco next add locations?
5. Wal-Mart executives expect Costco to enter China, soon, though Costco press releases suggest that France or Spain will be the next Costco destination. Why might drive Costco to locate in France or Spain next, instead of China? (List two possible reasons.)

B. Costco Nations' Differences

Wal-Mart managers believe that countries with Costco warehouses boast higher GDP per capita, higher urban populations, and more vehicles per person. Test their hypotheses.

1. Compare countries with Costco warehouses and countries without Costco warehouses. Identify the dimensions in which means between the two populations differ:

___ GDP per capita Evidence :_____________

___ urban population Evidence: ______________

___ vehicles per person Evidence: ______________

2. For any significant difference between Costco Nations and other countries' means construct approximate 95% confidence intervals to describe the extent of difference between means:

GDP per capita difference between means: ______________________

Urban population difference between means: ______________________

Vehicles per person difference between means: _____________________

3. Estimate conservatively the percent of 25 to 34 year olds who hold college degrees in

   countries with Costco warehouses: _____________

4. Estimate conservatively the percent of 25 to 34 year olds who hold college degrees in

   countries with no Costco warehouse: ___________

5. Wal-Mart executives conclude that

   (i) the majority of 25 to 34 year olds hold college degrees in countries with Costco locations
   (ii) the majority of 25 to 34 year olds do not hold college degrees in countries without Costco locations

   Are the executives' conclusions correct? ___Y ___N

Chapter 11 Indicator Variables

In this chapter, 0-1 *indicator* or "*dummy*" variables are used to incorporate segment differences, shocks, or structural shifts into models. With cross sectional data, indicators can be used to incorporate the unique responses of particular groups or segments. With time series data, indicators can be used to account for external shocks or structural shifts. Indicators also offer one option to account for seasonality or cyclicality in time series.

Analysis of variance sometimes is used as an alternative to regression when potential drivers are categorical, or when data are collected to assess the results of an experiment. In this case, the categorical drivers could be represented with indicators in regression, or analyzed directly with analysis of variance.

This chapter introduces the use of indicators to analyze data from conjoint analysis experiments. Conjoint analysis is used to quantify customer preferences for better design of new products and services.

Model variable selection begins with the choice of potential drivers from logic and experience. Indicators are added to account for segment differences, shocks, shifts or seasonality, and, in time series models, if autocorrelation remains, an indicator variable may be added to remedy the autocorrelation. The addition of indicators in the variable selection process is considered in this chapter.

11.1 Indicators Modify the Intercept to Account for Segment Differences

To compare two segments, a 0-1 indicator can be added to a model. One segment becomes the baseline, and the indicator represents the amount of difference from the base segment to the second segment. Indicators are like switches that turn on or off adjustments to a model intercept.

Example 11.1 Hybrid Fuel Economy. In a model of the impact of car characteristics on fuel economy:

$$M\hat{P}G = b_0(mpg) + b_1(mpg) \times hybrid + b_2\left(\frac{mpg}{ton}\right) \times emissions(tons) + b_3\left(\frac{mpg}{hp}\right) \times horsepower$$

$$= 48(mpg) + 8.8(mpg) \times hybrid - 2.3\left(\frac{mpg}{ton}\right) \times emissions(tons) - .025\left(\frac{mpg}{hp}\right) \times horsepower$$

The coefficient estimate of 8.8 (mpg) for the *hybrid* indicator modifies the intercept. For conventional cars, the *hybrid* indicator is 0, making the intercept for conventional cars 48 MPG:

C. Fraser, *Business Statistics for Competitive Advantage with Excel 2016*,
DOI 10.1007/978-3-319-32185-1_11

$$M\hat{P}G = 48(mpg) + 8.8(mpg) \times 0 - 2.3\left(\frac{mpg}{ton}\right) \times emissions(tons)$$
$$- .025\left(\frac{mpg}{hp}\right) \times horsepower$$

$$= 48(mpg) - 2.3\left(\frac{mpg}{ton}\right) \times emissions(tons) - .025\left(\frac{mpg}{hp}\right) \times horsepower$$

For hybrids in the sample, the *hybrid* indicator is 1, which adjusts the intercept for hybrids to 56.8 (mpg) by adding 8.8 (mpg) to the baseline 48 (mpg):

$$M\hat{P}G = 48(mpg) + 8.8(mpg) \times 1 - 2.3\left(\frac{mpg}{ton}\right) \times emissions(tons)$$
$$-- .025\left(\frac{mpg}{hp}\right) \times horsepower$$

$$= 56.8(mpg) - 2.3\left(\frac{mpg}{ton}\right) \times emissions(tons) \;-- .025\left(\frac{mpg}{hp}\right) \times horsepower$$

The adjustment is switched on when *hybrid* = 1, but remains switched off if *hybrid* = 0. The parameter estimate for the indicator tells us that on average, hybrid gas mileage is 8.8 (mpg) higher than conventional gas mileage.

Example 11.2 Yankees v Marlins Salaries[1]. The Yankees General Manager has discovered that the hot rookie whom the Yankees are hoping to sign is also considering an offer from the Marlins. The General Manager would like to know whether there is a difference in salaries between the two teams. He believes that, in addition to a possible difference between the two teams, *Runs* by players ought to affect salaries.

We will build a model of baseball salaries, including *Runs* and an indicator for Team. This variable, *Yankees,* will be equal to 1 if a player is on the Yankees Team, and equal to 0 if the player is a Marlin. The Marlins is the baseline team. Data are shown in Table 11.1, and regression results are shown in Table 11.2.

Table 11.1 Baseball team salaries

Player	*Team*	*Yankee*	*Runs*	*Salary (M$)*
Castillo	Marlin	0	72	5.2
Delgado	Marlin	0	81	4.0
Pierre	Marlin	0	96	3.7
Gonzalez	Marlin	0	45	3.4
Easley	Marlin	0	37	.8
Cabrera	Marlin	0	106	.4
Aguila	Marlin	0	11	.3
Treanor	Marlin	0	10	.3
Rodriguez	Yankee	1	111	21.7
Jeter	Yankee	1	110	19.6

[1] This example is a hypothetical scenario based on actual data

Sheffield	Yankee	1	94	13.0
Williams	Yankee	1	48	12.4
Posada	Yankee	1	60	11.0
Matsui	Yankee	1	97	8.0
Martinez	Yankee	1	41	2.8
Womack	Yankee	1	46	2.0
Sierra	Yankee	1	13	1.5
Giambi	Yankee	1	66	1.3
Flaherty	Yankee	1	8	.8
Crosby	Yankee	1	10	.3
Phillips	Yankee	1	7	.3

Table 11.2 Multiple regression of baseball salaries

SUMMARY OUTPUT						
Regression Statistics						
R Square	0.57					
Standard Error	4.2					
Observations	35					
ANOVA	*df*	*SS*	*MS*	*F*	*Significance F*	
Regression	2	754	377	21.3	.0000	
Residual	32	566	18			
Total	34	1320				
	Coefficients	*Standard Error*	*t Stat*	*p value*	*Lower 95%*	*Upper 95%*
Intercept	−3.90	1.56	−2.5	.02	−7.06	−0.73
Yankee	6.31	1.43	4.4	.0001	3.40	9.22
Runs	.104	.020	5.1	.0000	.062	.15

From the regression output, the model is:

$$\hat{Salary}(\$M) = -3.90(\$M)^a + 6.31(\$M)^b \times Yankee + .104\left(\frac{\$M}{run}\right)^b Runs$$

RSquare: $.57^b$
^{a}*Significant* at .05
^{b}*Significant* at .01

The coefficient estimate for the Yankee indicator is \$6.3M. The intercept for Yankees is \$6.3M greater than the intercept for Marlins. The rookie can expect to earn \$6.3 million more if he signs with the Yankees.

His expected salary, with 40 runs last season, is:

- As a Marlin, setting the *Yankee* indicator to zero:

$$
\begin{aligned}
Sa\hat{l}ary(\$M) &= -3.90(\$M) + 6.31(\$M) \times 0 + .104\left(\frac{\$M}{run}\right) \times 40(runs) \\
&= -3.90(\$M) + 4.16(\$M) = .26(\$M) = \$260{,}000
\end{aligned}
$$

- As a Yankee, setting the *Yankee* indicator to one:

$$
\begin{aligned}
Sa\hat{l}ary(\$M) &= -3.90(\$M) + 6.31(\$M) \times 1 + .104\left(\frac{\$M}{run}\right) \times 40(runs) \\
&= 2.41(\$M) + 4.16(\$M) = 6.57(\$M) = \$6{,}570{,}000
\end{aligned}
$$

The *Yankee* indicator modifies the intercept of the regression line, increasing it by $6.31 M.

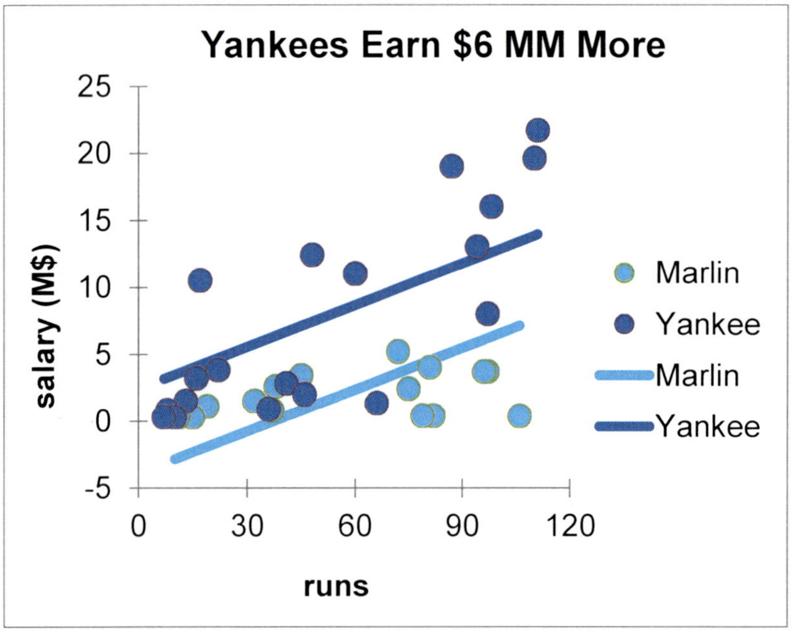

Figure 11.1 Yankees expect to earn $6 million more

In Figure 11.1, the intercept represents the baseline Marlins segment; the indicator adjusts the intercept to reflect the difference between Yankees and Marlins.

It does not matter which team is the designated baseline. The model will provide identical results either way.

11.2 Indicators Estimate the Value of Product Attributes

New product development managers sometimes use *conjoint analysis* to identify potential customers' most preferred new product design and to estimate the relative importance of product attributes. The conjoint analysis concept assumes that customers' preferences for a product are the sum of the values of each of the product's attributes, and that customers *trade off* features. A customer will give up a desired feature if another, more desired feature is offered. The offer of a more desired feature compensates for the lack of a second, less desired feature.

Example 11.3 New Smartphone Design. As an example, consider preferences for smartphones. Management believes that customers choose smartphones based on desired size, design, keypad, and price. For a new smartphone design, they are considering

- three sizes: bigger than shirt-pocket, shirt-pocket, and ultra thin shirt-pocket
- three designs: single unit, clamshell, and slider
- three keypads: standard, touch screen, and QWERTY
- three prices: \$150, \$250 and \$350

Management believes that price is a quality signal, and that customers suspect the quality of less expensive smartphones.

The least desirable, baseline configuration is expected to be:

shirt pocket size, with standard keypad at the lowest price.

It is not clear whether a single unit, clam shell or slider will be preferred. To find the *part worth utilities,* or the value of each cell phone feature, indicators are used to represent features that differ from the baseline. The conjoint analysis regression model is:

$$\begin{aligned} \textit{Smartphone}\ \widehat{\textit{preference}}_i = b_0 &+ b_1 \times \textit{bigger than shirt pocket size}_i \\ &+ b_2 \times \textit{ultra thin shirt size}_i \\ &+ b_3 \times \textit{clam shell}_i \quad + b_4 \times \textit{slider}_i \\ &+ b_5 \times \textit{touch screen}_i + b_6 \times \textit{QWERTY}_i \\ &+ b_7 \times \textit{\$250}_i \quad + b_8 \times \textit{\$350}_i \end{aligned}$$

for the *i*th smartphone configuration, where

b_0 is the intercept, which reflects preference for the baseline configuration,
b_1, b_2, b_3, b_4, b_5, b_6, b_7, and b_8 are estimates of the *part worth utilities* of features.

The conjoint analysis process assumes that it is easier for customers to rank or rate products or brands, rather than estimating the value of each feature. For price preferences, this may be particularly true. It will be easier to customers to rate hypothetical smartphone designs than it would be for customers to estimate the value of a \$250 smartphone, relative to a \$150 smartphone.

The four smartphone attributes could be combined in 81 (=3^4) unique ways. 81 hypothetical smartphones would be too many for customers to accurately evaluate. From the 81, a set of nine are carefully chosen so that the chance of each feature is equally likely (33%), and each feature is uncorrelated with other features. Slider designs, for example, are equally likely to be paired with each of the three sizes, each of the three keypads, and each of the three prices. This will eliminate multicollinearity among the indicators used in the regression of the conjoint model. Such a subset of hypothetical combinations is an *orthogonal array* and is shown in Table 11.3.

Table 11.3 Nine hypothetical smartphone designs in an orthogonal array

Size	*Shape*	*Keypad*	*Price*
Bigger than shirt-pocket	Single unit	Standard	$150
Bigger than shirt pocket	Clamshell	Touch screen	$250
Bigger than shirt pocket	Slider	QWERTY	$350
Shirt pocket	Single unit	Touch screen	$350
Shirt pocket	Clamshell	QWERTY	$150
Shirt pocket	Slider	Standard	$250
Ultra thin shirt pocket	Single unit	QWERTY	$250
Ultra thin shirt pocket	Clamshell	Standard	$350
Ultra thin shirt pocket	Slider	Touch screen	$150

Three customers rated the nine hypothetical smartphones after viewing concept descriptions with sketches. The configurations judged extremely attractive were rated 9 and those judged not at all attractive were rated 1. The regression with eight indicators is shown in Table 11.4.

Table 11.4 Regression of PDA preferences

R Square	.747						
Standard Error	1.644						
Observations	27						
ANOVA	*df*	*SS*	*MS*	*F*	*Significance F*		
Regression	8	143	17.9	6.6	.0004		
Residual	18	49	2.7				
Total	26	192					
	Coefficients	*Standard Error*	*t Stat*	*p value*	*Lower 95%*	*Upper 95%*	*1 tail p value*
Intercept	1.00	.95	1.1	.31	−.99	2.99	
Bigger than shirt pocket	1.89	.78	2.4	.03	.26	3.52	.02
ultra thin shirt pocket	.78	.78	1.0	.33	−.85	2.41	.17
clamshell	−1.56	.78	−2.0	.06	−3.18	.07	
slider	−1.44	.78	−1.9	.08	−3.07	.18	
touch screen	4.22	.78	5.4	.00	2.59	5.85	.00
QWERTY	3.78	.78	4.9	.00	2.15	5.41	.00
$250	1.67	.78	2.2	.05	.04	3.30	.03
$350	1.67	.78	2.2	.05	.04	3.30	.03

Smartphone size, keypad, and price features influence preferences, while design options do not. The preferred smartphone is *bigger than shirt pocket* or *ultra thin shirt pocket* size, features a *touch screen* or *QWERTY keypad,* and is priced at $250 or $350.

The *coefficients* estimate the part worth *utilities* of the smartphone features. Expected preference for the ideal design is the sum of the part worth utilities for features included. Design does not affect preferences, so the least expensive option would be used, and the two higher prices are equivalent to customers, so the higher, more profitable price would be charged. The expected preference rating for a hypothetical smartphone that is bigger than shirtpocket, single unit, with touch screen at $350 would be 8.8 on the 9 point scale:

$$\begin{aligned}
\widehat{\textit{Smartphone preference}}_i = 1.00 &+ 1.89 \times \textit{bigger than shirtpocket}_i \\
&+ .78 \times \textit{ultra thin shirtpocket}_i \\
&- 1.56 \times \textit{clamshell}_i \quad - 1.44 \times \textit{slider}_i \\
&+ 4.22 \times \textit{touch screen}_i + 3.78 \times \textit{QWERTY}_i \\
&+ 1.67 \times \textit{\$250}_i \quad + 1.67 \times \textit{\$350}_i
\end{aligned}$$

$$\begin{aligned}
= 1.00 &+ 1.89\,(1) \\
&+ .78\,(0) \\
&- 1.56\,(0) \qquad - 1.44\,(0) \\
&+ .22\,(1) \qquad + 3.78\,(0) \\
&+ 1.67\,(0) \qquad + 1.67\,(1) \\
= 1.00 &+ 1.89 + 4.22 + 1.67 \\
= 8.78 &
\end{aligned}$$

The part worth utilities from coefficient estimates are shown in Figure 11.2.

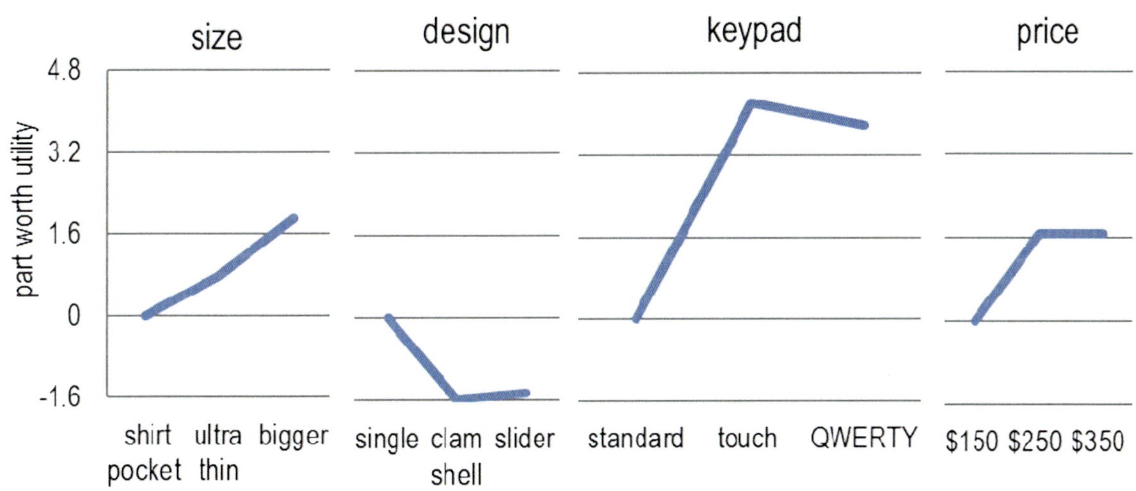

Figure 11.2 Smartphone part worth utilities

Preferred *larger than shirt pocket* size adds an expected 1.89 (= 1.89 − 0) to the preference rating, a *touch screen* adds an expected 4.22 (= 4.22 − 0), and a price of *$350* adds an expected 1.67 (= 1.67 − 0). The preferred design makes no significant difference, since the population coefficients for both *clamshell* and *slider* could be positive, zero, or negative.

The range in part worth utilities for each attribute is an indication of that attribute's importance. Preference depends most on the keypad configuration, which is more than twice as important as size or price, shown in Table 11.5.

Table 11.5 Relative importance of smartphone attributes

Attribute	*Part worth utility of least preferred*	*Part worth utility of most preferred*	*Part worth utility range*	*Attribute importance*
Size	0	1.9	1.9	1.9 / 7.8 = .24
keypad	0	4.2	4.2	4.2 / 7.8 = .54
price	0	1.7	1.7	1.7 / 7.8 = .22
	Sum of part worth utility ranges:		7.8	

Conjoint analysis been used to improve the designs of a wide range of products and services, including:

- seating, food service, scheduling and prices of airline flights
- offer of outpatient services and prices for a hospital
- container design, fragrance and design of a aerosol rug cleaner,
- digital camera pixels, features and prices

Conjoint analysis is versatile and the attributes studied can include characteristics that are difficult to describe, such as fragrance, sound, feel, or taste. It is difficult for customers to tell us how important color, package design, or brand name is in shaping preferences, and conjoint analysis often provides believable, valid estimates.

11.3 Indicators Estimate Segment Mean Differences

Indicators are used in regression to test hypotheses regarding equivalence of segment, group, or category means. With indicators, managers can compare mean performance across categories. The following are questions that managers might use indicators to address:

Does job satisfaction differ across divisions?
Does per capita demand differ across global regions?
Do preferences differ across flavors?
Do rates of return differ across portfolios?
Does customer loyalty differ across brands?

Where differences exist, regression with indicators enables estimation of the extent of those differences.

In each of these scenarios, the question concerns performance differences across categories or groups: divisions, global regions, flavors, portfolios, or brands.

Regression with indicators compares performance variation across groups with performance variation within groups, and more across group variation is evidence that the group performance levels differ.

Example 11.4 Background Music to Create Brand Interest. A brand manager suspects that the background music featured in a brand's advertising may affect the level of interest in the advertised brand. Several background options are being considered, and those options differ along two categories, or *factors*.

Three vocals options are:

(i) backgrounds which feature vocals,
(ii) backgrounds with brand related vocals substituted for original vocals, and
(iii) backgrounds with vocals removed.

Three orchestration options are:

(i) saxophone,
(ii) saxophone and percussion, and
(iii) saxophone and piano.

The hypotheses that the brand manager would like to test are:

$H_{vocals0}$: Mean interest ratings following exposure to ads with alternate vocals options are equivalent.

$$\mu_{original} = \mu_{brand\_specific} = \mu_{no\_vocals}$$

Versus

$H_{vocals1}$: At least one mean interest rating following exposure to ads with alternate vocals differs.

And

$H_{orchestration0}$: Mean interest ratings following exposure to ads with alternate orchestrations are equivalent.

$$\mu_{saxophone} = \mu_{saxophone+percussion} = \mu_{saxophone+piano}$$

Versus

$H_{orchestration1}$: At least one mean interest rating following exposure to ads with alternate orchestrations differs.

To determine whether vocals and orchestration of backgrounds affect brand interest ratings, the ad agency creative team designed nine backgrounds for a brand ad. Since the ad message, visuals, and length of ad could also influence interest, the agency creatives were careful to make those ad features identical across the nine versions. By using ads that were identical, except for their musical backgrounds, any difference in resulting brand interest could be attributed to the difference in backgrounds.

Nine consumers were randomly selected and then randomly assigned to one of the nine background *treatments,* or combination of *vocals* and *orchestration.* Each viewed the brand advertisement with one of the nine backgrounds, and then rated their interest in the brand using a scale from 1 ("not at all interested") to 9 ("very interested"). The data are shown in Table 11.6 and Figure 11.3.

Table 11.6 Brand interest ratings by vocals and orchestration levels

Brand Interest Ratings by Ad Background Music				
		orchestration		
vocals option	sax	sax & percussion	sax & piano	mean
None	9	6	7	7.3
original	6	4	5	5.0
brand specific	5	4	3	4.0
mean	6.7	4.7	5.0	5.4

To set up the data for regression analysis, zeros and ones are used to distinguish *levels* of both *factors*. Each factor in this experiment has three levels. Two indicator variables, shown in Tables 11.7 and 11.8, are needed to distinguish two of the three levels from the third *baseline* level. Regression results do not depend on which level is designated as the *baseline*.

Only the two indicator variables, *original* and *brand*, are included in the regression, since together with the baseline, the three form an identity matrix. The value of the baseline will be reflected in the intercept.

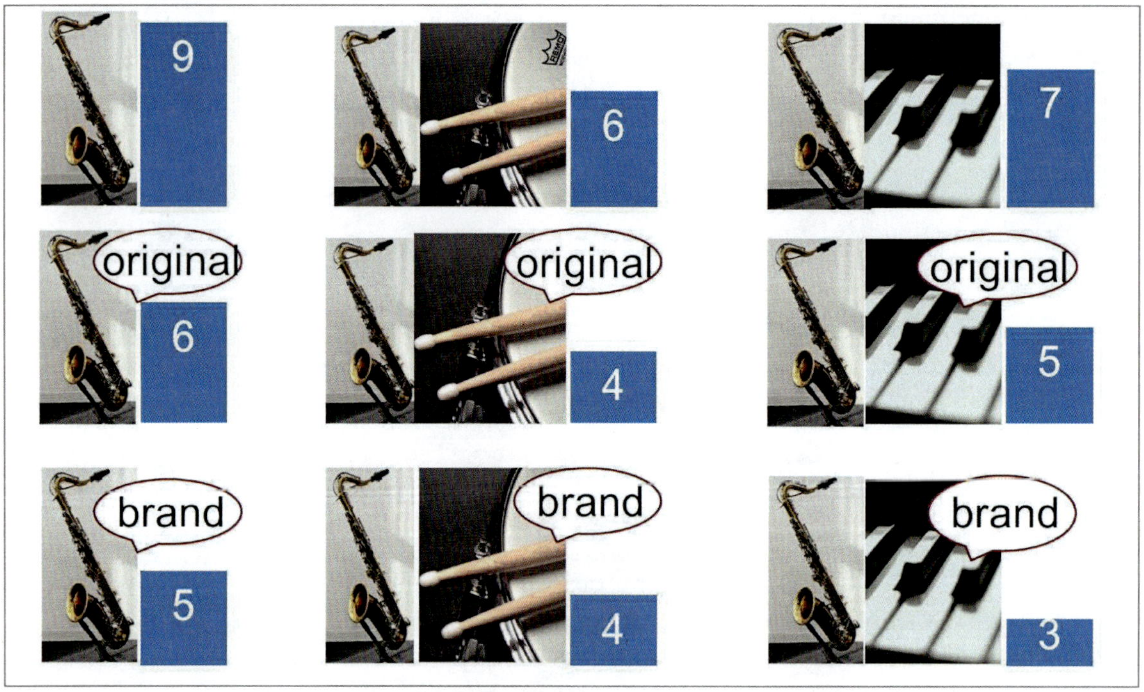

Figure 11.3 Brand interest by background instrumentation and vocals

Table 11.7 Vocals indicators

	Baseline	*Indicator variables*	
Vocals factor levels	*No vocals*	*Original*	*Brand*
No vocals	1	0	0
Original	0	1	0
Brand	0	0	1

Table 11.8 Orchestration indicators

	Baseline	*Indicator variables*	
Orchestration factor levels	*Sax*	*Saxperc*	*Saxpiano*
Sax	1	0	0
Sax + percussion	0	1	0
Sax + piano	0	0	1

Regression enables us to determine whether at least one of the factors, either one or both, matters. Regression also identifies particular levels which produce higher or lower expected performance relative to the baseline. To illustrate, a regression model of *vocals* and *orchestration* background influences on *brand interest* is shown below.

$$\widehat{Interest} = b_0 + b_{original} \times original + b_{brand} \times brand + b_{sax+perc} \times saxperc + b_{sax+piano} \times saxpiano$$

where *no vocals* with orchestration for *saxophone* are the baseline levels.

Regression results are below in Table 11.9:

Table 11.9 Multiple regression with indicators

SUMMARY OUTPUT						
Regression Statistics						
R Square	.932					
Standard Error	.667					
Observations	9					
ANOVA	*df*	*SS*	*MS*	*F*	*Significance F*	
Regression	4	24.4	6.1	13.8	.01	
Residual	4	1.8	0.4			
Total	8	26.2				
	Coefficients	*Standard Error*	*t Stat*	*P-value*	*Lower 95%*	*Upper 95%*
Intercept	8.6	.50	17.2	.0001	7.2	9.9
original	−2.3	.54	−4.3	.01	−3.8	−.8
brand	−3.3	.54	−6.1	.004	−4.8	−1.8
saxperc	−2.0	.54	−3.7	.02	−3.5	−.5
saxpiano	−1.7	.54	−3.1	.04	−3.2	−.2

The model *F* statistic, 13.8, has a *p value* (=.01) less than the *critical p value* of .05. Sample evidence allows the conclusion that at least one of the *vocals* or *orchestration* options is driving the level of brand *interest*. From the model *RSquare,* we learn that differences in *vocals* and *orchestration* together account for 93% of the variation in brand *interest* ratings.

Regression enables identification of indicators which differ from the baseline. The coefficient estimates for *original vocals* and *brand vocals* are significant. *Original* vocals reduces *interest* by 1 to 4 points, and *brand* vocals reduces *interest* by 2 to 5 points.

Both the coefficient estimates for *saxperc* and *saxpiano* are significant. Adding percussion to the background reduces expected brand *interest* ratings by 1 to 4 rating scale points. Adding piano to the background reduces *interest* by as much as 3 rating scale points.

When a regression model is built using indicators, part worth graphs can be used to illustrate results. In the background music example, *part worth interest ratings* can be compared, as Figure 11.4 illustrates:

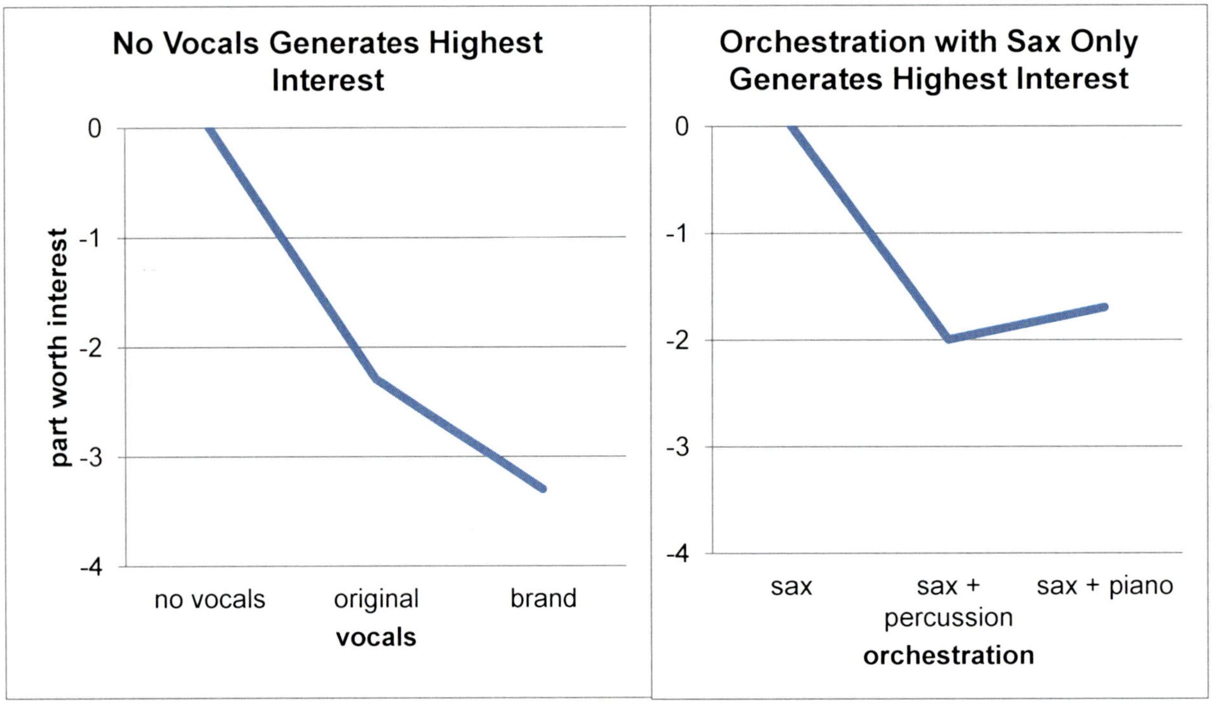

Figure 11.4 Interest part worths

From regression, we learn that together, *vocals* and *orchestration* options account for 93% of the variation in *interest* ratings, and that backgrounds with *no vocals* instead of *original* or *brand* vocals, and *saxophone* alone, instead of a combination with either piano or percussion, are expected to generate the highest ratings, as much as eight scale points higher than backgrounds with vocals and either piano or percussion.

11.4 Analysis of Variance Offers an Alternative to Regression with Indicators

ANalysis Of VAriance is an alternative to regression with indicators for situations in which all of the drivers are categorical. ANOVA also tests hypotheses regarding factor level means and provides individual F statistics for each factor but splitting the variation explained by factors, SSR in regression, into pieces explained by each factor.

In the background music example, variation across *vocals* levels, ignoring *orchestration* levels, is:

$$SSB_{vocals} = n_{no\ vocals} \times (\bar{X}_{no\ vocals} - \bar{\bar{X}})^2$$

$$+ n_{original} \times (\bar{X}_{original} - \bar{\bar{X}})^2$$

$$+ n_{brand} \times (\bar{X}_{brand} - \bar{\bar{X}})^2$$

$$= 3 \times (7.3 - 5.4)^2 + 3 \times (5.0 - 5.4)^2 + 3 \times (4.0 - 5.4)^2$$

$$= 17.6$$

There are three *vocals* levels. The degrees of freedom for variation across *vocals* levels is two, comparing two of the levels to the third baseline. Mean variation is:

$$MSB_{vocals} = SSB_{vocals} / df_{vocals}$$

$$= 17.6 / 2$$

$$= 8.8$$

Variation across *orchestration* levels is

$$SSB_{orchestration} = n_{sax} \times (\bar{X}_{sax} - \bar{\bar{X}})^2$$

$$+ n_{sax+perc} \times (\bar{X}_{sax+perc} - \bar{\bar{X}})^2$$

$$+ n_{sax+piano} \times (\bar{X}_{sax+piano} - \bar{\bar{X}})^2$$

$$= 3 \times (6.7 - 5.4)^2 + 3 \times (4.7 - 5.4)^2 + 3 \times (5.0 - 5.4)^2$$

$$= 6.9$$

And mean variation between *orchestration* levels is:

$$MSB_{orchestration} = SSB_{orchestration} / df_{orchestration}$$

$$= 6.9 / 2$$

$$= 3.4$$

To compare mean variation across *vocals* levels and *orchestration* levels with mean variation within *vocals* and *orchestration* levels, the variation within levels is calculated by subtracting SSB_{vocals} and $SSB_{orchestration}$ from total variation, SST:

$$\begin{aligned} SST &= (9-5.4)^2+(6-5.4)^2+(7-5.4)^2 \\ &+(6-5.4)^2+(4-5.4)^2+(5-5.4)^2 \\ &+(5-5.4)^2+(4-5.4)^2+(3-5.4)^2 \\ &= 26.2 \end{aligned}$$

Of the total variation of 26.2, 17.6 has been explained by differences across *vocals* levels, and 6.9 has been explained by differences across *orchestration* levels, leaving 1.8 unexplained from variation within levels:

$$\begin{aligned} SSW &= SST - SSB_{vocals} - SSB_{orchestration} \\ &= 26.2 - 17.6 - 6.9 \\ &= 1.8 \end{aligned}$$

Mean unexplained variation is:

$$\begin{aligned} MSW &= SSW \,/\, (N - df_{vocals} - df_{orchestration} - 1) \\ &= 1.8 \,/\, 4 \\ &= .4 \end{aligned}$$

To test each of the two sets of hypotheses, the corresponding F statistic is calculated from the ratio of *mean squares between*, MSB_{vocals} or $MSB_{orchestration}$, and *mean square within*, MSW:

$$\begin{aligned} F_{vocals_{2,4}} &= MSB_{vocals} \,/\, MSW \\ &= 8.8 \,/\, .4 \\ &= 19.8 \end{aligned}$$

$$\begin{aligned} F_{orchestration_{2,4}} &= MSB_{orchestration} \,/\, MSW \\ &= 3.4 \,/\, .4 \\ &= 7.8 \end{aligned}$$

With 2 and 4 degrees of freedom, the *critical F* for 95% confidence is 6.9. Both *F* statistics exceed the *critical F*, and have *p values* of .008 and .04. Based on the sample data, there is evidence that the *vocals* alternatives are not equally effective in backgrounds, and that the *orchestration* alternatives are also not equally effective. Both null hypotheses are rejected.

Excel provides the *F* statistics and their *p values,* as well as factor level means, shown below in Table 11.10:

Table 11.10 ANOVA results from Excel

Anova: Two Factor Without Replication						
SUMMARY	*Count*	*Sum*	*Average*	*Variance*		
No vocals	3	22	7.3	2.3		
original	3	15	5.0	1.0		
brand	3	12	4.0	1.0		
Sax	3	20	6.7	4.3		
sax & percussion	3	14	4.7	1.3		
sax & piano	3	15	5.0	4.0		
ANOVA						
Source of Variation	*SS*	*df*	*MS*	*F*	*p value*	*F crit*
Rows	17.6	2	8.8	19.8	.008	6.9
Columns	6.9	2	3.4	7.8	.04	6.9
Error	1.8	4	.4			
Total	26.2	8				

In the sample, ads with *no vocals* produced highest average brand *interest ratings,* $\bar{X}_{no\ vocals} = 7.3$, and ads with *brand* vocals, produced lowest average *interest ratings,* $\bar{X}_{brand} = 4.0$.. The F_{vocals} test allows the conclusion that at least one of the *vocals* factor levels differs. Therefore, it is possible that (i) *no vocals* is more effective than either option with vocals, (ii) *brand vocals* are less effective than either *original* or *no vocals,* or (iii) all three levels may differ. To determine which of the three levels differ, *multiple comparisons,* which resemble *t* tests, would be used, though Excel does not offer this ability. (Other more specialized software packages, such as SPSS and SAS, do offer multiple comparisons.)

Ads with *sax* produced highest average brand *interest ratings,* $\bar{X}_{sax} = 6.7$, and ads with *sax+percussion orchestration* produced the lowest average *interest ratings,* $\bar{X}_{saxperc} = 4.7$. The $F_{orchestration}$ test allows the conclusion that at least one of the *orchestration* factor levels differs; however, from analysis of variance results, it is not possible to determine whether any of the three levels are statistically unique.

11.5 ANOVA and Regression with Indicators Are Complementary Substitutes

The *F* statistics used to test hypotheses with analysis of variance and with regression are similar. Both compare variation explained by model drivers or factors with unexplained variation. Analysis of variance enables us to determine whether each factor matters. For example, both *vocals* and *instrumentation* in ad backgrounds matter and at least one *vocal* option and at least one *instrumental* option are more effective in generating *brand interest* following ad exposure. Regression enables us to determine whether at least one of the factors, either one or both, matters. Regression also identifies particular indicators which produce higher or lower expected performance relative to the baseline.

Regression *RSquare* provides a measure of the power of the model: differences in *vocals* and *orchestration* together account for 93% of the variation in brand *interest* ratings. While analysis of variance does not explicitly provide *RSquare,* it can be easily found from analysis of variance output as the ratio of explained variation, the sum of squares due to the factors, and total variation. Variation explained by the two factors in analysis of variance is equivalent to variation explained by the model in regression:

$$SSB_{vocals} + SSB_{orchestration} = SSR$$
$$17.6 + 6.9 = 24.4$$

And

$$RSquare = (SSB_{vocals} + SSB_{orchestration})/SST$$

$$= (17.6 + 6.9)/26.2$$

$$= .932$$

From analysis of variance, we learn that both *vocals* and *orchestration* influence *interest* ratings. From regression, we learn that together, *vocals* and *orchestration* options account for 93% of the variation in *interest* ratings, and that backgrounds with *no vocals* instead of *original* or *brand vocals,* and *saxophone* alone, instead of a combination with either piano or percussion, are expected to generate the highest ratings, as much as eight scale points higher than backgrounds with vocals and either piano or percussion.

Multiple regression with indicators and analysis of variance are substitutes, though they are each offer particular advantages. Multiple regression is designed to accommodate both categorical and continuous drivers, and interest is twofold: (i) identify performance drivers, including differences across groups, and (ii) forecast performance under alternate scenarios. Regression accounts for the impact of continuous drivers by building them into a model. Analysis of variance is designed to identify performance differences across groups. Where possible, continuous drivers are controlled by choosing groups that have equivalent profiles, often in the context of an experiment.

11.6 ANOVA and Regression in Excel

Regression's dual goals of (i) identification of drivers and quantification of their influence, plus (ii) forecasting performance under alternate scenarios, provides more information than analysis of variance in Excel, where output is primarily geared toward hypothesis tests of the factors. However, other, more specialized software packages, such as *SAS, JMP,* and *SPSS,* offer more powerful and versatile analysis of variance features, including multiple comparisons. Marketing researchers and psychometricians sometimes use *analysis of covariance* to account for variation in experiments that has not been controlled, and to compare factor levels to identify those that differ.

Analysis of variance is particularly well suited for use with experimental data, and, since experiments tend not to be routinely conducted by managers, experimental data collection and analysis are often outsourced to marketing research firms. Because Excel is targeted for use by managers, analysis of variance in Excel is basic. In Excel, there is the additional limitation that *replications,* the number of datapoints for each combination of factor levels, must be equivalent. In the background music experiment, for example, had fifteen consumers been randomly selected to view one of the nine ads, data from only nine consumers could be used in analysis of variance with Excel. Six of the ads would have been viewed by two consumers each, and three of the ads would have been viewed by only one consumer. Data from six consumers would have to be ignored in order to use analysis of variance in Excel. Since all of the data could be used in regression with indicators, regression is a more useful choice in Excel, and allows both hypothesis tests and forecasts under alternate scenarios.

Excel 11.1 Use Indicators to Find Part Worths and Attribute Importances

Nine consumers rated their interest in a brand after viewing and ad with one of nine music backgrounds configurations. A 9 point rating scale from 1 ("not at all interested") to 9 ("very interested") was used. The data are in **Excel 11.1 Backgrounds.**

In columns D and E, create *vocals* indicators, and in columns F and G, create *orchestration* indicators.

	A	B	C	D	E	F	G
1	*vocals*	*orchestration*	*rating*	*none*	*original*	*sax*	*sax & piano*
2	original	sax	6	0	1	1	0
3	brand specific	sax	5	0	0	1	0
4	none	sax	9	1	0	1	0
5	original	sax & percussion	4	0	1	0	0
6	brand specific	sax & percussion	4	0	0	0	0
7	none	sax & percussion	6	1	0	0	0
8	original	sax & piano	5	0	1	0	1
9	brand specific	sax & piano	3	0	0	0	1
10	none	sax & piano	7	1	0	0	1

Baseline hypothetical. The baseline background, with indicators for *none* and *original vocals* and *sax* and *sax & piano orchestration* is *brand specific vocals* with *sax & percussion orchestration,* the fifth combination out of nine. Notice that in that row 6, all indicator values in columns D, E, F and G are zero.

Run a regression of *rating*, with the four indicators:

	A	B	C	D	E	F	G
1	SUMMARY OUTPUT						
2							
3	*Regression Statistics*						
4	Multiple F	0.965507					
5	R Square	0.932203					
6	Adjusted I	0.864407					
7	Standard E	0.666667					
8	Observati	9					
9							
10	ANOVA						
11		*df*	*SS*	*MS*	*F*	*gnificance F*	
12	Regressio	4	24.44444	6.111111	13.75	0.013166	
13	Residual	4	1.777778	0.444444			
14	Total	8	26.22222				
15							
16		*Coefficient*	*andard Err*	*t Stat*	*P-value*	*Lower 95%*	*Upper 95%*
17	Intercept	3.222222	0.496904	6.484597	0.002916	1.842596	4.601849
18	none	3.333333	0.544331	6.123724	0.003602	1.822028	4.844639
19	original	1	0.544331	1.837117	0.140066	-0.51131	2.511305
20	sax	2	0.544331	3.674235	0.021312	0.488695	3.511305
21	sax & pian	0.333333	0.544331	0.612372	0.573392	-1.17797	1.844639

Original vocals do not influence interest *ratings* relative to *brand* specific *vocals*, though no vocals improve interest. *Sax & piano orchestration* generates interest equivalent to *sax & percussion,* the baseline *orchestration*, though *sax* alone improves ratings.

Part worths. The *coefficients* are estimates of the part worths, the impact of each background feature on interest rating. Copy the data in columns A through G in the data sheet and paste below regression results. Then, add the intercept and part worths to find predicted interest ratings:

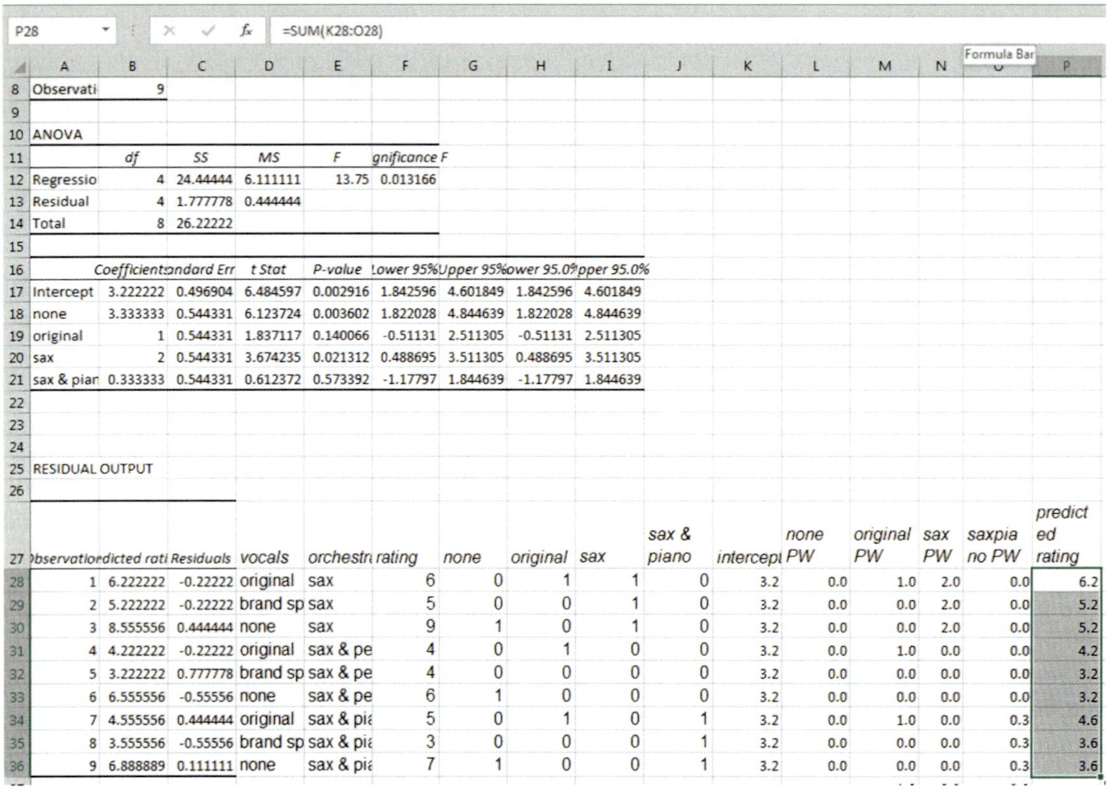

P28 =SUM(K28:O28)

	A	B	C	D	E	F	G	H	I
8	Observati	9							
9									
10	ANOVA								
11		df	SS	MS	F	gnificance F			
12	Regressio	4	24.44444	6.111111	13.75	0.013166			
13	Residual	4	1.777778	0.444444					
14	Total	8	26.22222						
15									
16		Coefficient	:andard Err	t Stat	P-value	Lower 95%	Upper 95%	ower 95.0%	pper 95.0%
17	Intercept	3.222222	0.496904	6.484597	0.002916	1.842596	4.601849	1.842596	4.601849
18	none	3.333333	0.544331	6.123724	0.003602	1.822028	4.844639	1.822028	4.844639
19	original	1	0.544331	1.837117	0.140066	-0.51131	2.511305	-0.51131	2.511305
20	sax	2	0.544331	3.674235	0.021312	0.488695	3.511305	0.488695	3.511305
21	sax & piar	0.333333	0.544331	0.612372	0.573392	-1.17797	1.844639	-1.17797	1.844639
22									
23									
24									
25	RESIDUAL OUTPUT								
26									

	A	B	C	D	E	F	G	H	I	J	K	L	M	N	O	P
27	Observatio	edicted rati	Residuals	vocals	orchestr	rating	none	original	sax	sax & piano	intercept	none PW	original PW	sax PW	saxpia no PW	predicted rating
28	1	6.222222	-0.22222	original	sax	6	0	1	1	0	3.2	0.0	1.0	2.0	0.0	6.2
29	2	5.222222	-0.22222	brand sp	sax	5	0	0	1	0	3.2	0.0	0.0	2.0	0.0	5.2
30	3	8.555556	0.444444	none	sax	9	1	0	1	0	3.2	0.0	0.0	2.0	0.0	5.2
31	4	4.222222	-0.22222	original	sax & pe	4	0	1	0	0	3.2	0.0	1.0	0.0	0.0	4.2
32	5	3.222222	0.777778	brand sp	sax & pe	4	0	0	0	0	3.2	0.0	0.0	0.0	0.0	3.2
33	6	6.555556	-0.55556	none	sax & pe	6	1	0	0	0	3.2	0.0	0.0	0.0	0.0	3.2
34	7	4.555556	0.444444	original	sax & pi	5	0	1	0	1	3.2	0.0	1.0	0.0	0.3	4.6
35	8	3.555556	-0.55556	brand sp	sax & pi	3	0	0	0	1	3.2	0.0	0.0	0.0	0.3	3.6
36	9	6.888889	0.111111	none	sax & pi	7	1	0	0	1	3.2	0.0	0.0	0.0	0.3	3.6

To see the difference that each background feature makes, plot the part worth utilities for each attribute. Insert a row above the two *vocals* indicators in rows 18 and 19 for *brand* with baseline part worth of zero, then select the six cells with *vocals* labels and part worths and request a line plot.

In row 18,
Alt HIR
In A18,
brand
In B18,
0
In A18,
Shift+down down
Shift+right
Alt NN

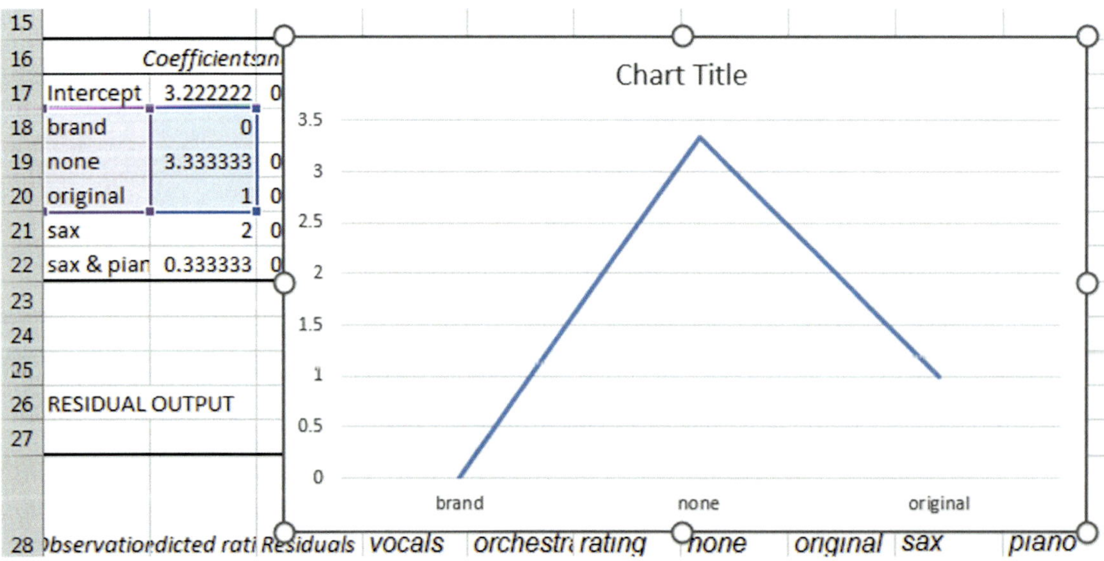

Insert a row below the two *orchestration* indicators in rows 21 and 22 for *sax & perceussion* with baseline part worth of zero, then select the six cells with *orchestration* labels and part worths and request a line plot.	In A23, Sax & percussion In B23, 0 In A21, **Shift+down down** **Shift+right** **Alt NN**

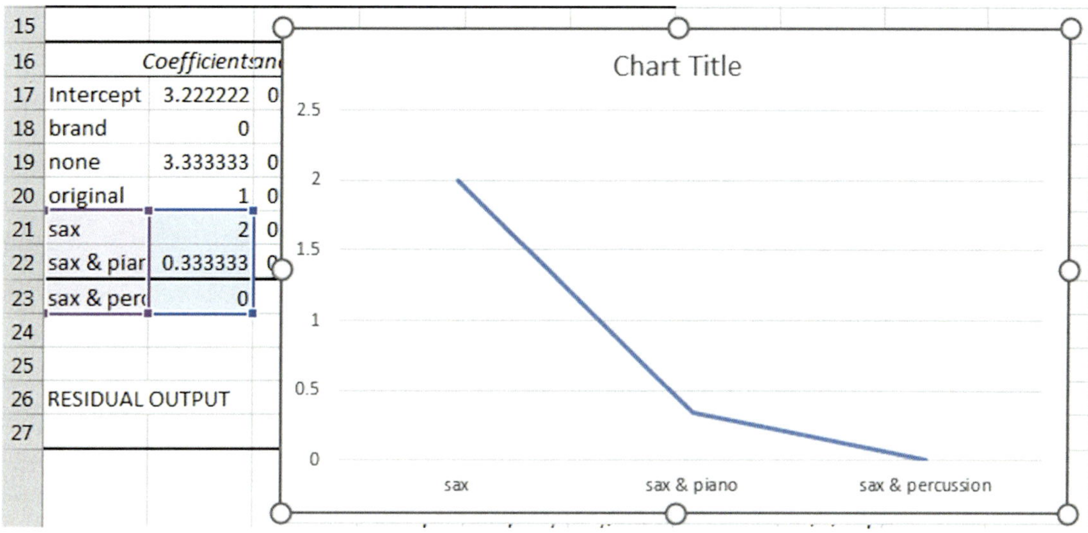

So that attributes can be compared, reformat the vertical axis range, from the most negative to the most positive part worth, 0 to 3.5, choosing a value for major unit, such as .5:

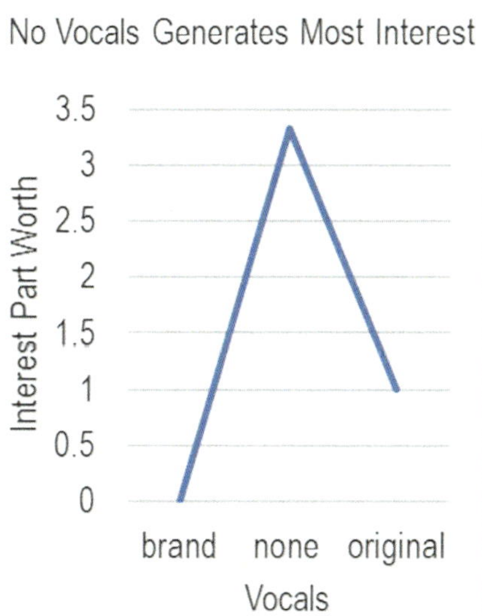

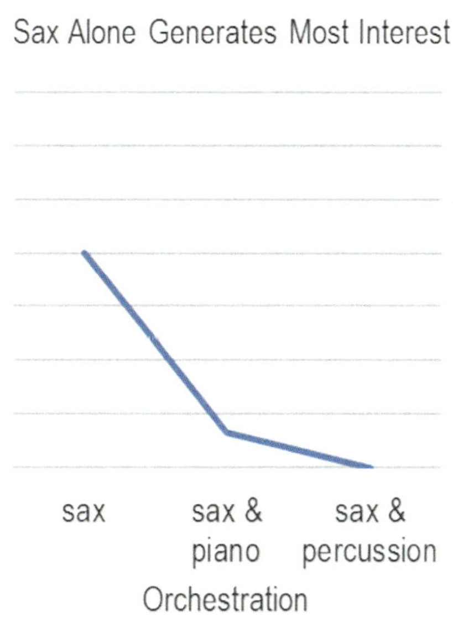

Attribute importances. To find the *attribute importances*, first find the range of the two *vocals* part worths and the range of *orchestration* part worths.

In L37,
Range
In M37,
=max(L28:m36)-min(L28:M36)
In N37,
=max(N27:O36)-min(N27:O36)

N37 =MAX(N28:O36)-MIN(N28:O36)

	I	J	K	L	M	N	O	
26								
27	sax	sax & piano	intercept	none PW	original PW	sax PW	saxpia no PW	pre ed rat
28	1	0	3.2	0.0	1.0	2.0	0.0	
29	1	0	3.2	0.0	0.0	2.0	0.0	
30	1	0	3.2	0.0	0.0	2.0	0.0	
31	0	0	3.2	0.0	1.0	0.0	0.0	
32	0	0	3.2	0.0	0.0	0.0	0.0	
33	0	0	3.2	0.0	0.0	0.0	0.0	
34	0	1	3.2	0.0	1.0	0.0	0.3	
35	0	1	3.2	0.0	0.0	0.0	0.3	
36	0	1	3.2	0.0	0.0	0.0	0.3	
37				range	1.0	2.0		

Sum the two ranges, and find the percent of the sum contributed by each of the two attributes, the importances.

In O37,
=sum(M37:N37)
In L38,
Importance
In M38,
=M37/O37 fn4
In M38,
Shift+right
Cntl+R

M38 =M37/$O$37

	J	K	L	M	N	O
27	*sax & piano*	*intercept*	*none PW*	*original PW*	*sax PW*	*saxpia no PW*
28	0	3.2	0.0	1.0	2.0	0.0
29	0	3.2	0.0	0.0	2.0	0.0
30	0	3.2	0.0	0.0	2.0	0.0
31	0	3.2	0.0	1.0	0.0	0.0
32	0	3.2	0.0	0.0	0.0	0.0
33	0	3.2	0.0	0.0	0.0	0.0
34	1	3.2	0.0	1.0	0.0	0.3
35	1	3.2	0.0	0.0	0.0	0.3
36	1	3.2	0.0	0.0	0.0	0.3
37			range	1.0	2.0	3.0
38			importan	0.3	0.7	

Excel 11.2 Use ANOVA to Test Equivalence of Mean Interest Ratings

Test the equality of mean interest ratings by *vocals* and *orchestration* options with ANOVA. In columns H through K, rows 1 through 4, reformat the data into a matrix of *ratings* by *vocals* (in rows) by *orchestration* (in columns):

	A	B	C	D	E	F	G	H	I	J	K
1	*vocals*	*orchestration*	*rating*	*none*	*original*	*sax*	*sax & piano*	*vocals option*	sax	sax & percussion	sax & piano
2	original	sax	6	0	1	1	0	original	6	4	5
3	brand specific	sax	5	0	0	1	0	brand specific	5	4	3
4	none	sax	9	1	0	1	0	none	9	6	7
5	original	sax & percussion	4	0	1	0	0				
6	brand specific	sax & percussion	4	0	0	0	0				
7	none	sax & percussion	6	1	0	0	0				
8	original	sax & piano	5	0	1	0	1				
9	brand specific	sax & piano	3	0	0	0	1				
10	none	sax & piano	7	1	0	0	1				

Run analysis of variance (two factor without replication).	**Alt AY*n* A down down H1:k4 tab L**

	A	B	C	D	E	F	G
1	Anova: Two-Factor Without Replication						
2							
3	*SUMMARY*	*Count*	*Sum*	*Average*	*Variance*		
4	original	3	15	5	1		
5	brand spe	3	12	4	1		
6	none	3	22	7.333333	2.333333		
7							
8	sax	3	20	6.666667	4.333333		
9	sax & perc	3	14	4.666667	1.333333		
10	sax & pian	3	15	5	4		
11							
12							
13	ANOVA						
14	*ce of Varic*	*SS*	*df*	*MS*	*F*	*P-value*	*F crit*
15	Rows	17.55556	2	8.777778	19.75	0.008456	6.944272
16	Columns	6.888889	2	3.444444	7.75	0.042078	6.944272
17	Error	1.777778	4	0.444444			
18							
19	Total	26.22222	8				

To see the plot of *vocals* part worths, move the average interest ratings in column D to Column B, next to the background options (or factor levels), and then select the labels and average ratings in rows 4 through 6 and columns A and B and request a line plot.	In D, **Cntl+spacebar** **Cntl+X** In B, **Cntl+spacebar** **Alt HIE** In A4, **Shift+down down** **Shift+right** **Alt NN**

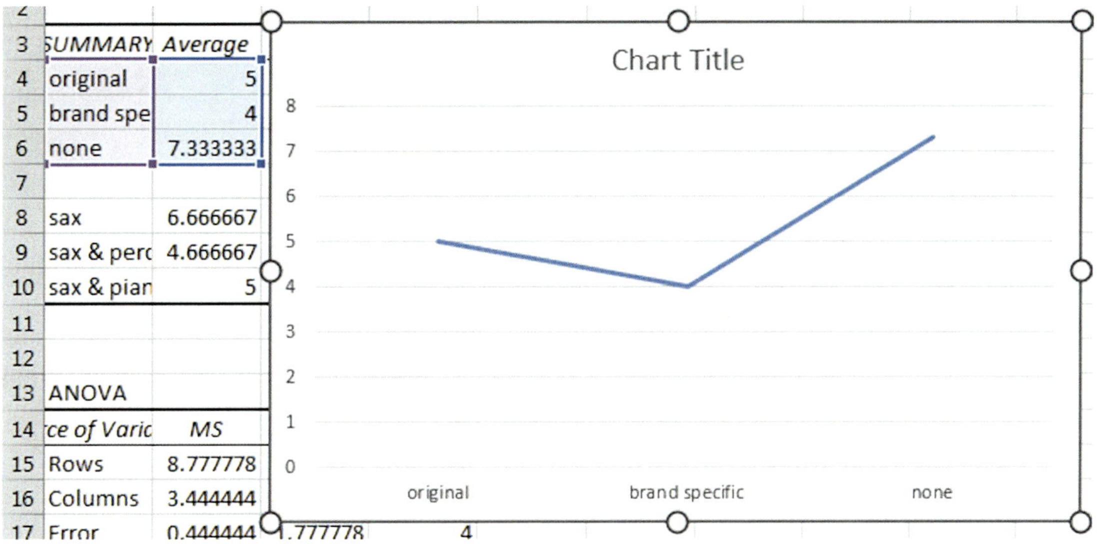

To see the plot of *orchestration* part worths, select the labels and average ratings in rows 8 through 10 in columns A and B and request a line plot.

In A8,
Shift+down down
Shift+right
Alt NN

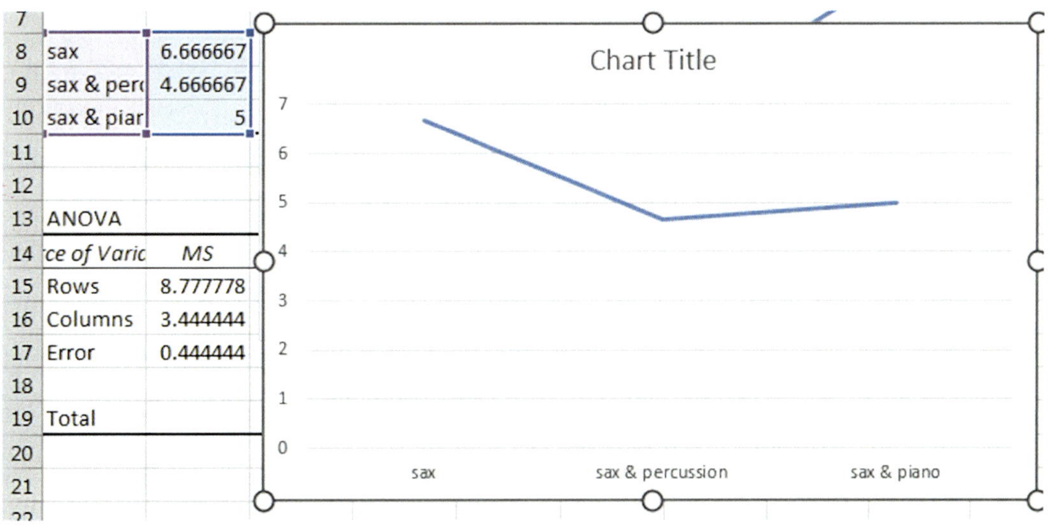

To compare the relative influence of *vocals* and *orchestration*, reformat axes to reflect the minimum and maximum average ratings, 4 to 7.5, add a vertical axis title and chart titles.

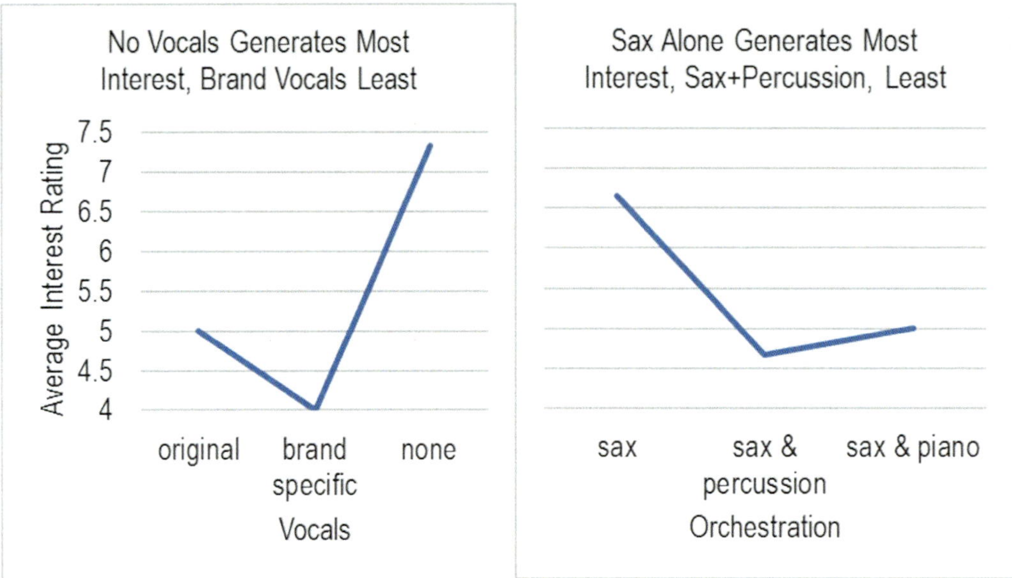
No Vocals Generates Most Interest, Brand Vocals Least
Average Interest Rating
7.5
7
6.5
6
5.5
5
4.5
4
original
brand specific
none
Vocals
Sax Alone Generates Most Interest, Sax+Percussion, Least
sax
sax & percussion
sax & piano
Orchestration

Lab 11.1 Revere Bank Profits

Revere Bank management would like to increase profitability per customer, targeting relatively more profitable customer segments, and convincing existing customers to use online services and online billpay, if they would increase profits. In addition, management wonders whether customer tenure influences customer profitability. Are loyal customers more profitable than new customers? Or are new customers more profitable than long time customers?

If tenure is positively linked to profitability, how much more would it be worth to incentivize existing customers to remain loyal one more year?

They require a demographic profile of relatively more profitable customers, and they would like to know how much the use of online services and online billpay drive annual customer profitability, if at all. Data are in **Lab 11.1 Revere Bank Profits**.

Build a model of customer profitability.

1. Present your regression model equation, including RSquare and the significance levels of each coefficient and RSquare. *If your model does not use all of the potential drivers, remove extra terms.* Substitute
   (i) variable names for y, I1, I2, X1 through X3,
   (ii) units for uy and u1 through u3,
   (iii) coefficients for b0 through b5,
   (iv) significance levels for L0 through L5 and LR.

$$\hat{y}(uy) = b0(uy)^{L0} + b1(uy)^{L1} \times I1 + b2(uy)^{L2} \times I2$$

$$+b3(\frac{uy}{u1})^{L3} \times X1(u1) + b4(\frac{uy}{u2})^{L4} \times X2(u2) + b5(\frac{uy}{u3})^{:5} \times X3(u3)$$

RSquare: ___ LR

2. Which are drivers of customer profitability?

___ income ___ tenure ___ age ___ online services ___ online billpay

3. How much does use of online services influence customer profitability?

____

4. How much does use of online billpay influence customer profitability?

____

5. Describe customers who are the most profitable: ______________________________

Lab 11.2 Power PowerPoints

Corporate Relations management is considering alternate designs for the firm's PowerPoint presentation. Content will focus on performance in firm divisions and new ventures by the firm in several global markets. Ideally, audience members would remember twelve key points made in the presentation.

Graphics. Some managers believe that graphics illustrating performance and market potential are easier for audience members to digest and remember. Others argue that tables with precise numbers are more effective. A third group prefers to use photographs with visual images that support conclusions from the numbers stated directly in text.

Text. There is also difference of opinion regarding the amount of text to include in each slide. Some favor a single sentence. Others prefer to use bullet points to remind presenters what material should be covered.

To identify the most effective design, Corporate Relations personnel created six PowerPoint sets:

Graphs, single sentence	Photographs, single sentence	Tables, single sentence
Graphs, bullet points	Photographs, bullet points	Tables, bullet points

A random sample of 12 shareholders not employed for the firm viewed one of the six PowerPoint sets, then answered questions about the content. For each of the twelve key points mentioned, one point was added, making scores from zero to twelve possible.
To account for the influence of experience, the shareholders were asked to report their years of professional experience. **Lab 11.2 Power PowerPoints** contains this data.

One of the participants had to leave early, leaving responses incomplete, reducing the sample to eleven.

Use regression to test whether visuals or text influence audience content recall and to identify the most effective combination(s).

1. State the hypotheses that you are testing:

2. Report your conclusions, noting the statistical tests that you used:

3. Find the part worth recall scores and attribute importances:

Attibute level		*Part worth recall score*	*Attribute importance*
Visuals	Graphics		
	Photographs		
	Tables		
Text	Single sentence		
	Bullet points		
		Total:	

4. Plot the part worth recall scores by level for visuals and text using the same scale for *y* axes.

5. Write regression equations for PowerPoints with

   Graphics and single sentences:

   Graphics and bullet points:

   Photographs and single sentences:

   Photographs and bullet points:

   Tables and single sentences:

   Tables and bullet points:

6. Identify the most effective visual, text combination(s): ________________________

7. Identify the least effective visual, text combination(s): ________________________

8. Find the importance of experience:

Coefficient b (1)	*Max* (2)	*Min* (3)	*Range* (4) =(2)−(3)	*importance* = (1) * (4)

9. Find the standardized importances of attributes and experience:

Attribute	*Importance* *(1)*	*Standardized importance* = *(1) / Total*
Visuals		
Text		
Experience		
Total		

10. Which matters more: ___visuals ___text ___ audience experience

Lab 11.3 ANOVA and Regression with Indicators: Powerful PowerPoints

I. Use ANOVA to test whether type of visual or text format influences audience content recall. Data are in **Lab 11.2 Power PowerPoints**.

1. State the hypotheses that you are testing.

2. Report your conclusions, noting the statistical tests that you used.

3. Illustrate average recall scores by type of visual and by text format, choosing the same y axis scale for each.

4. What proportion of variation in recall scores can be accounted for by PowerPoint design differences? ______________

5. Which visual type is most effective? ______________________

6. Which visual type should be avoided? ____________________

7. Which text format is most effective? _____________________

8. Which makes a bigger difference, the visual type or the text format? _____________

II. Compare ANOVA with Regression

Use the same sample that you used for ANOVA, ignoring the influence of experience to investigate the impact of PowerPoint design on content recall.

1. Do visuals and/or text influence recall scores? Y N

   Evidence: p value from F test: ____

2. Report your conclusions, noting the statistical tests that you used:

Assignment 11 Forecasting Chipotle Revenue in the Long Range

Chipotle executives desire a long range forecast of revenues through 2020, and they are particularly interested in learning the impacts of the expansion with the introduction of Asian style Shophouses in 2012 and the "No GMOs" product enhancement introduced in 2014. Since "No GMOs" has been in effect a little more than one year, you will not be able to validate your model; however, management believes that a forecast without validation will be insightful.

Chipotle long range contains annual revenues, 2001 through 2014. Build a naïve model to forecast revenue in 2020 and quantify the impacts of the Shophouse expansion and the "No GMOs" enhancement.

1. Assess residuals for presence of positive autocorrelation and report your conclusion, citing the appropriate statistic:
2. Present your naïve model equation:
3. Illustrate your fit and forecast:
4. What is the margin of error in your 2020 forecast? ____
5. What is the expected contribution to revenue of the Shophouse expansion in 2020? __
6. What is the expected contribution to revenue of the "No GMOs" enhancement in

   2020? ___

Case 11 Store24 (A): Managing Employee Retention* and Store24 (B): Service Quality and Employee Skills**

I. Shortcut challenge

Repeat the steps used in our Lab 11 Revere Bank to build a model of bank profits and then to find predicted profit at a baseline store. Record your time. If your time is more than 5 minutes, repeat. Repeat up to 4 times (5 times, total), if your times are more than 5 minutes. Submit times here:

II. Store24 (A) and (B)

Download and read the HBS cases before you begin your analysis. You may work with a partner.

Problem. Management needs to know what controllable "people" factors are driving store sales and profits. If management or crew tenure, management or crew skill, or service quality is driving performance, programs to increase tenure, skill or quality will be created, focusing on the stores performing below median.

A number of uncontrollable factors probably influence store sales and profits, in addition to "people" influences. While these cannot be changed by management, at least in the short term, managers are aware that those influences cannot be ignored.

There has been some grumbling from managers of stores not located in residential neighborhoods, as well as from managers of stores not open 24 hours, since sales or profit potential may be limited in those stores. While location and hours open cannot be changed, at least in the short term, management compensation could be adjusted if some stores are found to have lower performance potential.

The case hints that some store performance responses to drivers may be nonlinear. Please assume that responses are approximately linear.

While both sales and profits matter, *focus here on sales response*. Build a model of Store24 sales, removing insignificant variables, one at a time, until all remaining are significant. Use one tail t tests for any potential driver for which the direction of influence is known.

Data are in **Case 11 Store24.**

1. Present your regression model equation, including RSquare and the significance levels of each coefficient and RSquare. *If your model does not use all of the potential drivers, remove extra terms.* Substitute

    (v) variable names for y, I1, I2, X1 through X9,
    (vi) units for uy and u1 through u9,
    (vii) coefficients for b0 through b11,
    (viii) significance levels for L0 through L11 and LR.

* Harvard Business School case 9602096
** Harvard Business School case 9602097

$$\hat{y}(uy) = b0(uy)^{L0} + b1(uy)^{L1} \times I1 + b2(uy)^{L2} \times I2$$

$$+b3(\frac{uy}{u1})^{L3} \times X1(u1) + b4(\frac{uy}{u2})^{L4} \times X2(u2) + b5(\frac{uy}{u3})^{:5} \times X3(u3)$$

$$+b6(\frac{uy}{u4})^{L6} \times X4(u4) + b7(\frac{uy}{u5})^{L7} \times X5(u5) + b8^{L8} \times X6(u6)$$

$$+b9(\frac{uy}{u7})^{L9} \times X7(u7) + b10(\frac{uy}{u8})^{L10} \times X8(u8) + b11(\frac{uy}{u9})^{L11} \times X9(u9)$$

RSquare: ___ LR

2. Identify sales drivers *among the controllable factors*, so that resources can be directed to programs to improve those driver levels:

___ Mgmt tenure ___ Crew tenure ___ Mgmt skill _____Crew skill ___ Service quality

3. How much does nonresidential location reduce sales, on average? $______

4. How much does not being open 24 hours reduce sales, on average? $_____

Chapter 12
Model Building and Forecasting with Multicollinear Time Series

An explanatory regression model from time series data allows us to identify performance drivers and forecast performance given specific driver values, just as regression models from cross sectional data do. When decision makers want to forecast *future* performance in the shorter term, a time series of past performance is used to identify drivers and fit a model. A time series model can be used to identify drivers whose variation over time is associated with later variation in performance over time.

Naïve models based on trend are *naïve*, and offer no explanation for the stable trend. In time series models with dual goals of forecasting in the shorter term and explaining performance over time, we seek links between past variation in drivers with later variation in performance. These links between drivers and performance require that changes in the drivers precede change in performance. Therefore, lagged predictor variables are used. Patterns of change in drivers that also occur in the dependent variable in later time periods are identified to choose driver lags. Time series models are built using predictor values from past periods to explain and forecast later performance. Figure 12.1 illustrates the differences in model building processes between naïve time series models based on trend, for longer term forecasts, and explanatory time series models for shorter term forecasts.

Three differences in the model building process distinguish explanatory cross sectional and time series models:

- the use of lagged predictors,
- addition seasonality and cyclical variables, and
- the model validation process.

Figure 12.2 illustrates the differences in model building processes between explanatory cross sectional and time series models.

Most business performance variables and economic indicators are cyclical. Economies cycle through expansion and recession, and performance in most businesses fluctuates following economic fluctuation. Business and economic variables are also often seasonal. Cyclicality and seasonality are accounted for by adding cyclical and seasonal predictors.

Before a time series model is used to forecast future performance, whether for short term or longer term forecasts, it is validated:

- the two most recent observations are hidden while the model is built,
- the model equation is used to forecast performance in those two most recent periods,
- model prediction intervals are compared with actual performance values in those two most recent periods, and if the prediction intervals contain actual performance values, this is evidence that the model has *predictive validity* and can be reliably used to forecast unknown performance in future periods.

C. Fraser, *Business Statistics for Competitive Advantage with Excel 2016*,
DOI 10.1007/978-3-319-32185-1_12

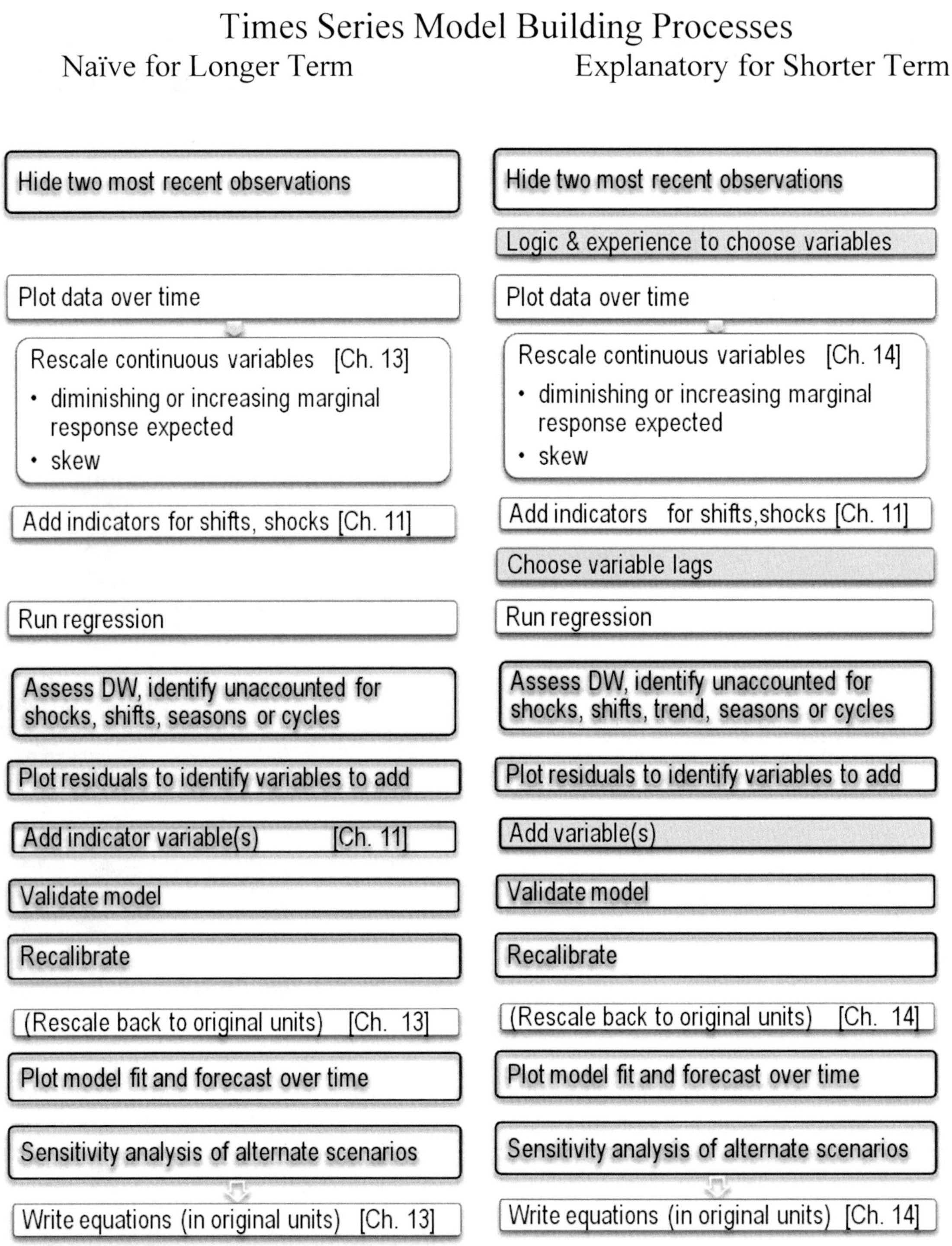

Figure 12.1 Model building processes with time series data

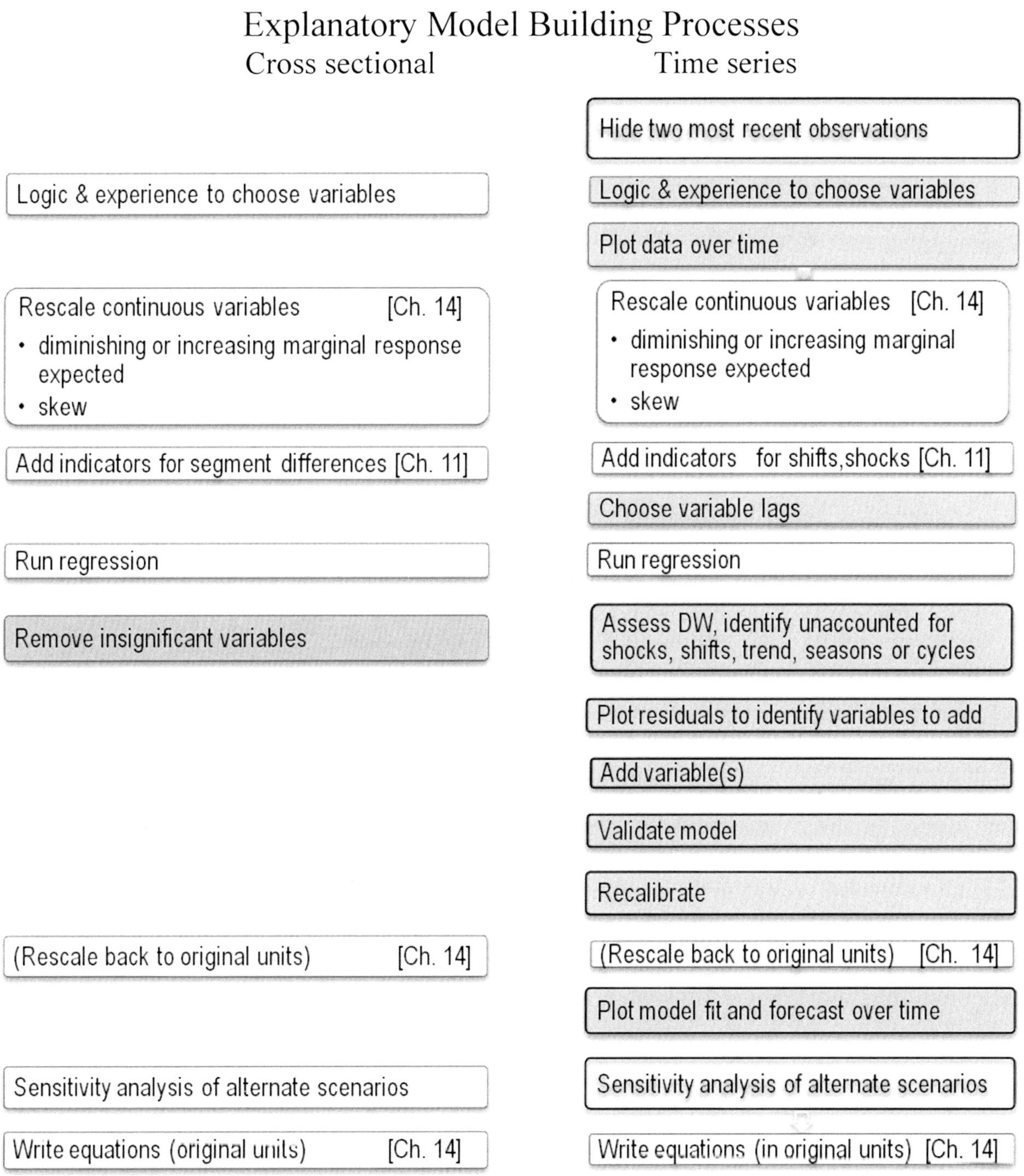

Figure 12.2 Explanatory model building processes

12.1 Time Series Models Include Decision Variables, External Forces, and Leading Indicators

Most successful forecasting models logically assume that performance in a period, y_t, depends upon

- decision variables under the management control,
- external forces, including
  - shocks such as 9/11, Hurricane Katrina, change in Presidential Party
  - market variables,
  - competitive variables,
- Leading indicators of the economy, industry or the market
- Seasonality
- Cyclicality

Ultimately, the multiple regression explanatory forecasting models contain several of these components, which together account for variation in performance. This chapter introduces leading indicator components of regression models built from time series.

Performance across time depends on decision variables and the economy. Decision variables, such as spending on advertising, sales effort and research and development tend to move together. In periods of prosperity, spending in all three areas may increase; in periods where performance is sluggish, spending in all three areas may be cut. Firm strategy guides resource allocation to the various firm functions. As a result, it is common for spending and investment variables to be correlated in time series data.

Many economic indicators also move together across time. In times of economic prosperity, GDP is growing faster, consumer expectations increase, and investments increase. Increasing wealth filters down from the economy to consumers and stock holders, where some proportion of gains are channeled back into consumption of investments.

It is common for decision variables, past performance, and leading indicators to be correlated in time series data. This inherent correlation of performance drivers in time series data makes logical choice of drivers a critical component of good model building.

It is also often more promising to build models by adding variables, one at a time, looking at residuals for indications of the most promising variables to add next. Multicollinearity, including its consequences, diagnosis and alternate remedies, is further considered in this chapter.

Example 12.1 Home Depot Revenues[1]. Home Depot executives were concerned in early 2015 that new home sales, a leading indicator of Home Depot revenues, were not robustly recovering from the economic recession of 2008–2010. The U.S. economy had recovered, though growth was slow. The financial crisis had reduced lending, and *new home sales* had slowed. Traditionally, Home Depot Revenues have grown following growth in *New Home Sales*, since builders and homeowners buy construction materials, flooring, and appliances at Home Depot.

---

[1] This example is a hypothetical scenario based on actual data

Lowes' business was similar to Home Depot's business. Both firms' revenues were seasonal and linked to the housing market, though Lowes offered installation services and Home Depot had not. Lowes revenues had not reacted strongly to the financial crisis, slowing, but not declining. Amanda was not sure whether Lowes revenues had a positive or a negative impact on Home Depot revenues. Whenever either firm advertised or promoted home improvement items, later sales at both tended to be higher. Nonetheless, the two firms were competing for the business of many of the same customers. It was not clear whether *Lowes*, Home Depot's major competitor, was helping to expand the home improvement market, or taking business from Home Depot.

12.2 Indicators of Economic Prosperity Lead Business Performance

A *leading indicator* model links changes in a leading indicator, such as *new home sales,* and later performance:

$$re\widehat{v}enue(\$B)_q = b_0(\$B) + b_1\left(\frac{\$B}{K\ new\ home\ sales}\right) \times New\ Home\ Sales(K)_{q-l}$$

where l denotes the length of lag, or delay from change in new home sales to change in revenues.

Amanda, a recent business school graduate with modeling expertise, was asked to build a model of Home Depot Revenues, which would both explain revenue fluctuations and forecast revenues in the next four quarters.

Home Depot executives wanted to know how strongly

- past *New Home Sales*
- past *Lowes revenues*

influenced revenues.

After being briefed by the executives, Amanda created a model reflecting their logic. She considered as possible drivers in her model:

- *new home sales*$(K)_{q-l}$
- *Lowes revenues*$_{q-l}$

12.3 Hide the Two Most Recent Datapoints to Validate a Time Series Model

Amanda used datapoints for quarterly revenue in 2001 through 2015, including quarters before, during and after the financial crisis and recession. Before Amanda proceeded further, she excluded the two most recent observations from first and second quarter 2015. These *hold out* observations would allow her to compare forecasts for the two most recent periods with actual revenues to *validate* her model. If the 95% prediction intervals from the model contained the actual revenues for both quarters, she would be able to conclude that her model is valid. She could then use the model to forecast with confidence.

12.4 Compare Scatterplots to Choose Driver Lags: Visual Inspection

The potential drivers each reflect economic conditions and move together over time. Consequently, they are highly correlated. Including all of the drivers in a multiple regression model at once would introduce a high degree of multicollinearity and make it difficult to identify each of their marginal impacts. To most effectively build a time series model, start with one driver, and then add additional drivers, one at a time.

Amanda began by plotting *Home Depot revenues* by quarter. She focused more heavily on pattern in recent quarters, since her goal was to build a model which produced valid forecasts. She added a trend line for reference. (The trend is the average linear growth over the series.) She noted that revenues grew through mid 2006, then began declining through 2009. Revenue losses seemed to be linked to the housing bubble of 2006–2008 and to the recession of 2008–2009. In recent quarters, from 2010, revenues had again begun growing. Revenues were also seasonal. Her scatterplot is shown in Figure 12.3.

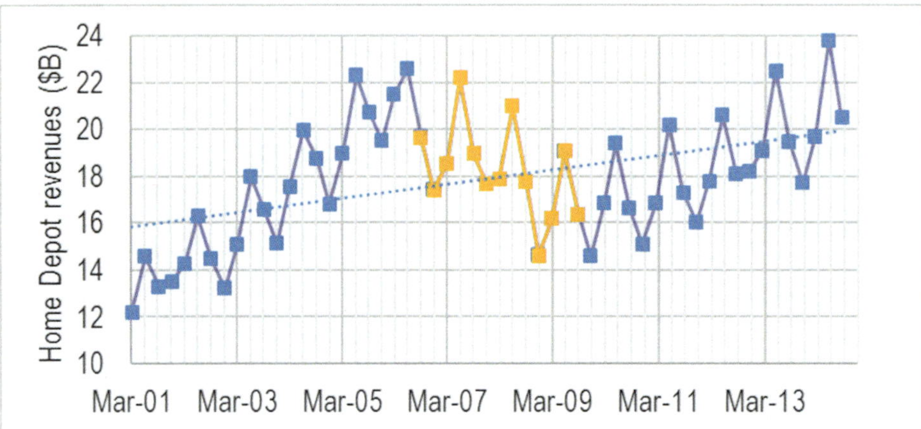

Figure 12.3 Home Depot revenues by quarter

To choose lags for the two potential drivers, Amanda plotted *new home sales* and *Lowes revenues*. Both were seasonal, and so multiples of four were considered, four quarter lags and eight quarter lags, to line up seasonality in the two leading indicators with seasonality in *Home Depot revenues*. *New home sales* began declining early in 2006, while *Home Depot revenues* began declining later in 2006, and so a four quarter lag made sense. This correspondence is shown in Figure 12.4.

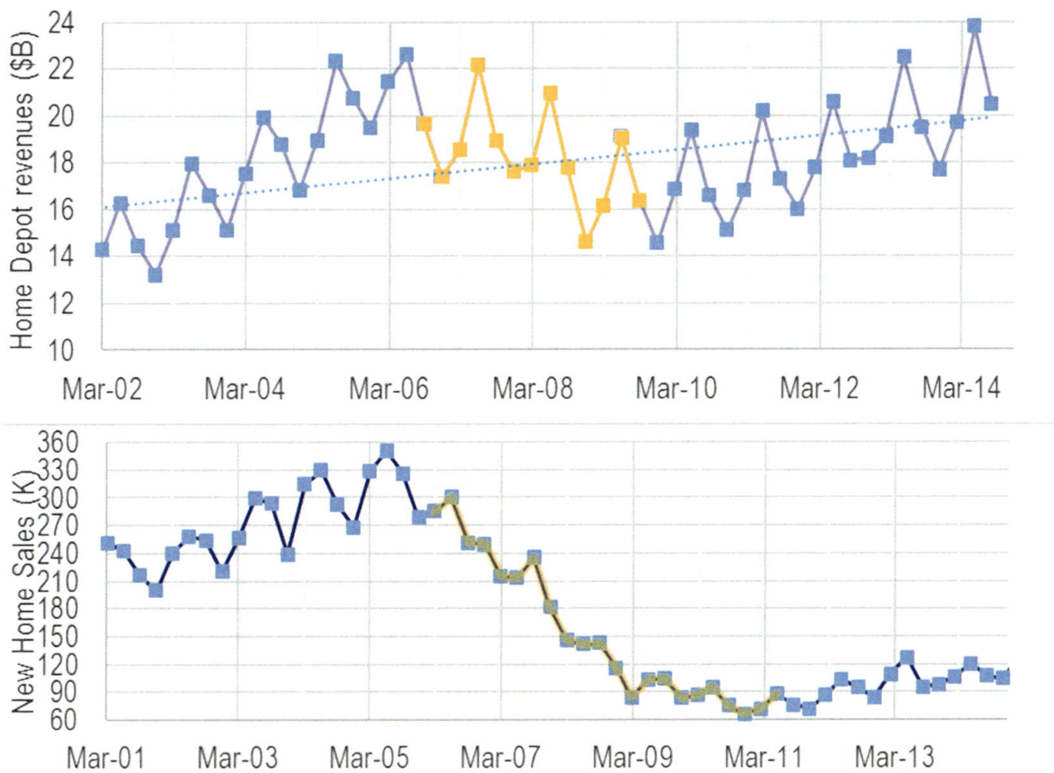

Figure 12.4 Home Depot revenues with past new home sales by quarter

Lowes revenues had been relatively resilient from 2006 through 2009, slowing, but not declining. Consequently, both four quarter and eight quarter lags made sense, shown in Figure 12.5.

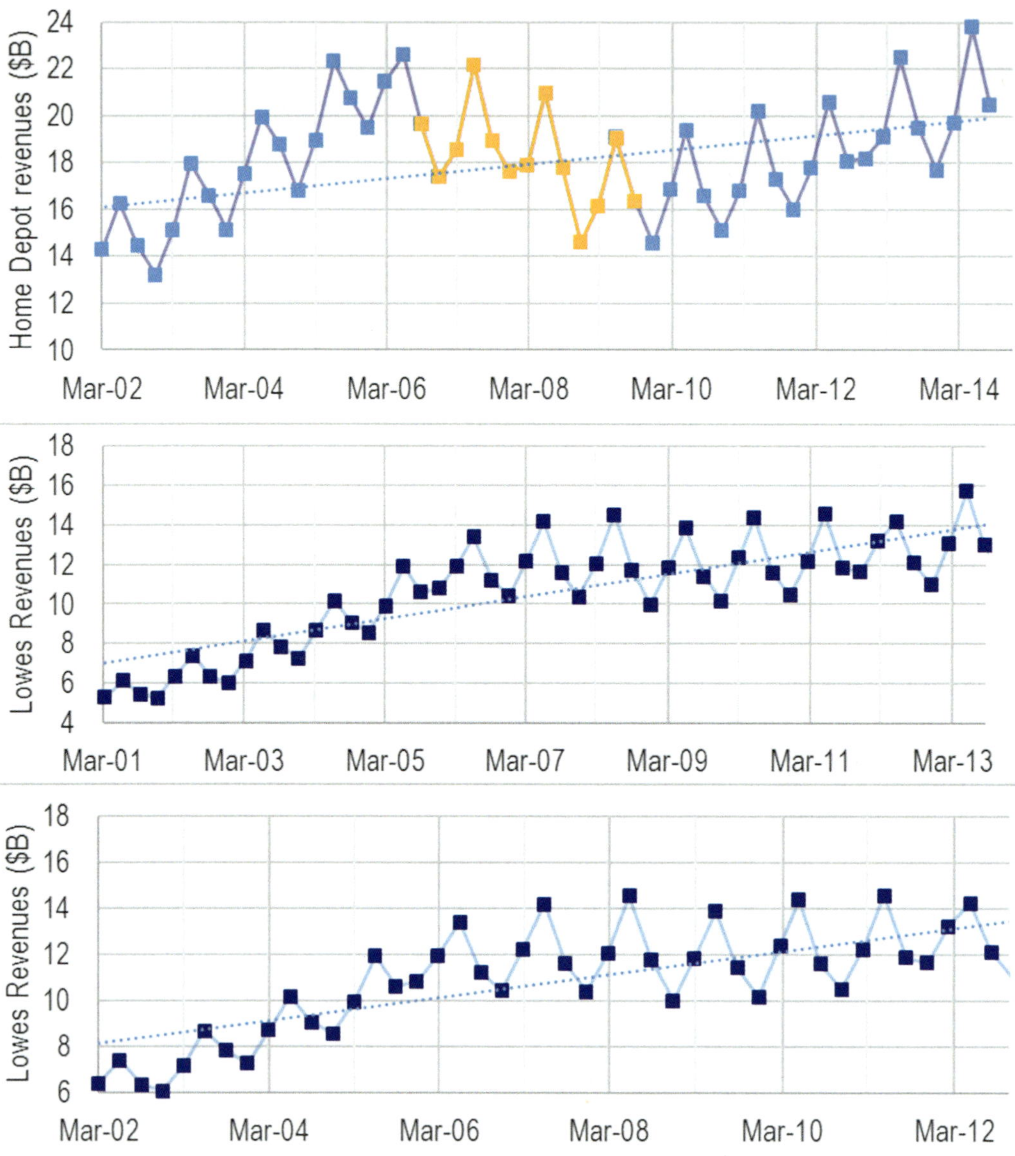

Figure 12.5 Home Depot revenues with past Lowes revenues by quarter

To choose between the three potential drivers, Amanda compared correlations, shown in Table 12.1. The shorter, four quarter *Lowes revenue* lag was strongest, and she selected it for her first regression, shown in Table 12.2.

Table 12.1 Correlations between potential drivers and Home Depot revenues

	Home Depot revenues (B$) q
new home sales(K) q-4	0.26
Lowes Revenues (B$) q-4	0.48
Lowes Revenues (B$) q-8	0.32

Table 12.2 Regression of Home Depot revenues with past Lowes revenues

SUMMARY OUTPUT

Regression Statistics	
R Square	.239
Standard Error	2.00
Observations	47

ANOVA

	df	*SS*	*MS*	*F*	*Significance F*
Regression	1	54	53.6	13.5	.0006
Residual	45	179	4.0		
Total	46	233			

	Coefficients	*Standard Error*	*t Stat*	*P-value*	*Lower 95%*	*Upper 95%*
Intercept	13.6	1.39	9.8	.00	10.8	16.3
Lowes Revenues (B$) q-4	.45	.12	3.7	.0006	.20	.70

Past *Lowes revenues* is a significant driver. The model is significant. However, with RSquare are just 24 percent, there are clearly other drivers, as well. In order to identify the driver to add next, Amanda plotted the residuals and assessed Durbin Watson.

12.5 Assess Residuals to Identify Unaccounted for Trend or Cycles

Amanda's initial model, with one driver, plus intercept, and a sample size of 47, has *DW* critical values of dL=1.49 and dU=1.57. The model *DW* statistic is .66, leading to the conclusion that the residuals are exhibit positive autocorrelation. The a trend, cycle, shock or shift remains to be accounted for. There remains variation in revenues to be explained. Visual inspection of the residuals can provide clues which may suggest which potential driver to add to the model. Figure 12.6 contains a scatterplot of residuals by quarter, with the vertical axis units set at the standard error, 2.0 ($B).

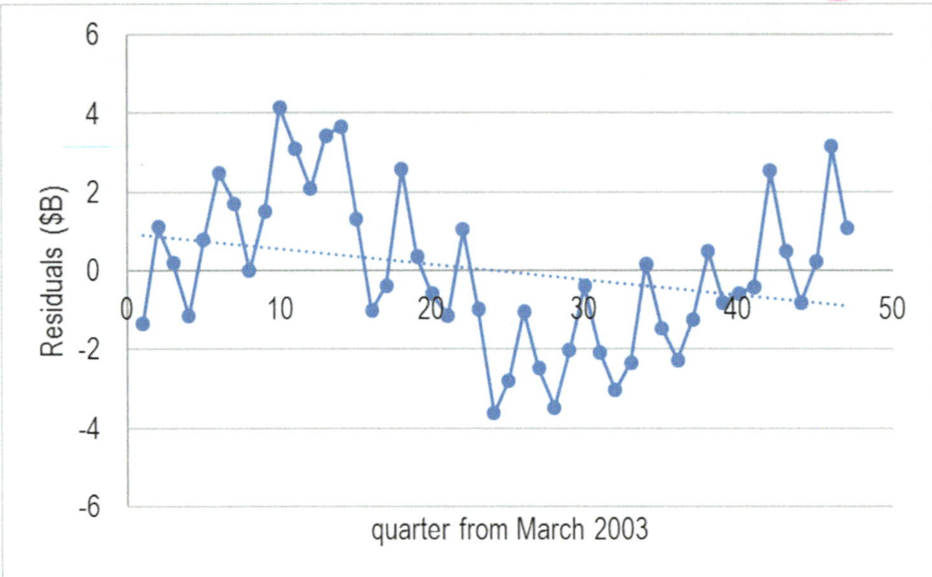

Figure 12.6 Residuals reveal variation to be explained

A negative trend remains, as well as both a cycle and seasonality. *New home sales* may improve the model. Regression results with past *new home sales* are shown in Table 12.3.

Table 12.3 Regression with past new home sales

SUMMARY OUTPUT	
Regression Statistics	
R Square	.573
Standard Error	1.50
Observations	47

ANOVA

	df	*SS*	*MS*	*F*	*Significance F*
Regression	2	133	66.7	29.5	.0000
Residual	44	99	2.3		
Total	46	233			

	Coefficients	*Standard Error*	*t Stat*	*P-value*	*Lower 95%*	*Upper 95%*
Intercept	6.9	1.53	4.6	.0000	3.9	10.0
Lowes Revenues (B$) q-4	.75	.11	7.2	.0000	.56	.99
New home sales (K) q-4	.016	.0027	5.9	.0000	.011	.022

Results confirm that both past *Lowes revenues* and past *new home sales* drive *Home Depot revenues.* The model is significant, and both coefficients are significant with correct signs. However, RSquare, though now larger by 34%, at just 57%, suggests that other drivers remain to be identified. Assessing Durbin Watson and the residuals plot will provide clues concerning potential drivers to add to improve the model.

Amanda found that the Durbin Watson statistic was .47, even lower than the simple regression Durbin Watson statistic at .66, and also lower than the lower critical value, 1.44, for a two driver model with 47 quarters. Since seasonality, which is negative autocorrelation, would increase Durbin Watson values, it is likely that the addition of past *new home sales* to the model accounted for some remaining seasonality. The plot of residuals from the two driver model is shown in Figure 12.7, with the vertical axis units set at the new standard error of 1.5($B).

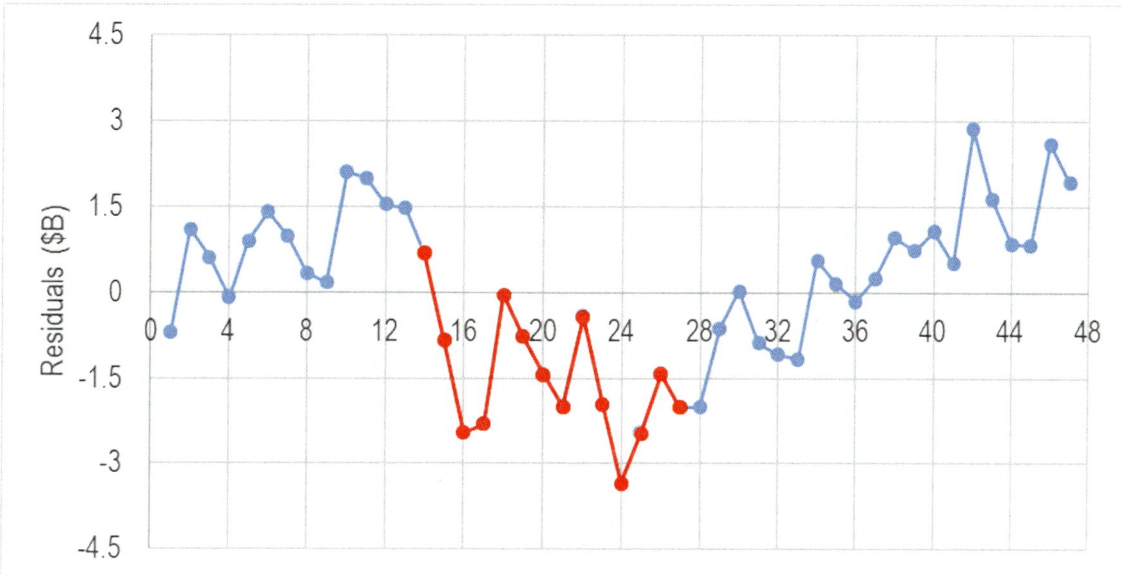

Figure 12.7 Residuals ($B) from the two driver model

The residual plot suggests that the *housing bubble* impact, followed by the global recession induced by the *housing bubble*, reduced *Home Depot revenues* for close to four years, from June 2006 through September 2009. In quarter 24 of the series, residuals are more than two standard errors from the expected value of zero. The model is not fitting well. To account for this shock, Amanda added a *housing bubble* indicator, equal to one in quarters from June 2006 through September 2009, and equal to zero in other quarters. Results are shown in Table 12.4.

Table 12.4 Three driver regression

SUMMARY OUTPUT	
Regression Statistics	
R Square	.884
Standard Error	.82
Observations	47

ANOVA

	df	*SS*	*MS*	*F*	*Significance F*
Regression	3	204	67.9	100.3	.0000
Residual	43	29	.68		
Total	46	233			

	Coefficients	*Standard Error*	*t Stat*	*P-value*	*Lower 95%*	*Upper 95%*
Intercept	1.8	.97	1.8	.07	−.13	3.8
Housing bubble q	−3.4	.33	−10.2	.0000	−4.0	−2.7
Lowes Revenues (B$) q-4	1.15	.069	16.6	.0000	1.01	1.29
New home sales (K) q-4	.027	.0018	14.8	.0000	.023	.031

The *housing bubble* shifted revenues down by an average of $3.4B each quarter. This shift accounts for a large proportion of variation in revenues. RSquare is now 88%, 31% greater.

To decide whether or not another driver ought to be added, Amanda again assess Durbin Watson, now 1.25, below the lower critical value dL=1.40. The residual plot, with vertical axis units set to the new and smaller standard error of .82, is shown in Figure 12.8.

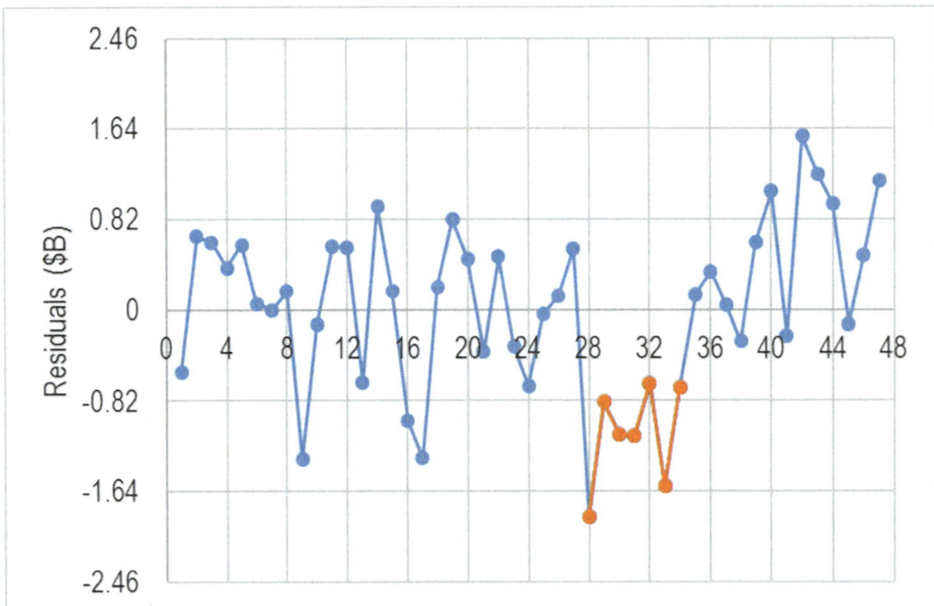

Figure 12.8 Residuals from the three driver model

There is a noticeable shock downwards in quarters 28 through 34, December 2009 through June 2011. Amanda added a *recession* indicator turned on in those quarters. Results are shown in Table 12.5.

Table 12.5 Four driver model

SUMMARY OUTPUT	
Regression Statistics	
R Square	.931
Standard Error	.62
Observations	47

ANOVA

	df	*SS*	*MS*	*F*	*Significance F*
Regression	4	217	54.1	140.7	.0000
Residual	42	16	.38		
Total	46	233			

	Coefficients	*Standard Error*	*t Stat*	*P-value*	*Lower 95%*	*Upper 95%*
Intercept	2.4	.74	3.2	.003	.88	3.9
Housing bubble q	−3.6	.25	−14.2	.0000	−4.1	−3.1
Recession q	−1.6	.28	−5.8	.0000	−2.2	−1.1
Lowes Revenues (B$) q-4	1.16	.052	22.2	.0000	1.05	1.27
New home sales (K) q-4	.025	.0014	17.6	.0000	.022	.028

The recession shifted revenues down by $1.6B each quarter, on average. With the recession indicator, RSquare is now 93%, up from 88%.

Durbin Watson has increased to 1.95, and is well above the upper critical value, dU=1.72. The residuals are free of unaccounted for trend, cycles, shifts and shocks. The residuals plot is shown below in Figure 12.9.

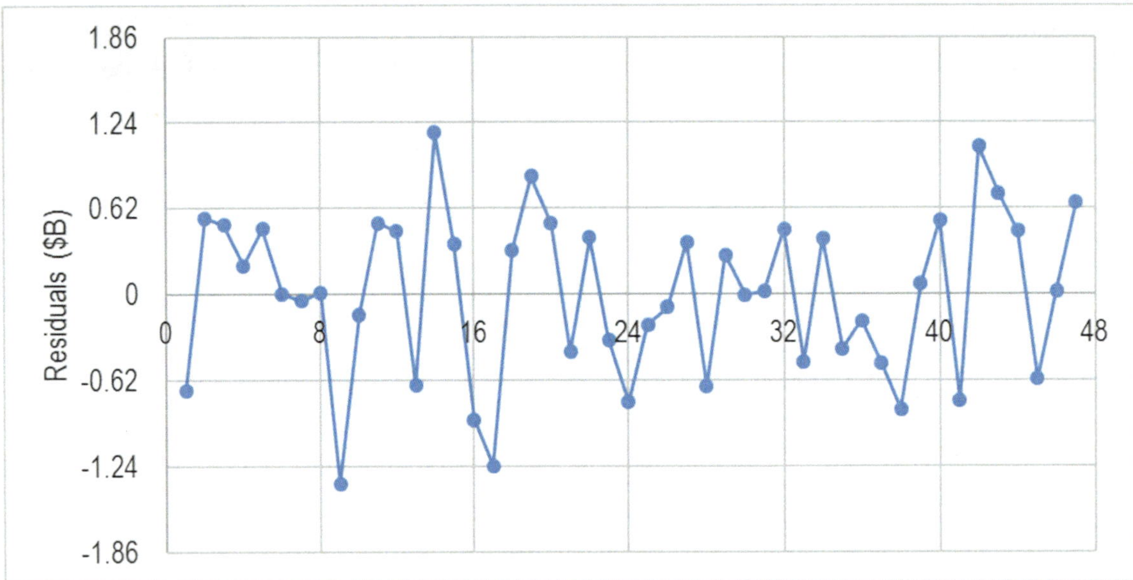

Figure 12.9 Residuals from the three driver model

12.6 Forecast the Recent, Hidden Points to Assess Predictive Validity

With a significant model, significant drivers, a Durbin Watson statistic above the upper critical value, and no apparent residual pattern, Amanda decided to test the predictive validity of her model. Comparing actual *revenues* in the two most recent, hidden quarters, with model predictions provided evidence of predictive validity. Both fell within the 95% prediction interval, shown in Table 12.6.

Table 12.6 Evidence of predictive validity of the model

q	*Home Depot revenues (B$) q*	*lower 95% prediction interval bound*	*upper 95% prediction interval bound*
Dec-14	19.2	17.2	19.7
Mar-15	20.9	19.3	21.8

12.7 Add the Most Recent Datapoints to Recalibrate

With evidence of predictive validity, Amanda used the model to forecast revenues in the next four quarters. Before making the forecast, she added the two most recent observations that were hidden to validate. The recalibrated model became:

$$\begin{aligned} r\hat{e}venues(\$B)_q = {} & 2.4(\$B)^a - 3.6(\$B)^a \times housing\ bubble_q \\ & -1.7(\$B)^a \times recession_q \\ & +1.1\left(\frac{\$B}{rev(\$B)}\right)^u \times Lowes\ revenue(\$B)_{q-4} \end{aligned}$$

$$+ .025\left(\frac{\$B}{sales(K)}\right)^{a} \times New\ Home\ Sales(K)_{q-4}$$

RSquare: .93[a]
[a]*Significant* at .01.

Variation in past *Lowes revenue,* and *new home sales*, with the *housing bubble* and *recession,* together account for 93% of the quarterly variation in *Home Depot revenues*. Using this multiple linear regression model, quarterly revenues in the next four quarters are expected to fall within \$1.2B of predictions with 95% confidence, shown in Figure 12.10.

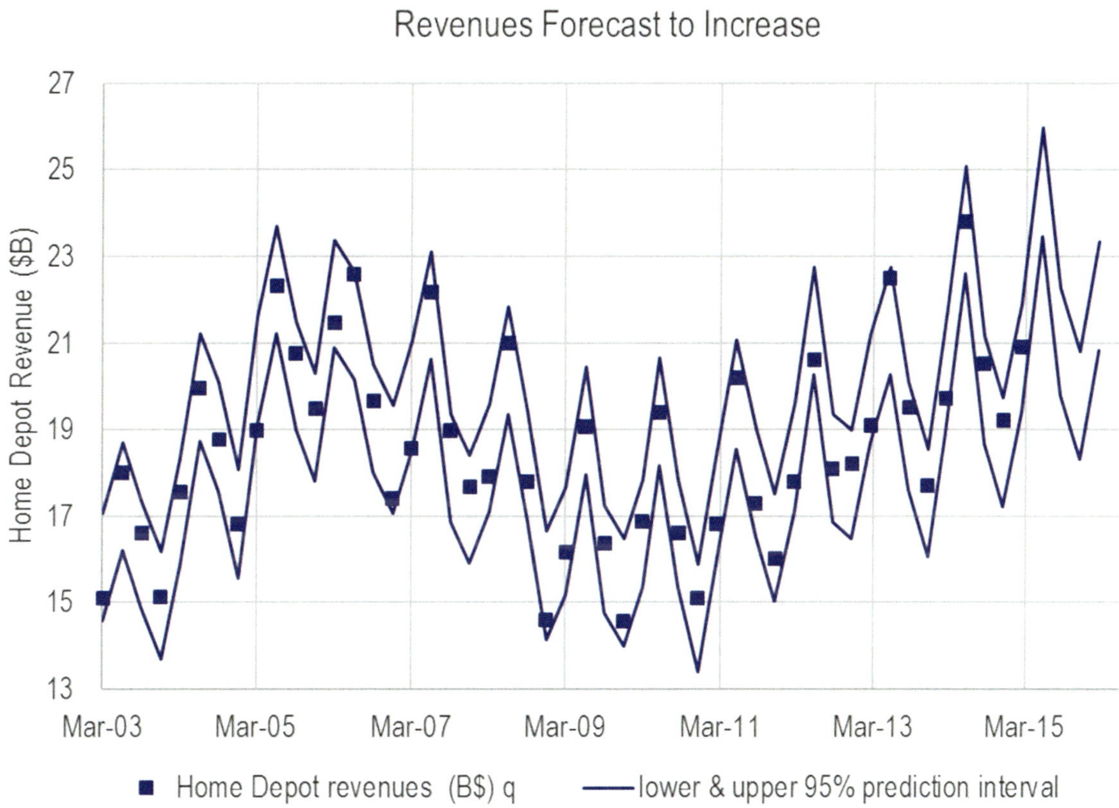

Figure 12.10 Model fit and forecast

Annual quarterly growth (from same quarter in the past year) has been 5 to 8% in the past four quarters, averaging 6.4%. Home Depot revenues are expected to grow less in the next four quarters, from 2 to 6%, with average quarterly growth of 3.5% over same quarter in the past year, on average. In all four quarters, there is the chance that revenues could be lower than same quarter revenues in the past year. Table 12.7 contains the four quarter forecast with annual growth projections.

Table 12.7 Quarterly revenue forecast

quarter	*95% lower prediction (B$)*	*95% upper prediction ($B)*	*Prior year Revenues ($B)*	*Forecast annual growth for quarter*
Jun-15	23.5	26.0	23.8	4.9%
Sep-15	19.8	22.3	20.5	2.5%
Dec-15	18.3	20.8	19.2	1.8%
Mar-16	20.8	23.3	20.9	5.7%

12.8 Compare Part Worths to Assess Driver Importances

Comparing driver part worths, the product of each driver's value in a quarter and its coeffiicient estimate, reveals the importance of *Lowes Revenues* in driving *Home Depot revenues*. Stacked part worths are shown in Figure 12.11, showing the driver contributions to predicted *Home Depot revenues*.

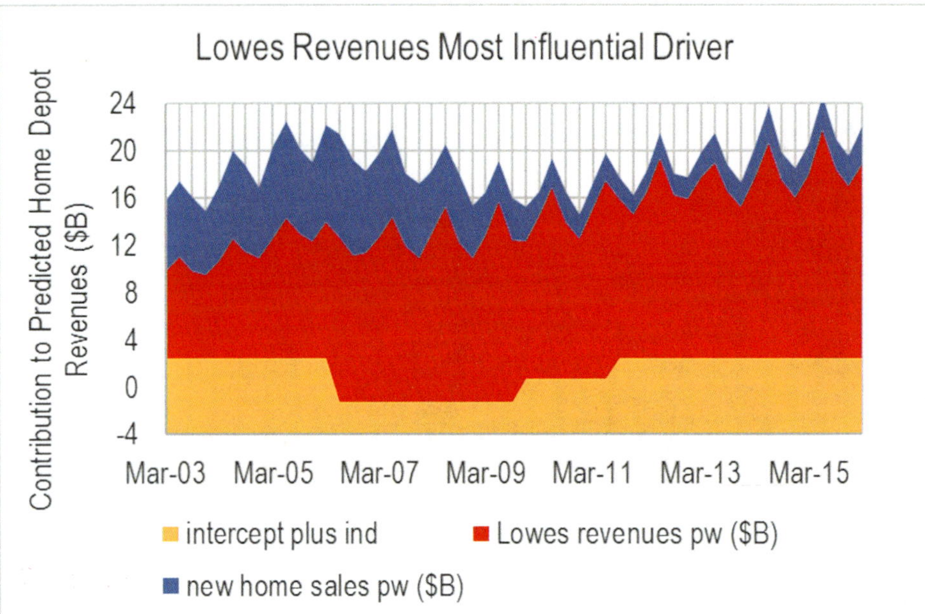

Figure 12.11 Stacked driver part worths

Lowes, while a competitor, provides advertising and promotion to drive home improvement projects, from which Home Depot benefits. Here, *Lowes Revenues* represent the broader home improvement industry, as well. Lowes' marginal contribution to predicted *Home Depot revenues*, shown in Figure 12.10 as change in the height of the red area, is \$12.3B (=\$19.3B, at the highest - \$7.0B, at the lowest).

Past *new home sales* is an important driver, though less influential in post recession quarters. This suggests that, post recession, home owners may be renovating, rather than buying new homes. The marginal contribution of past *new home sales* to *Home Depot revenues*, shown in Figure 12.10 as the change in height of the blue area, is \$7.1B (=\$8.8B, at the highest - \$1.6B, at the lowest).

Amanda summarized her model results for Management:

MEMO

Re: Slow, Stable Growth Forecast in Next Four Quarters
To: Home Depot Management
From: Amanda Chanel
Date: June 2015

Following past growth Lowes revenues, new home sales, quarterly revenues are expected to increase an average of 3.5% annually over the next four quarters.

A regression model of quarterly revenues was built from past new home sales, and Lowes revenues, accounting for housing bubble and recession shift. The model accounts for 93% of the variation in revenues and produces valid forecasts within \$1.2 billion of actual revenues.

Revenues are driven by growth in home improvement

Revenues have benefitted from growth in the home improvement sector, represented by past Lowes revenues, and from new home sales, to a smaller degree. The stacked contributions to predictions of these two drivers are shown in Exhibit 1. Revenues have improved from downshifts during the housing bubble and consequent global recession, a major influence.

Recovering growth is forecast

Revenues in the next four quarters are expected to exceed those in the past four quarters, though could fall below same quarter past year. Growth over same quarter past year Is forecast from 3.9% to 5.7%.

Quarter	*Forecast (\$B)*	*Forecast growth*
15-II	22.5 to 25.7	3.9%
15-III	18.9 to 22.2	2.5%
15-IV	17.5 to 20.8	1.8%
16-I	20.0 to 23.2	5.7%

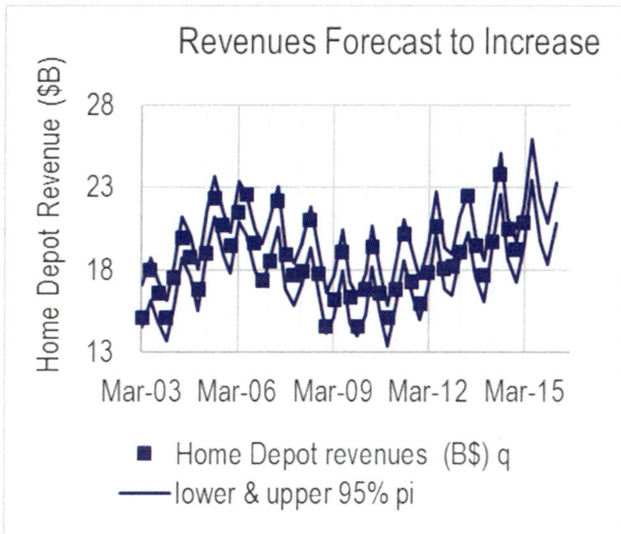

$$\widehat{revenues}(\$B)_q = 2.4 - 3.6(\$B)^a \times HBubble_q - 1.7 \times recession_q + 1.2\left(\frac{\$B}{rev(\$B)}\right)^a \times Lowes\ rev(\$B)_{q-4} - .025\left(\frac{\$B}{homes(K)}\right)^a \times New\ Homes(K)_{q-4}$$

RSquare: $.93^a$ $\quad$ ^{a}*Significant* at .01.

Revenue response may be nonlinear This model assumes constant response, though nonlinear response is possible.

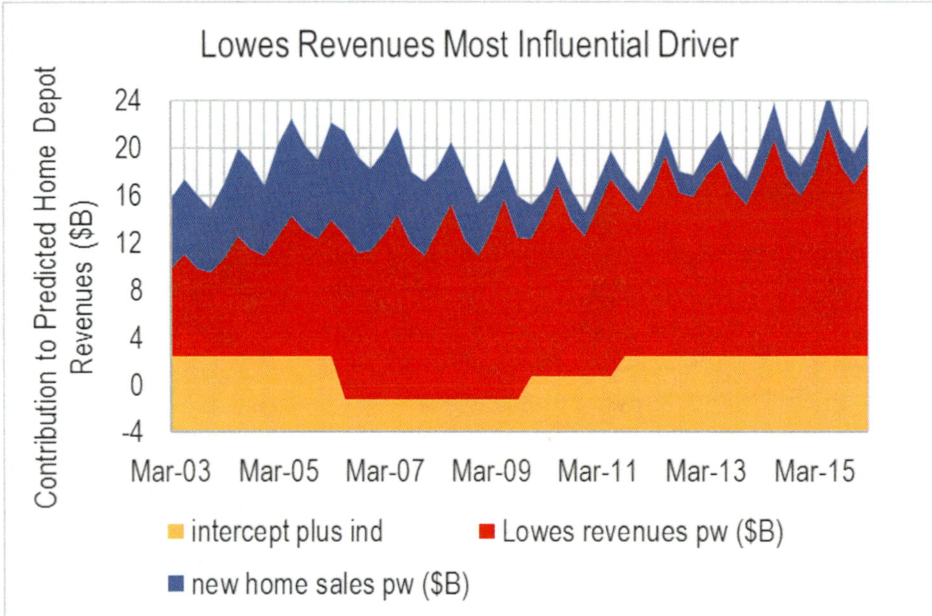

Exhibit 1. Stacked contributions, part worths (pw) to predicted revenue

12.9 Leading Indicator Components Are Powerful Drivers and Often Multicollinear

Like cross sectional models, explanatory time series models allow identification of performance drivers. However, time series models differ from cross sectional models, with forecasts as a dual goal, and the model building process with time series contains additional steps.

- Often lagged predictors are used to make driver identification more certain and to enable forecasts.
- Lagged predictors tend to move together across time and are often highly correlated. Consequently, to minimize multicollinearity issues, model building begins with one predictor, and then others are added, considering their joint influence and incremental model improvement.

Like naïve time series models based on trend, where the singular focus is on forecasting, explanatory time series models are also validated. Forecasting accuracy of time series models is tested, or validated, before they are used for prediction of future performance.

Predictors in time series models tend to be highly correlated, since most move with economic variables and most exhibit predictable growth (*trend*). Model building with time series begins with the strongest among logical predictors, and additional predictors are added which improve the model.

Time series typically contain trend, business cycles, and seasonality that are captured with these components. Unaccounted for trend, cycles, or seasonality are detected through inspection of the residual plot and the Durbin Watson statistic. Leading indicators are often stable and predictable performance drivers. Competitive variables may account for trend, seasonality or cycles common to a market.

Useful forecasting models must be valid. Holding out the two most recent performance observations allows a test of the model's forecasting capability. With successful prediction of the most recent performance, the model is validated and the recalibrated model can be used with confidence to forecast performance in future periods.

Excel 12.1 Build and Fit a Multiple Regression Model with Multicollinear Time Series

Home Depot Revenues. Build a model of Home Depot quarterly revenues which potentially includes past growth in the housing market and the home improvement sector, represented by a competitor's revenues. The data are in **12 Home Depot quarterly revenues**.

Plot *Home Depot revenues* by quarter to see the pattern of movement over time. (Hide or ignore the two most recent datapoints, December 2014 and March 2015.)

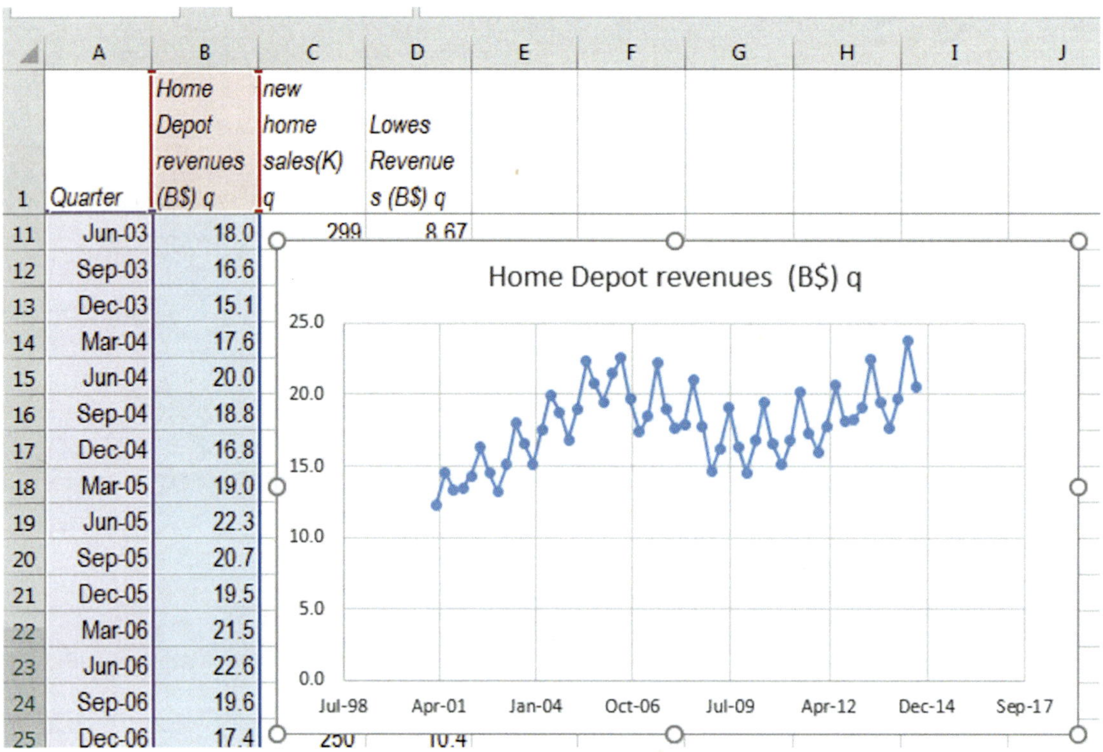

In Excel scatterplots, time is measured in days. To set the quarter axis beginning and end points, format the axis, with minimum 38060, maximum 42000, major units 732 (two years), and minor units 366 (one year).

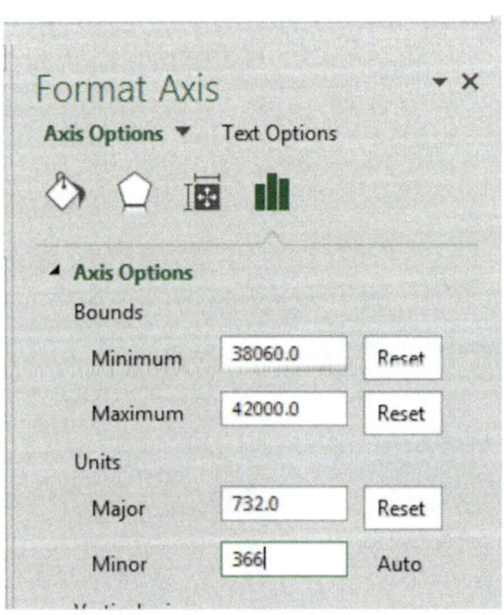

Add a trendline. | **Alt JCATL**

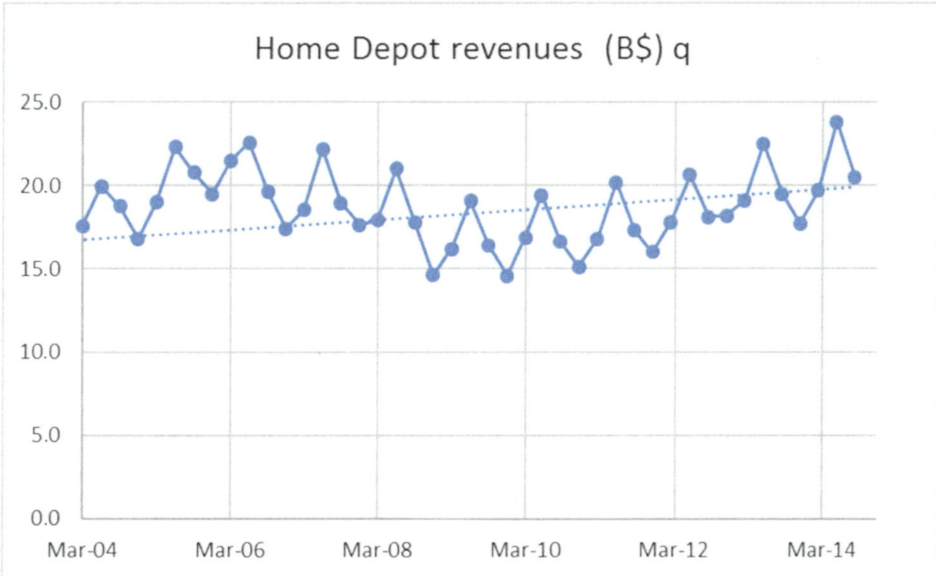

Plot both potential drivers, looking for trend, seasonality or a cycle to explain the trend, seasonality and cycle in the Home Depot revenues data.

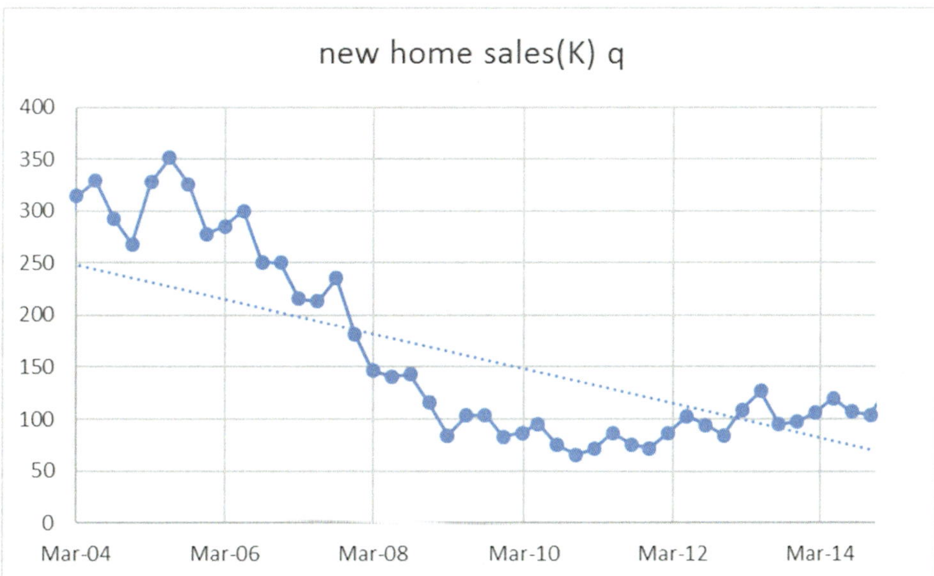

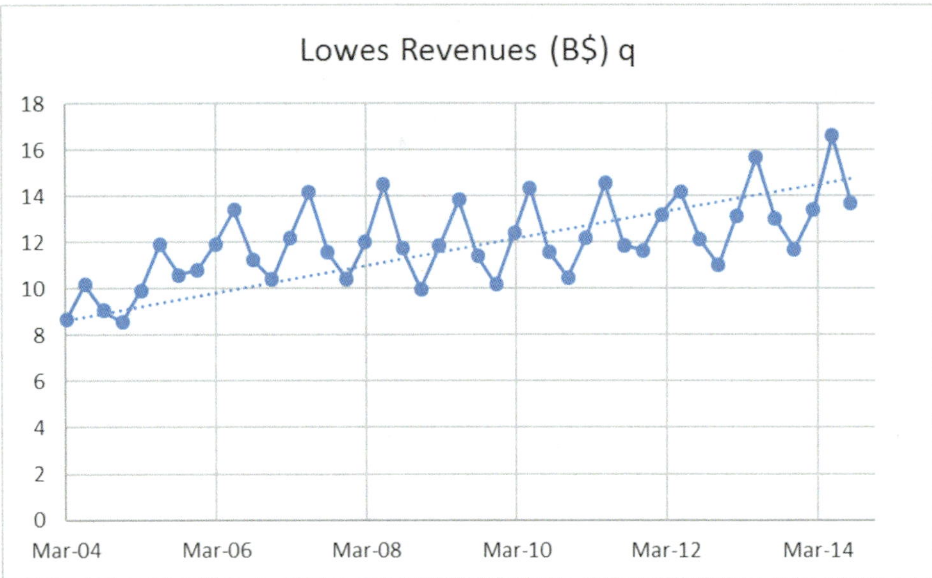

Excel 12.2 Create Potential Driver Lags

Create working columns that will contain the lagged drivers by copying columns A through D and then pasting into column E:

	A	B	C	D	E	F	G	H
1	Quarter	Home Depot revenues (B$) q	Lowes Revenues (B$) q	new home sales(K) q	Quarter	Home Depot revenues (B$) q	Lowes Revenues (B$) q	new home sales(K) q
32	Sep-08	17.8	11.7	143	Sep-08	17.8	11.7	143
33	Dec-08	14.6	10.0	116	Dec-08	14.6	10.0	116
34	Mar-09	16.2	11.8	84	Mar-09	16.2	11.8	84
35	Jun-09	19.1	13.8	104	Jun-09	19.1	13.8	104

Both Lowes revenues and new home sales are seasonal, and so lags that are multiples of four are preferred, to align with the seasonality in Home Depot revenues. Consider both four and eight quarter lags.

To make two alternate lags of the two potential drivers, copy columns G and H and paste into columns I and J. Make four quarter lags in columns G and H by inserting four blank cells in rows 2 through 5.

In G1,
Lowes Revenues ($B) q-4
In H2,
New home sales (K) q-4
In G2,
Shift+right
Shift+down down down
Alt HIID

	G	H
1	Lowes Revenues (B$) q-4	new home sales(K) q-4
2		
3		
4		
5		
6	5.28	251
7	6.13	243
8	5.45	216

Make eight quarter lags in columns I and J by inserting eight blanks cells in rows 2 through 9.

In I1,
Lowes revenues ($B) q-8
In J1,
New home sales ($B) q-8,
In I2,
Shift+right
Shift+down down down down down down down
Alt HIID

	I	J
1	Lowes Revenues ($B) q-8	new home sales(K) q-8
2		
3		
4		
5		
6		
7		
8		
9		
10	5.28	251
11	6.13	243
12	5.45	216

Finish creating the potential driver lags by selecting cells in rows of columns E through J that contain blanks and delete those cells.	In E2, **Shift+down down down down down down down** **Shift+right right right right right** **Alt HDDU**

	E	F	G	H	I	J
1	*Quarter*	*Home Depot revenues (B$) q*	*Lowes Revenues (B$) q-4*	*new home sales(K) q-4*	*Lowes Revenues ($B) q-8*	*new home sales(K) q-8*
2	Mar-03	15.1	6.37	240	5.28	251
3	Jun-03	18.0	7.39	258	6.13	243
4	Sep-03	16.6	6.32	254	5.45	216
5	Dec-03	15.1	6.04	220	5.25	199
6	Mar-04	17.6	7.12	256	6.37	240
7	Jun-04	20.0	8.67	299	7.39	258
8	Sep-04	18.8	7.8	294	6.32	254
9	Dec-04	16.8	7.25	239	6.04	220
10	Mar-05	19.0	8.68	314	7.12	256

Excel 12.3 Select the Most Promising Driver

To increase the chances that the model will be valid, identify the potential driver(s) that show a matching pattern in the most recent quarters (excluding the two most recent quarters). Highlight cells in which *Home Depot revenues* declined from the prior quarter

	E	F	G	H	I	J
1	*Quarter*	*Home Depot revenues (B$) q*	*Lowes Revenues (B$) q-4*	*new home sales(K) q-4*	*Lowes Revenues ($B) q-8*	*new home sales(K) q-8*
43	Jun-13	22.5	14.2	103	14.5	87
44	Sep-13	19.5	12.1	94	11.9	76
45	Dec-13	17.7	11	84	11.6	72
46	Mar-14	19.7	13.1	109	13.2	87
47	Jun-14	23.8	15.7	127	14.2	103
48	Sep-14	20.5	13	95	12.1	94
49	Dec-14	19.2	11.7	98	11	84

Since all four potential drivers show a matching pattern, compare their correlations with *Home Depot revenues*:

	A	B	C	D	E	F
1		*Home Depot revenues (B$) q*	*Lowes Revenues (B$) q-4*	*new home sales(K) q-4*	*Lowes Revenues ($B) q-8*	*new home sales(K) q-8*
2	Home Depot revenues (B$) q	1				
3	Lowes Revenues (B$) q-4	0.480018	1			
4	new home sales(K) q-4	0.261383	-0.50698	1		
5	Lowes Revenues ($B) q-8	0.324013	0.964778	-0.64024	1	
6	new home sales(K) q-8	0.134226	-0.36013	0.910275	-0.46134	1

The shorter, four quarter *Lowes revenue* lag is most highly correlated with *Home Depot revenues.* Run a simple regression with *Lowes revenue*$_{q-4}$.

	A	B	C	D	E	F	G
1	SUMMARY OUTPUT						
2							
3	*Regression Statistics*						
4	Multiple F	0.480018					
5	R Square	0.230418					
6	Adjusted I	0.213316					
7	Standard E	1.995041					
8	Observati	47					
9							
10	ANOVA						
11		*df*	*SS*	*MS*	*F*	*gnificance F*	
12	Regressio	1	53.62614	53.62614	13.47327	0.000639	
13	Residual	45	179.1084	3.980187			
14	Total	46	232.7346				
15							
16		*Coefficients*	*andard Err*	*t Stat*	*P-value*	*Lower 95%*	*Upper 95%*
17	Intercept	13.55153	1.388267	9.761473	1.1E-12	10.75542	16.34764
18	Lowes Rev	0.451528	0.123012	3.670595	0.000639	0.203769	0.699287

Excel 12.4 Plot Residuals to Identify Unaccounted for Trend, Cycles, or Seasonality and Assess Autocorrelation

Plot residuals, setting the vertical axis units to the standard error, 2.0, and add a trendline.

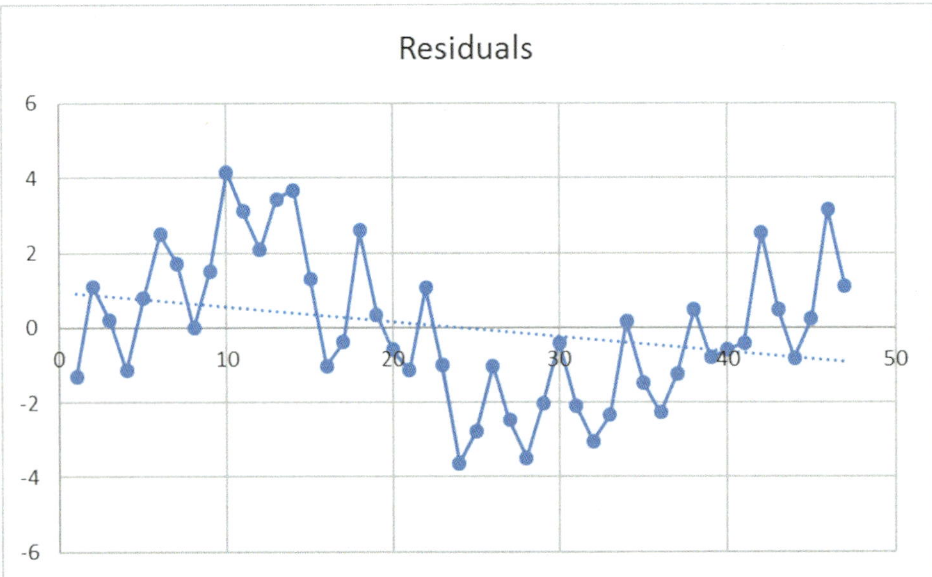

The residual plot shows trend, a cycle and seasonality. Adding one of the *new home sales* lags will likely improve the regression.

Assess Durbin Watson to confirm that trend, cycles, shifts or shocks remain.

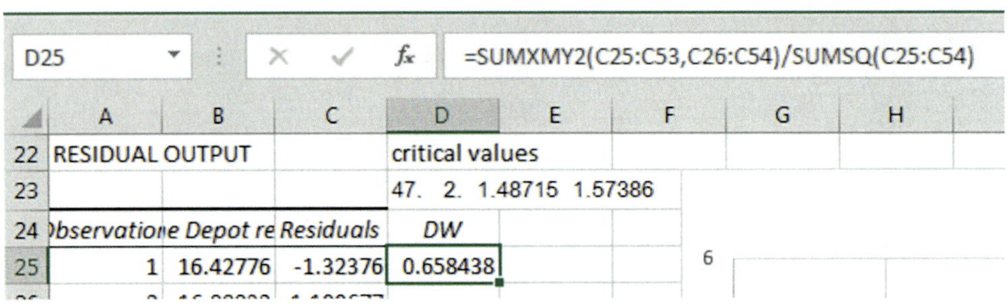

D25 =SUMXMY2(C25:C53,C26:C54)/SUMSQ(C25:C54)

	A	B	C	D	E	F	G	H
22	RESIDUAL OUTPUT			critical values				
23				47. 2. 1.48715 1.57386				
24	*)bservatio*	*e Depot re*	*Residuals*	*DW*				
25	1	16.42776	-1.32376	0.658438				

DW is below the lower critical value, 1.49, confirming that positive autocorrelation from unaccounted for trend, cycles, shifts or shocks is present.

Copy the residuals, return to the data sheet, and paste residuals next to the data. Since *Lowes revenues* is already in the model, the longer *Lowes revenues* lag can be moved right of the residuals.

E	F	G	H	I	J
Quarter	Home Depot revenues (B$) q	Lowes Revenues (B$) q-4	new home sales(K) q-4	new home sales(K) q-8	Residuals
Jun-10	19.4	13.8	104	141	-0.40249
Sep-10	16.6	11.4	104	143	-2.09895
Dec-10	15.1	10.2	83	116	-3.04267

Highlight declines from past quarter in the residuals in the most recent quarters:

	E	F	G	H	I	J	K	L
1	Quarter	Home Depot revenues (B$) q	Lowes Revenues (B$) q-4	new home sales(K) q-4	new home sales(K) q-8	Residuals	Lowes Revenues ($B) q-8	Residuals r
44	Sep-13	19.5	12.1	94	76	0.48498	11.9	1.636713
45	Dec-13	17.7	11	84	72	-0.81834	11.6	0.853103
46	Mar-14	19.7	13.1	109	87	0.233452	13.2	0.816206
47	Jun-14	23.8	15.7	127	103	3.159479	14.2	2.605891
48	Sep-14	20.5	13	95	94	1.078605	12.1	1.922425

Since pattern in both of the *new home sales* lags match the residual pattern, compare correlations between the two *new home sales* lags and the residuals to choose one to add to the regression.

	A	B	C	D
1		new home sales(K) q-4	new home sales(K) q-8	Residuals
2	new home sales(K) q-4	1		
3	new home sales(K) q-8	0.910275	1	
4	Residuals	0.575363	0.350063	1
5				

Add *new home* $sales_{q-4}$ to the regression:

	A	B	C	D	E	F	G
1	SUMMARY OUTPUT						
2							
3	*Regression Statistics*						
4	Multiple F	0.757177					
5	R Square	0.573317					
6	Adjusted I	0.553922					
7	Standard E	1.5023					
8	Observati	47					
9							
10	ANOVA						
11		*df*	*SS*	*MS*	*F*	*gnificance F*	
12	Regressio	2	133.4307	66.71535	29.56054	7.28E-09	
13	Residual	44	99.30386	2.256906			
14	Total	46	232.7346				
15							
16		*Coefficient*	*andard Err*	*t Stat*	*P-value*	*Lower 95%*	*Upper 95*
17	Intercept	6.94433	1.525589	4.5519	4.17E-05	3.869707	10.0189
18	Lowes Rev	0.775505	0.107465	7.216352	5.51E-09	0.558924	0.99208
19	new home	0.016333	0.002747	5.946441	4.05E-07	0.010798	0.02186

Again plot residuals, setting the vertical axis units to the standard error, 1.5.
Assess Durbin Watson to decide whether additional drivers are needed. (Reuse the Durbin Watson formula from the simple regression. Copy and paste into the same cells in your new regression.

	A	B
23		critical values
24		47. 3. 1.44352 1.62038
25	*Residuals*	*DW*
26	-0.70033	0.472577
27	1.099655	
28	0.603779	
29	-0.09674	
30	0.902708	
31	1.408336	
32	0.976693	
33	0.341561	
34	0.168579	
35	2.100075	
36	2.004223	
37	1.536734	
38	1.47404	

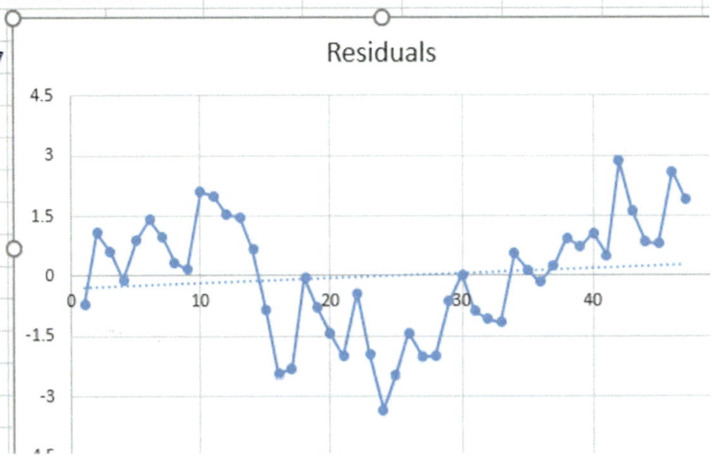

Residuals in quarters 1 through 13 show positive trend. Residuals in quarters 14 through 27 suggest a downward shift. This downward shift includes a residual more than two standard errors below the expected value of zero. Quarters 14 through 27 correspond to June 2006 through September 2009, a period during which the housing bubble reduced real estate values and ultimately precipitated the global recession. Add an indicator in the data sheet, to the right of *Home Depot revenues*, which is equal to 1 in June 2006 through September 2009, and equal to 0 elsewhere.

	E	F	G
1	*Quarter*	*Home Depot revenues (B$) q*	*housing bubble*
14	Mar-06	21.5	0
15	Jun-06	22.6	1
16	Sep-06	19.6	1
17	Dec-06	17.4	1
18	Mar-07	18.5	1
19	Jun-07	22.2	1
20	Sep-07	19.0	1
21	Dec-07	17.7	1
22	Mar-08	17.9	1

Add the *housing bubble* indicator to the regression model.

	A	B	C	D	E	F	G
1	SUMMARY OUTPUT						
2							
3	*Regression Statistics*						
4	Multiple F	0.935442					
5	R Square	0.875052					
6	Adjusted I	0.866335					
7	Standard E	0.822358	2.016692	1.658442			
8	Observati	47					
9							
10	ANOVA						
11		*df*	*SS*	*MS*	*F*	*gnificance F*	
12	Regressio	3	203.6549	67.88496	100.3812	1.9E-19	
13	Residual	43	29.07969	0.676272			
14	Total	46	232.7346				
15							
16		*Coefficient*	*andard Err*	*t Stat*	*P-value*	*Lower 95%*	*Upper 95%*
17	Intercept	1.83122	0.974255	1.879611	0.066948	-0.13355	3.795992
18	housing b	-3.37299	0.331003	-10.1902	4.86E-13	-4.04052	-2.70546
19	Lowes Re\	1.147682	0.069242	16.57494	2.74E-20	1.008042	1.287321

Plot the residuals and assess Durbin Watson.

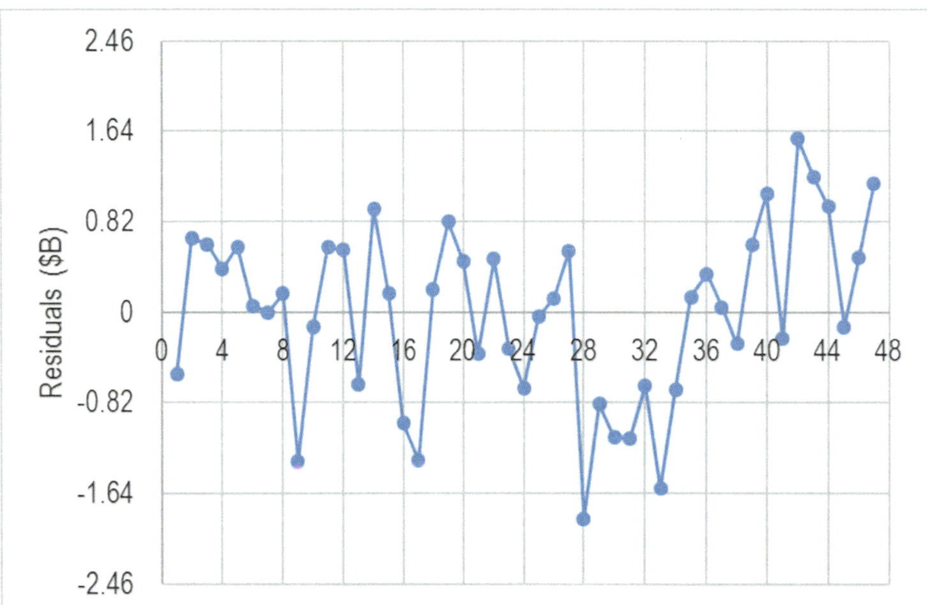

Residuals in quarters 28 through 34 are each negative, signaling an unaccounted for downward shock.

	D
24	T k dL dU
25	47. 4. 1.39894 1.66923
26	*DW*
27	1.245177735

DW is below the lower critical value, confirming that the residuals are not free of unaccounted for trend, cycles, shifts or shocks.

Return to the data sheet and add a *recession* indicator, equal to 1 in quarters 28 through 34, and equal to 0 elsewhere.

	E	F	G	H
1	*Quarter*	*Home Depot revenues (B$) q*	*housing bubble*	*recession*
28	Sep-09	16.4	1	0
29	Dec-09	14.6	0	1
30	Mar-10	16.9	0	1
31	Jun-10	19.4	0	1
32	Sep-10	16.6	0	1

Run regression, adding the *recession* indicator.

	A	B	C	D	E	F	G
1	SUMMARY OUTPUT						
2							
3	*Regression Statistics*						
4	Multiple F	0.964644					
5	R Square	0.930537					
6	Adjusted I	0.923922					
7	Standard E	0.620413					
8	Observati	47					
9							
10	ANOVA						
11		*df*	*SS*	*MS*	*F*	*gnificance F*	
12	Regressio	4	216.5682	54.14206	140.6606	9.76E-24	
13	Residual	42	16.16634	0.384913			
14	Total	46	232.7346				
15							
16		*Coefficient*	*andard Err*	*t Stat*	*P-value*	*Lower 95%*	*Upper 95%*
17	Intercept	2.376322	0.74101	3.206868	0.002567	0.880903	3.871741
18	housing b	-3.5877	0.252456	-14.2112	1.21E-17	-4.09718	-3.07823
19	recession	-1.62958	0.281345	-5.79213	7.9E-07	-2.19736	-1.06181
20	Lowes Rev	1.159728	0.05228	22.18311	8.18E-25	1.054224	1.265233
21	new home	0.025165	0.001431	17.57965	5.62E-21	0.022276	0.028054

Plot the residuals and assess Durbin Watson, with T=47 observations and k=5 coefficients.

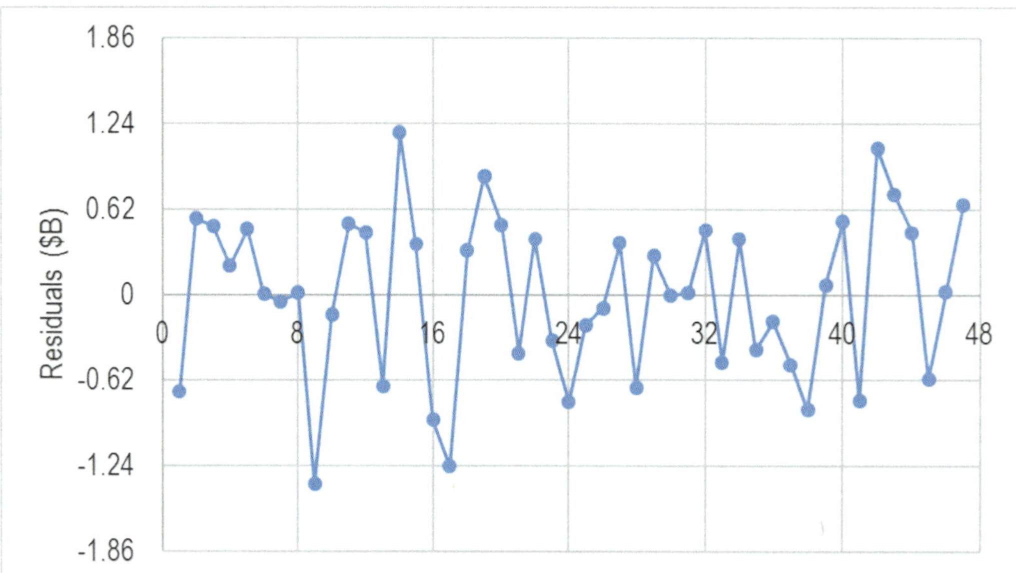

D28 =SUMXMY2(C28:C73,C29:C74)/SUMSQ(C28:C74)

	A	B	C	D	E
25	RESIDUAL OUTPUT			T k dL dU	
26				47. 5. 1.35350 1.72033	
27	*Observation*	*Home Depot re*	*Residuals*	*DW*	
28	1	15.80345	-0.69945	1.951128993	

Excel 12.5 Test the Model's Forecasting Validity

Given a significant model with high RSquare, significant drivers with correct signs, and DW above the upper critical level, test the model's validity.

Copy Quarter, Home Depot revenues, the indicators and the two drivers in rows 1 through 54 from the data sheet and paste next to residuals.

	D	E	F	G	H	I	J
27	*DW*	*Quarter*	*Home Depot revenues (B$) q*	*housing bubble*	*recession*	*Lowes Revenues (B$) q-4*	*new home sales(K) q-4*
28	1.951129	Mar-03	15.1	0	0	6.37	240
29		Jun-03	18.0	0	0	7.39	258
30		Sep-03	16.6	0	0	6.32	254
31		Dec-03	15.1	0	0	6.04	220
32		Mar-04	17.6	0	0	7.12	256

Use the regression equation to make the five part worths in columns K through O, and then sum to find predicted *Home Depot revenues* in column P.

	K	L	M	N	O	P
27	*intercept ($B)*	*housing bubble pw ($B)*	*recession pw ($B)*	*Lowes revenues pw ($B)*	*new home sales pw ($B)*	*predicted Home Depot revenues ($B)*
28	2.376322	0	0	7.38747	6.0396534	15.80344524
29	2.376322	0	0	8.570393	6.4926274	17.43934224
30	2.376322	0	0	7.329484	6.3919665	16.09777193
31	2.376322	0	0	7.00476	5.5363489	14.91743041

Find the *margin of error* from *critical t* and the *standard error*:

D7 =B7*C7

	A	B	C	D
3	*Regression Statistics*			
4	Multiple F	0.964644		
5	R Square	0.930537		
6	Adjusted I	0.923922	critical t	me
7	Standard E	0.620413	2.018082	1.252045
8	Observati	47		

Add and subtract the *margin of error* from *predicted* values to find the *lower* and *upper 95% prediction interval bounds* in columns Q and R:

Q28 =P28-$D$7

	P	Q	R
27	*predicted Home Depot revenues ($B)*	*lower 95% pi ($B)*	*upper 95% pi ($B)*
28	15.80344524	14.6	17.1
29	17.43934224	16.2	18.7

Compare actual revenues with the 95% prediction interval in the two most recent quarters to assess the model's validity for forecasting:

	E	F	G	H	I	J	K	L	M	N	O	P	Q	R
75	Quarter	*Home Depot revenues (B$) q*	*housing bubble*	*recession*	*Lowes Revenues (B$) q-4*	*new home sales(K) q-4*	*intercept ($B)*	*housing bubble pw ($B)*	*recession pw ($B)*	*Lowes revenues pw ($B)*	*new home sales pw ($B)*	*predicted Home Depot revenues ($B)*	*lower 95% pi ($B)*	*upper 95% pi ($B)*
76	Dec-14	19.2	0	0	11.7	98	2.376322	0	0	13.56882	2.4661918	18.41133619	17.2	19.7
77	Mar-15	20.9	0	0	13.4	106	2.376322	0	0	15.54036	2.6675136	20.5841963	19.3	21.8
78	Jun-15		0	0	16.6	120	2.376322	0	0	19.25149	3.0198267	24.64764038	23.4	25.9

The model correctly forecast the two hidden datapoints from the most recent quarters, providing confidence in its validity for forecasting.

Excel 12.6 Recalibrate to Forecast

Recalibrate the model to update the coefficients by rerunning regression, this time with all 50 rows of data.

	A	B	C	D	E	F	G
1	SUMMARY OUTPUT						
2							
3	*Regression Statistics*						
4	Multiple F	0.964104					
5	R Square	0.929497					
6	Adjusted I	0.923087					
7	Standard E	0.618304					
8	Observati	49					
9							
10	ANOVA						
11		*df*	*SS*	*MS*	*F*	*gnificance F*	
12	Regressio	4	221.7668	55.44169	145.0214	9.82E-25	
13	Residual	44	16.82121	0.3823			
14	Total	48	238.588				
15							
16		*Coefficient*	*andard Err*	*t Stat*	*P-value*	*Lower 95%*	*Upper 95%o*
17	Intercept	2.421827	0.734334	3.297992	0.001933	0.941874	3.90178
18	housing b	-3.61875	0.248997	-14.5333	2.09E-18	-4.12057	-3.11693
19	recession	-1.69045	0.27569	-6.1317	2.16E-07	-2.24606	-1.13483
20	Lowes Rev	1.162689	0.051547	22.55571	8.22E-26	1.058802	1.266576
21	new home	0.024955	0.001417	17.61367	1.5E-21	0.0221	0.027811

Reuse your formulas for *critical t, margin of error, part worths, predicted revenues,* and *95% prediction interval bounds.*

	K	L	M	N	O	P	Q	R
27	*intercept ($B)*	*housing bubble pw ($B)*	*recession pw ($B)*	*Lowes revenues pw ($B)*	*new home sales pw ($B)*	*predicted Home Depot revenues ($B)*	*lower 95% pi ($B)*	*upper 95% pi ($B)*
28	2.421827	0	0	7.406326	5.989279	15.81743	14.6	17.1
29	2.421827	0	0	8.592268	6.438475	17.45257	16.2	18.7
30	2.421827	0	0	7.348192	6.338654	16.10867	14.9	17.4
31	2.421827	0	0	7.022639	5.490172	14.93464	13.7	16.2
32	2.421827	0	0	8.278342	6.388564	17.08873	15.8	18.3
33	2.421827	0	0	10.08051	7.461643	19.96398	18.7	21.2

Excel 12.7 Illustrate the Fit and Forecast

To see the model fit and forecast, move the *lower* and *upper 95% prediction interval bounds* next to *Home Depot revenues ($B)* and plot the prediction intervals with actual revenues through March 2016:

	E	F	G	H
27	*Quarter*	*Home Depot revenues (B$) q*	*lower 95% pi ($B)*	*upper 95% pi ($B)*
28	Mar-03	15.1	14.6	17.1
29	Jun-03	18.0	16.2	18.7
30	Sep-03	16.6	14.9	17.4
31	Dec-03	15.1	13.7	16.2
32	Mar-04	17.6	15.8	18.3
33	Jun-04	20.0	18.7	21.2
34	Sep-04	18.8	17.6	20.1
35	Dec-04	16.8	15.6	18.1

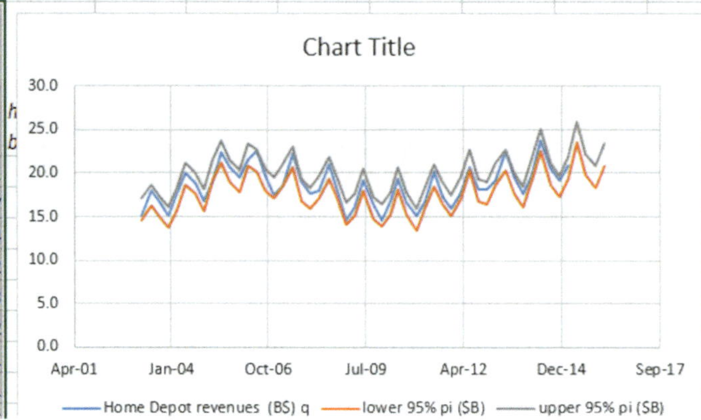

Recolor one of the 95% prediction intervals so that both are the same color.
Change the *Home Depot revenues ($B)* series from a line to markers.

Format axes to reduce white space, add a vertical axis label and a chart title that summarizes your conclusion, and set fontsize to 12:

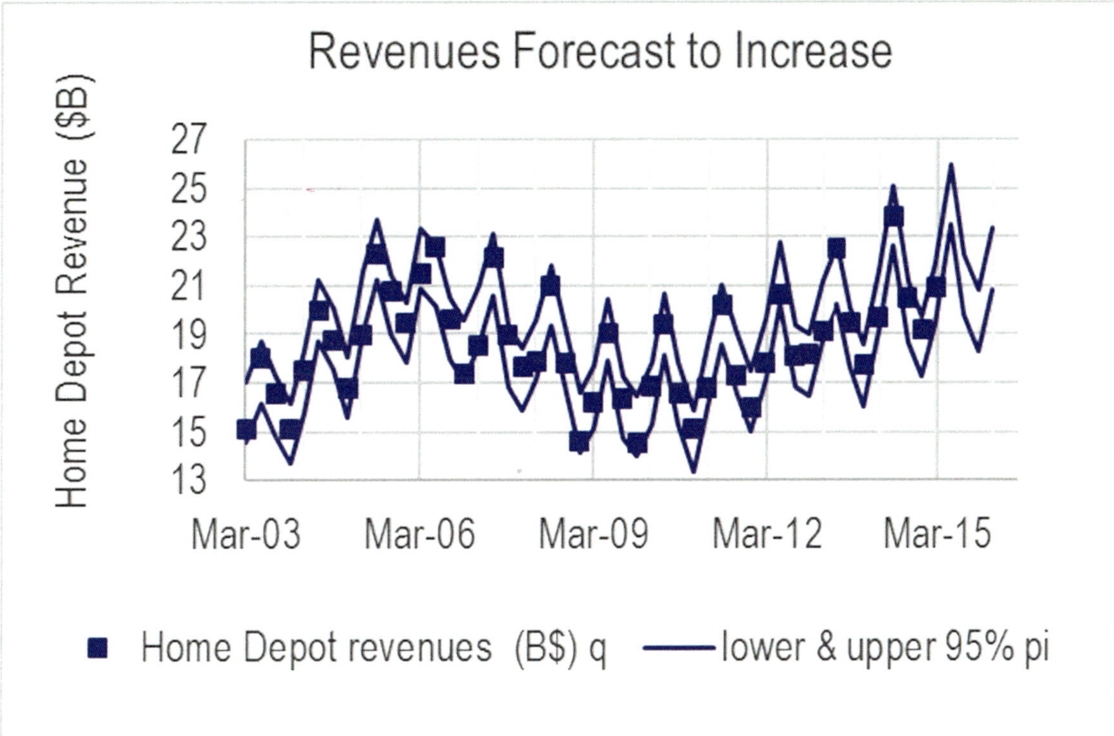

Excel 12.8 Assess the Impact of Drivers

Use the part worths to compare the impacts of each of the drivers on model forecasts. Find the cumulative part worths that produce predicted *Home Depot revenues* in columns S through V.

In column S, add to the intercept in M the *housing bubble pw* in N.
In column T, add to the sum in S the *recession part worth* in O.
In column U, add to the sum in T the *Lowes revenue part worth* in P, and
in column V, add to the sum in U the *new home sales part worth* in Q.

S28 =M28+N28

	M	N	O	P	Q	R	S	T	U	V
27	intercept ($B)	housing bubble pw ($B)	recession pw ($B)	Lowes revenues pw ($B)	new home sales pw ($B)	predicted Home Depot revenues ($B)	plus housing bubble pw	plus recession pw	plus Lowes pw	plus new home sales pw
28	2.421827	0	0	7.406326	5.989279	15.81743	2.4	2.4	9.8	15.8
29	2.421827	0	0	8.592268	6.438475	17.45257	2.4	2.4	11.0	17.5
30	2.421827	0	0	7.348192	6.338654	16.10867	2.4	2.4	9.8	16.1
31	2.421827	0	0	7.022639	5.490172	14.93464	2.4	2.4	9.4	14.9

To plot the cumulative part worths by quarter, move quarter, from column E, and the intercept, from column M, to the left of *plus housing buble pw.*

	Q	R	S	T	U	V
27	Quarter	intercept ($B)	plus housing bubble pw	plus recession pw	plus Lowes pw	plus new home sales pw
28	Mar-03	2.421827	2.4	2.4	9.8	15.8
29	Jun-03	2.421827	2.4	2.4	11.0	17.5
30	Sep-03	2.421827	2.4	2.4	9.8	16.1
31	Dec-03	2.421827	2.4	2.4	9.4	14.9
32	Mar-04	2.421827	2.4	2.4	10.7	17.1

Select *quarter, intercept, plus housing bubble, plus recession, plus Lowes* and *plus new home sales,* in columns Q through V, and request a scatterplot.

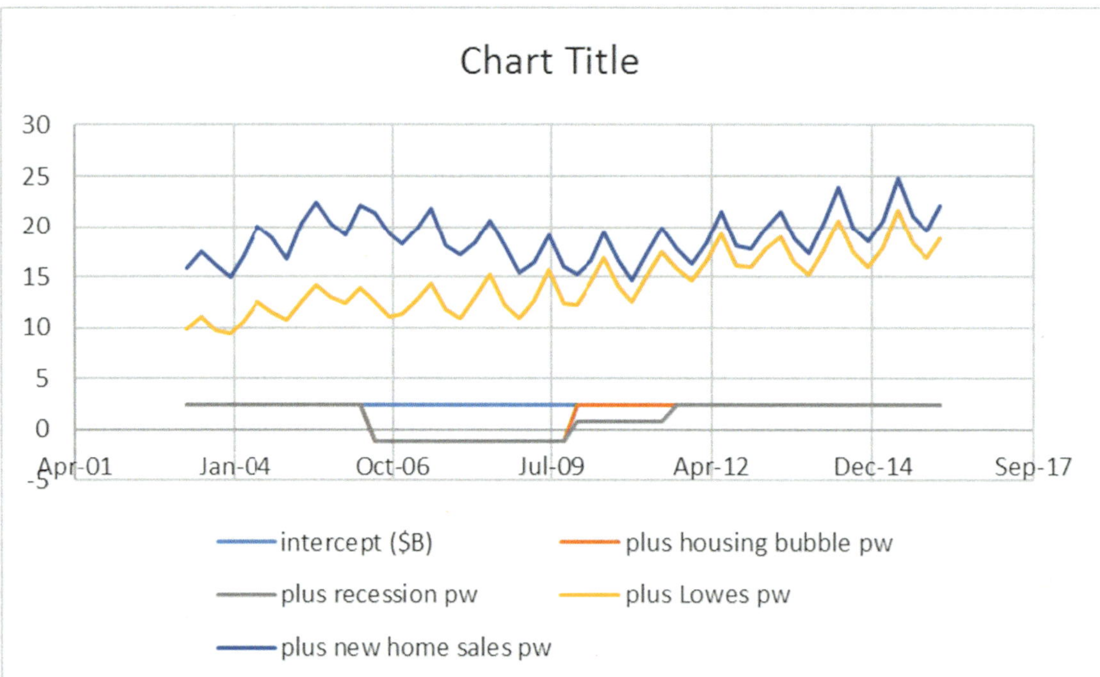

Adjust the vertical axis to make good use of space. Adjust the horizontal axis units to 8 quarters (733 days). Add a vertical axis title, chart title, and set font size to 12.

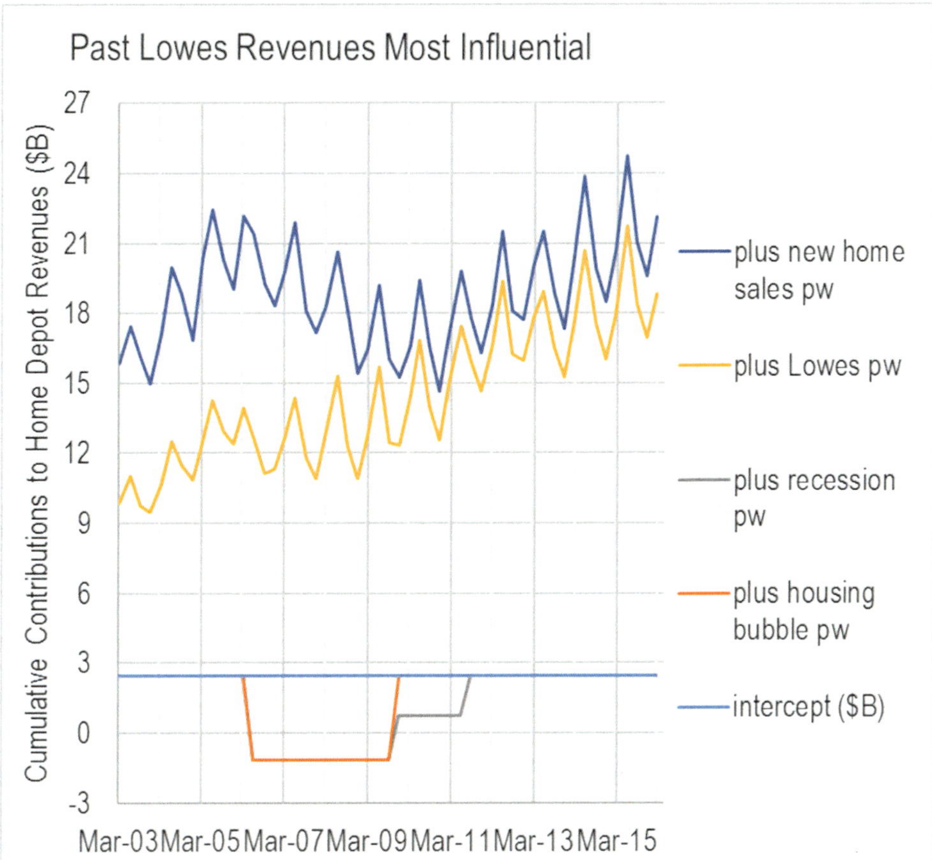

Lab 12.1 What Is Driving WFM Revenues... and What Revenues Can WFM Expect Next Year?

Management believes that some of WFM's success may be due to demographics. Is WFM successful because a growing number of WFM shoppers are younger?

Competition with the larger chain stores has intensified in recent years, as Wal-Mart, Safeway, Kroger, Costco, and others add organic lines. Revenues of competitors may be hurting WFMs. Or demand for organic products may simply be increasing, allowing revenues of WFM and competitors to increase together. When competitors promote organics, WFM revenues may benefit. **12 WFM revenues** contains a ten year time series of quarterly data on:

WFM Revenues, WFM stores, Populations by age group, Kroger revenues

I. Choose Candidate Predictor Lags

1. Plot WFM revenues by quarter. What do you see?

___ increasing trend ___ decreasing trend ___ seasonality ___other:_______________

2. Identify the potential driver lags which you could use to produce a forecast of at least *four* quarters:

Food revenues are seasonal. Choose a lag for Kroger revenues which aligns Kroger seasonality with Whole Food market seasonality. Set up this lag.

Kroger revenues: _____quarter lag

For each population variable, choose a lag which is a multiple of 4 quarters to correspond to age in years, and which allow a *four* quarter forecast. Set up these lags.

population 18 to 44: ____quarter lag population 45 to 64: ____quarter lag

population 65 plus: ____quarter lag

WFM stores increase in nearly all quarters, so there is little difference between lag choices. Set up an 8 quarter lag, which would enable an 8 quarter forecast.

II. Decide the direction of influence that you expect for each potential driver.

More people ought to be positively correlated with more revenue.

Population 18 44: ___+ ___− or ___? Population 45 to 64; ___+ ___− or ___?

Population 65 plus: ___+ ___− or ___? Kroger revenues: ___+ ___− or ___?

WFM stores: ___+ ___− or ___?

III. Find drivers with matching pattern. WFM revenues exhibit an increase followed by a decline in the two most recent quarters (ignoring the two hidden quarters). Which candidate predictor lags show a similar increase followed by a decline in the two most recent quarter?

___ WFM stores 8q lag ___ population 18 to 44 __q lag

___ population 45 to 64 __ q lag ___ population 65 plus __ q lag

___ Kroger revenues ______ q lag(s)

IV. Build your model.

A. Simple regression model. Choose the one of those four predictor lags which shows the largest positive correlation with WFM revenues and then build a simple regression model.

1. Is the predictor significant? Y or N Is the coefficient sign as expected? Y or N
2. Plot the residuals to see patterns in unaccounted for variation in McDonalds revenues.

   What do you see? ___ increasing trend ___decreasing trend ___seasonality

   ___other:_______________
3. Assess Durbin Watson. Residuals

   ___contain ___may contain ___are free of unaccounted for trend or cycles.

B. Two driver model. Among the candidate predictors, add to your model the predictor which has the highest positive correlation with residuals.

1. Are both predictors significant? Y or N
2. Are both coefficient signs as expected? Y or N
3. How much did the second predictor improve explanatory power (RSquare)? _____
4. How much did the second predictor reduce the margin of error in forecasts? _____
5. Plot the residuals to see patterns in unaccounted for variation in WFM revenues. What do you see?

   ___ increasing trend ___decreasing trend ___seasonality ___other:___________

6. Assess Durbin Watson. Residuals:

   ___contain ___may contain ___are free of unaccounted for trend or cycles,

C. Four driver model

WFM acquired Wild Oats late in 2007, which was divested in March 2009. Add a Wild Oats indicator, 1 in quarters March 2008 through March 2009, 0 elsewhere.

The recession affected purchases of relatively expensive organic food. Add a recession indicator, 1 in quarters December 2008 through December 2009, 0 elsewhere.
Build a four variable model, adding the two indicators.

1. Are all predictors significant? Y or N Are all coefficient signs as expected? Y or N
2. How much did the indicators improve explanatory power (RSquare)? _____
3. How much did the indicators reduce the margin of error in forecasts? _____
4. Plot the residuals to see patterns in unaccounted for variation in WFM revenues.

   What do you see? ___ increasing trend ___decreasing trend ___seasonality

   ___other:_________________
5. Assess Durbin Watson. Residuals:

   ___contain ___may contain ___are free of unaccounted for trend or cycles.

Lab 12.2 What Is Driving WFM Revenues… and What Revenues Can WFM Expect Next Year?

I. Validate, Recalibrate and Present Results

Test the validity of your model for forecasting.

1. Is your model valid? Y or N

Recalibrate your model and plot your fit and forecast through December 2012.

2. Describe your model precision:

   Model predictions will be no further from revenues than _____ ($M) with 95% confidence.

3. Present your forecast for WFM revenues in December, 2012: _____ to ____ ($M)

4. Present your model equation, substituting
   (i) your model coefficients for b0, b1, b2, b3 and b4,
   (ii) units which predictors are measured in for u1 and u2,
   (iii) significance levels, "a" for 95% confidence, "b" for 99% confidence, for s1, s2, s3, s4, and s5,
   (iv) the variable names for I1, I2, X1, X2, and
   (v) the lags which you used, L1 and L2, for X1 and X2.

$$W\hat{F}M\ rev(\$M)_q = b0(\$M) + b1(\$M)^{s1} \times I1_q + b2^{s2} \times I2_q$$
$$+b3^{s3}(\$M/u1) \times X1(u1)_{q-L1}$$
$$+b4^{s4}(\$M/u2) \times X2(u2)_{q-L2}$$

RSquare: _____ ^{s5}

5. Identify drivers of WFM revenues: ___*population 18 to 45 years*

   population 45 to 64 years ___*population over 65* ___*Kroger revenues*

   ___ *WFM stores* ___*recession* ___*Wild Oats acquisition*

II. Sensitivity Analysis

1. Make a graph of stacked part worths:

2. Which driver is the most influential?

___*population 18 to 45 years* ___*population 45 to 64 years* ___*population over 65*

___*Kroger revenues* ___ *WFM stores* ___*recession* ___*Wild Oats acquisition*

Case 12 McDonalds Revenue Drivers and Future Prospects

McDonalds executives believe that success depends on demographics, economic productivity, and the growing fast food industry. They require identification of drivers of revenues, as well as a forecast of revenues over the next four quarters.

McDs QR contains data on quarterly revenues, global GDP per capita among high and upper middle income countries, the global fast food market value, competitor Yum Foods (Taco Bell, Kentucky Fried Chicken and others) revenues, and global population of the youngest and oldest segments for the eight year period 2003 through 2011. Each potential driver differs in data availability, with data from some drivers being available as early as March 2003, and data for other drivers being available only since March 2004.

It is a widely held belief that people's incomes in developing global regions are growing, and more families are electing to dine out, motivating new restaurants to open. McDonalds is a favorite choice for these new restaurant diners. Experts believe that there is a lag of about twelve quarters before increases in incomes are felt in the fast food restaurant industry.

Yum Foods competes with McDonalds, though McDonalds is substantially larger, both in terms of revenues and number of restaurants. Following increases in Yum Foods revenues, executives have observed slower growth in McDonalds revenues, about 6 quarters later.

The global fast food restaurant business is growing. As new restaurant diners enter the market, new restaurants appear, and advertising and promotion levels increase. Following expansion in the industry, McDonalds benefits later, from this growth, and this delayed benefit is thought to occur 8 or 9 quarters later.

Executives are convinced that demographics are key drivers of revenues. There is particular interest in learning whether the youngest and oldest population segments are particularly strong drivers. Both global population segments are growing. Young children are thought to first become fast food restaurant diners between 18 and 24 months of age (e.g., a 6 to 8 quarter delay). The over 65, retired segment is thought to alter fast food dining habits about 18 to 24 months after retirement (e.g. a 6 to 8 quarter delay).

Build a valid model of quarterly revenues, identifying drivers, and forecasting revenues in the next seven quarters.

I. Choose Candidate Predictor Lags

1. Plot McDonalds revenues by quarter. What do you see?

___ increasing trend ___ decreasing trend ___ seasonality ___other:___________

2. Identify the potential driver lags which you could use to produce a forecast of at least *seven* quarters:

    a. Plot Yum revenues by quarter. What do you see?

    Yum revenues: ___ increasing trend ___ decreasing trend ____ seasonality

    Yum revenues are seasonal. Choose a lag for Yum revenues which aligns Yum seasonality with McDonalds seasonality. Set up this lag: ___ quarters

b. Plot global populations under 15 and 65+ , global fast food market and global GDP per capita by quarter. What do you see?

Global population under 15: ___ increasing trend ___ decreasing trend

____ seasonality

Global population 65+: ___ increasing trend ___ decreasing trend

____ seasonality

Global fast food market: ___ increasing trend ___ decreasing trend

____ seasonality

Global GDP per capita: ___ increasing trend ___ decreasing trend

____ seasonality

For each of these four potential drivers, set up lags which allow a forecast of at least *seven* quarters. For population variables, choose lags which are multiples of 4 to correspond to years:

Global population under 15: ____ q lag Global population 65+: ____ q lag

Global fast food market: ____ to ____ q lags

Global GDP per capita: ____ to ____ q lags

II. Specify Expected directions of influence

Specify the direction of influence which you expect for each of the potential drivers. You can assume that more people ought to be related to higher revenues. Higher GDP per capita allows families in developing countries to afford to dine out at fast food restaurants; higher GDP per capita allows families in developed countries to choose expensive restaurants. Yum is a competitor, which may divert diners from McDonalds; Yum revenues also reflect a similar fast food business, and their promotions probably benefit McDonalds.

Global population under 15: ___+ ___− or ___?

Global population 65+: ___+ ___− or ___?

Global GDP per capita: ___ + ___ − or ___?

Global fast food market: ___ + ___ − or ___?

Yum revenues: ___ + ___ − or ___ ?

III. Find driver(s) with matching pattern

McDonalds revenues exhibit two quarters of decrease, followed by an increase in the three most recent quarters (ignoring the two hidden quarters). Which candidate lags provide a matching pattern in the two of the three most recent quarters?

___Global population under 15 lag ___Global population 65+ lag

___ Yum revenues lag ___ Global fast food lag ___Global GDP per capita ___ q lag

IV. Build your model

A. Simple regression model

Choose the candidate lag which shows matching pattern in the two out of three most recent quarters (ignoring the two hidden quarters) and which has the strongest correlation (of correct sign) with McDonalds revenues, and then build a simple regression model.

1. Is the predictor significant? Y or N Is the coefficient sign as expected? Y or N
2. Plot the residuals to see patterns in unaccounted for variation in McDonalds revenues. What do you see?
   ___ increasing trend ___decreasing trend ___seasonality ___other:___________
3. Assess Durbin Watson to determine whether there is, or may be, unaccounted for trend or cycles in the residuals. Residuals

   ___contain ___may contain ___are free of unaccounted for trend or cycles.

B. Two driver model

Choose the candidate lag with the strongest correlation (of correct sign) with residuals to add to your model.

1. Are both predictors significant? Y or N
2. Are both coefficient signs as expected? Y or N

3. How much did the second predictor improve explanatory power (RSquare)? _____

4. How much did the second predictor reduce the margin of error in forecasts? ___($M)

5. Plot the residuals to see patterns in unaccounted for variation in McDonalds revenues. What do you see?

   ___ increasing trend ___decreasing trend ___seasonality __other:_____________

6. Assess Durbin Watson to determine whether there is, or may be, unaccounted for trend or cycles in the residuals. Residuals:

   ___contain ___may contain ___are free of unaccounted for trend or cycles.

C. Three driver model

The global recession may have affected McDonalds revenues. Add a recession indicator, equal to 1 in quarters December 2008 through March 2011, 0 elsewhere. (These are the quarters where residuals are negative.) Build a three variable model, adding the indicator.

1. Are all predictors significant? Y or N Are all coefficient signs as expected? Y or N

2. How much did the indicator improve explanatory power (RSquare)? _____

3. How much did the indicator reduce the margin of error in forecasts? _____($M)

4. Plot the residuals to see patterns in unaccounted for variation in McDonalds revenues. What do you see?

   ___ increasing trend ___decreasing trend ___seasonality __other:_____________

5. Assess Durbin Watson to determine whether there is, or may be, unaccounted for trend or cycles in the residuals. Residuals:

   ___contain ___may contain ___are free of unaccounted for trend or cycles.

D. Four driver model

Choose the candidate lag with the strongest correlation (or correct sign) with residuals to add to your model.

1. Are all predictors significant? Y or N Are all coefficient signs as expected? Y or N

2. How much did the indicator improve explanatory power (RSquare)? _____

3. How much did the indicator reduce the margin of error in forecasts? _____($M)

4. Plot the residuals to see patterns in unaccounted for variation in McDonalds revenues. What do you see?

   ___ increasing trend ___decreasing trend ___seasonality __other:_____________

5. Assess Durbin Watson to determine whether there is, or may be, unaccounted for trend or cycles in the residuals. Residuals:

   ___contain ___may contain ___are free of unaccounted for trend or cycles.

V. Validate your model

1. Find the margin of error in your model forecasts: ________($M)

2. Find predicted revenues and the lower and upper 95% prediction interval bounds through September 2013. Do the prediction intervals contain the two most recent, hidden, revenues? Y or N

VI. Recalibrate and Present your Model

1. Recalibrate your model and update your equation, substituting
   (i) your model coefficients for b0, b1, b2, b3 and b4
   (ii) units which predictors are measured in for u1, u2, and u3
   (iii) significance levels, "a" for 95% confidence, "b" for 99% confidence, for s1 through s5,
   (iv) the variable names for I, X1, X2, and X3,
   (v) the lags which you used, L1, L2, and L3.

$$McDo\tilde{n}alds\ rev(\$M)_q = b0(\$M) + b1(\$M)^{s1} \times I_q + b2^{s2}(\$M/u1) \times X1(u1)_{q-L1}$$

$$+b3^{s3}(\$M/u2) \times X2(u2)_{q-L2} + b4^{s4}(\frac{\$M}{u3}) \times X3(u3)_{q-L3}$$

   RSquare: _____^{s5}

   [a]Significant at .01. [b]Significant at .05.

2. Plot your fit and forecast, showing McDonalds revenues through December 2011 and the lower and upper prediction interval bounds through September 2013. Add a title that describes a conclusion which viewers can see.

3. What is the margin of error in your forecasts? _____($M)

4. What is your forecast for September 2013: _____($M) to _____($M)

5. Which variables drive McDonalds revenues?

___past Global per capita GDP ___past Global fast food market

___past Yum revenues ___past global population under 15

___past global population 65+ ___past global recession

V. Sensitivity Analysis

1. Produce a scatterplot showing the cumulative Part Worths that drivers contribute to predicted McDonalds revenue for quarters from March 2006 through September 2013
   Adjust the y (vertical) axis to make good use of space.
   Adjust the x (horizontal) axis to show quarters.

   Add a chart title which summarizes your conclusion regarding the most influential driver.

2. Which driver is most influential? ________________

Case 12.1 Chipotle Quarterly Revenues Model and Forecast

Nearing the end of 2014, Chipotle executives have been pleased with quarterly revenues. While the recession reduced business noticeably from the fourth quarter of 2008 through 2009, revenues recovered in 2010. Management believes that it will be important to continue opening new locations in order to drive revenue growth, and new locations have been added at a steady rate. The impact of new locations is thought to impact revenues about one year later, after new locations attract regular customers.

Executives believe that much of the Chipotle success hinges on appeal to both young students, ages 15 to 24, and to young professional customers, ages 25 to 34, who are believed to value local, organic food. Recently, in June 2014, Chipotle announced that their restaurants would serve only food free of GMOs, a product enhancement that management believes will particularly appeal to both student and young professional segments.

Promotion of local, organic food by Whole Foods (a complementary food business which provides local, organic food) is thought to have a positive impact on Chipotle's business about one year later. While both businesses are seasonal, Chipotle's business is typically greatest in the second quarter each year, and WFM's business tends to be greatest in the first quarter of each year.

Overall prosperity may also drive revenues, making U.S. GDP a potentially powerful driver, particularly for Chipotle's upscale chain. Increasing economic prosperity could be felt as soon as one year later, though the delayed impact could be as long as two years later.

Chipotle quarterly revenues contains a time series of data with Chipotle quarterly revenues, population ages 15 to 24, population ages 25 to 34, U.S. GDP, Whole Foods revenues, and the number of Chipotle locations, by quarter. Build a model to explain variation in Chipotle quarterly revenues and to forecast revenues in the next four quarters, through December 2015.

I. Plot the data

1. Plot Chipotle revenues and each of the potential drivers. What do you see?

Chipotle revenues show: ___ trend ___ seasonality ___ cycle ___shock

Chipotle locations show: ___ trend ___ seasonality ___ cycle ___shock

GDP shows: ___ trend ___ seasonality ___ cycle ___shock

Whole Foods revenues show: ___ trend ___ seasonality ___ cycle ___shock

Population 15 to 24 shows: ___ trend ___ seasonality ___ cycle ___shock

Population 25 to 34 shows: ___ trend ___ seasonality ___ cycle ___shock

II. Specify Expected directions of influence

Specify the direction of influence which you expect for each of the potential drivers. You can assume that more people ought to be related to higher revenues. Higher GDP allows families to choose more upscale restaurants. Whole Foods business is complementary, reflecting the value of local and organic produce, similar to the Chipotle business model.

Chipotle locations: ___ + ___ − or ___? GDP: ___+ ___− or ___?

Whole Foods revenues: ___ + ___− or ___? Population 15 to 24: ___ + ___− or ___?

Population 25 to 34: ___ + ___ − or ___ ?

III. Choose the first driver

Make driver lags that will enable a forecast of at least four quarters. For populations in age groups, choose (a) multiple(s) of four. For Whole Foods lags, choose lags that align Whole Foods seasonality with Chipotle seasonality.

Mark the declines from past quarter in the four most recent quarters (excluding the two hold out quarters).

1. Which potential driver lags match the pattern in Chipotle revenues in the four most recent quarters (excluding the two holdout quarters)?

____ Chipotle locations q-__ ___ GDP q-__ ___Whole Foods revenues q-__

___ Population 15 to 24 q-__ ___ Population 25 to 34 q-__

2. Of the potential drivers that have a matching pattern, which is most highly correlated with Chipotle revenues?

____ Chipotle locations q-__ ___ GDP q-__ ___Whole Foods revenues q-__

___ Population 15 to 24 q-__ ___ Population 25 to 34 q-__

Select one driver and run regression.

IV. Assess the one driver model

1. Is your model free from positive autocorrelation (due to trend, cycles, or shifts)?

   ___ yes __ maybe __ no. Explain how you know, citing the appropriate statistic: __

2. Describe the power of your model in a single sentence: ____
3. Describe the precision of your model in a single sentence: ____

Plot the residuals.

4. What do you see? ___ trend ___ seasonality ___ cycle ___shock

V. Choose a second driver and assess two driver model

Mark the pattern in residuals in the four most recent quarters.

1. Which potential driver matches at least three out of four of the most recent quarters?

____ Chipotle locations q-__ ___ GDP q-__ ___Whole Foods revenues q-__

___ Population 15 to 24 q-__ ___ Population 25 to 34 q-__

Add a second driver.

2. Is your model free from positive autocorrelation (due to trend, cycles, or shifts)?

   ___ yes __ maybe __ no. Explain how you know, citing the appropriate statistic: __

3. Describe the power of your model in a single sentence: ____

4. Describe the precision of your model in a single sentence: ____

Plot the residuals.

5. What do you see? ___ trend ___ seasonality ___ cycle ___shock

VI. Choose a third driver and assess three driver model

Run correlations between the residuals and remaining potential drivers.

1. Which potential driver has the strongest correlation with residuals?

____ Chipotle locations q-__ ___ GDP q-__ ___Whole Foods revenues q-__

___ Population 15 to 24 q-__ ___ Population 25 to 34 q-__

Add a third driver.

2. Is your model free from positive autocorrelation (due to trend, cycles, or shifts)?

   ___ yes __ maybe __ no. Explain how you know, citing the appropriate statistic: __

3. Describe the power of your model in a single sentence: ____

4. Describe the precision of your model in a single sentence: ____

Plot the residuals.

5. What do you see? ___ trend ___ seasonality ___ cycle ___shock

VII. Add indicators and assess five driver model

Add an indicator of the recession, December 2008 through March 2009, and add an indicator of No GMOs, June 2014 through December 2015.

1. Is your model free from positive autocorrelation (due to trend, cycles, or shifts)?

   ___ yes __ maybe __ no. Explain how you know, citing the appropriate statistic: __

2. Describe the power of your model in a single sentence: ____

3. Describe the precision of your model in a single sentence: ____

Plot the residuals.

4. What do you see? ___ trend ___ seasonality ___ cycle ___shock

VIII. Validate your model

Does your model correctly forecast the two hidden quarters? ___Y ___N

IX. Recalibrate and present your model

1. Recalibrate your model and update your equation, substituting
   - (i) your model coefficients for b0, b1, b2, b3, b4 and b5
   - (ii) units which predictors are measured in for u1, u2 and u3,
   - (iii) significance levels, "a" for 95% confidence, "b" for 99% confidence, for s1 through s6,
   - (iv) the variable names for I1, I2, X1, X2, and X3,
   - (v) the lags which you used, L1, L2, and L3.

$$Chi\hat{p}otle\ rev(\$M)_q = b0(\$M) + b1(\$M)^{s1} \times I1_q + b2(\$M)^{s2} \times I2_q$$

$$+b3^{s3}\left(\frac{\$M}{u1}\right) \times X1(u1)_{q-L1} + b4^{s4}\left(\frac{\$M}{u2}\right) \times X2(u2)_{q-L2}$$

$$+ b5^{s5}(\frac{\$M}{u3}) \times X3(u3)_{q-L3}$$

RSquare: ______ ^{s6}

[a]Significant at .01. [b]Significant at .05.

2. Illustrate your model fit and forecast, adjusting the vertical axis, removing unnecessary decimals, and including vertical axis and chart titles, all shown in at least 12 pt font.

3. What is your forecast for December 2015? _____

4. Which are drivers of quarterly revenues? ___*population ages 15 to 24*

___*population ages 23 to 49* ___ *U.S. GDP* ___ *Whole Foods revenues*

___ *number of Chipotle locations*

X. Sensitivity Analysis

1. Illustrate the impact of each of the quarterly revenue drivers in your model with a graph of stacked part worths::

2. Which driver is most important? _____

Chapter 13
Nonlinear Multiple Regression Models

In this chapter, nonlinear transformations are introduced that expand linear regression options to include situations in which marginal responses are either increasing or decreasing, rather than constant. We will explore Tukey's Ladder of Powers to identify particular ways to rescale variables to produce valid models with superior fit. An example will be offered in the context of naïve models built for forecasting, and in Chapter 14, examples with explanatory multiple regression models will be added.

13.1 Consider a Nonlinear Model When Response Is Not Constant

To decide whether or not to use a nonlinear model, first rely on your logic:

- Do you expect the response, or change in the dependent, performance variable, to be constant, regardless of whether a change in an independent variable is at minimum values or at maximum values? Linear models assume constant response.
- Is the dependent variable limited or unlimited?

Linear models are unlimited. If your dependent variable couldn't be negative, because it is measured in dollars, purchases, people, or uses, a nonlinear model is logically more appropriate. After consulting your logic, plot your data.

13.2 Skewness Signals Nonlinear Response

When a dependent, performance variable y is skewed, performance response to differences or changes in drivers is nonlinear. Rescaling a skewed variable to roots or logarithms will linearize response, allowing use of multiple linear regression which will link the rescaled variables.

Increasing marginal response, in which differences or changes in x produce larger and larger responses in y, occurs if the dependent variable y is positively skewed, as in the upper plot in Figure 13.1.

C. Fraser, *Business Statistics for Competitive Advantage with Excel 2016*,
DOI 10.1007/978-3-319-32185-1_13

Figure 13.1 Nonconstant marginal response

Decreasing marginal response, in which differences or changes in *x* produce smaller and smaller responses in *y*, occurs if the dependent variable *y* is negatively skewed.

This possibility is shown in the lower plot in Figure 13.1.

To linearize response so that linear regression can be used, the goal is to rescale toward the center of Figure 13.1. Tukey offered a simple heuristic to quickly suggest ways to rescale variables when residuals from linear regression would be either skewed or heteroskedastic. A scale is chosen which reduces skewness of the dependent variable. A model built with a dependent variable which has been rescaled to reduce skewness will be nonlinear.

If a variable is positively skewed, as is the variable on the left in Figure 13.2, shrinking it by rescaling in roots, or natural logarithms will *Normalize*. Square roots are lower power, .5, than cube roots, .33, and are less radical. Natural logarithms make a bigger difference than square roots.

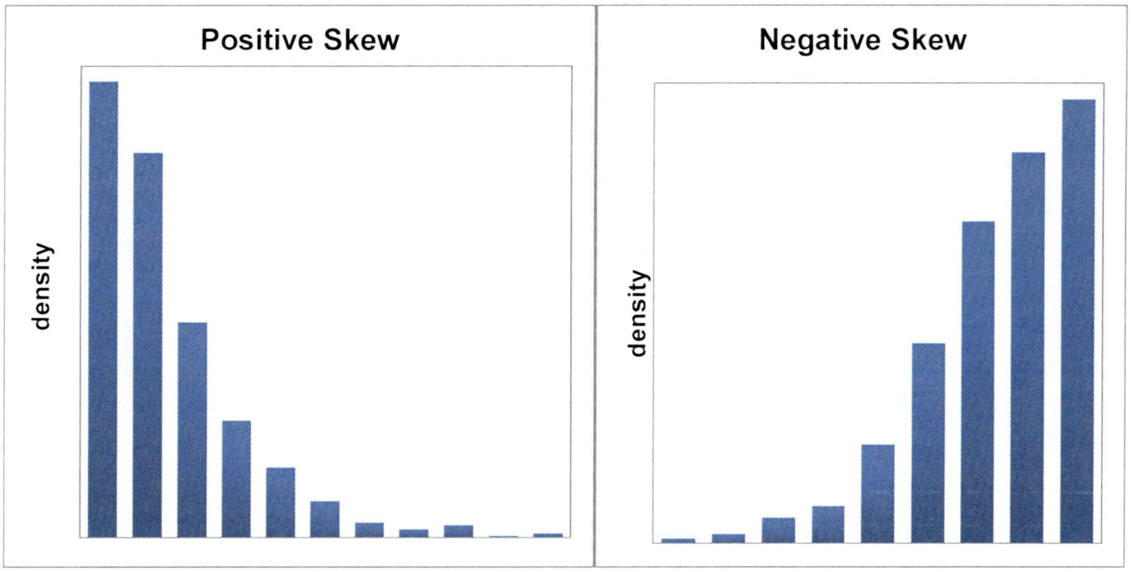

Figure 13.2 Positively and negatively skewed variables

When a variable is negatively skewed, as is the variable on the right in Figure 13.2, expanding it by rescaling to squares or cubes will *Normalize*. A higher power, such as cubes, will make a bigger difference.

Moving from the center up or down the *Ladder of Powers,* Figure 13.3, changing the power more, changes the data and its skewness more. More skewness calls for adjusting more.

The upper, blue plot in Figure 13.1 reflects a positively skewed dependent variable. Rescaling values of that dependent variable to their natural logarithms or roots will linearize the relationship, moving the plot down to Normal skewness in Figure 13.4.

The lower, gold plot in Figure 13.1 features a negatively skewed dependent variable. Rescaling dependent variable values to their squares or cubes will linearize the plot, moving it up to Normal skewness in Figure 13.4.

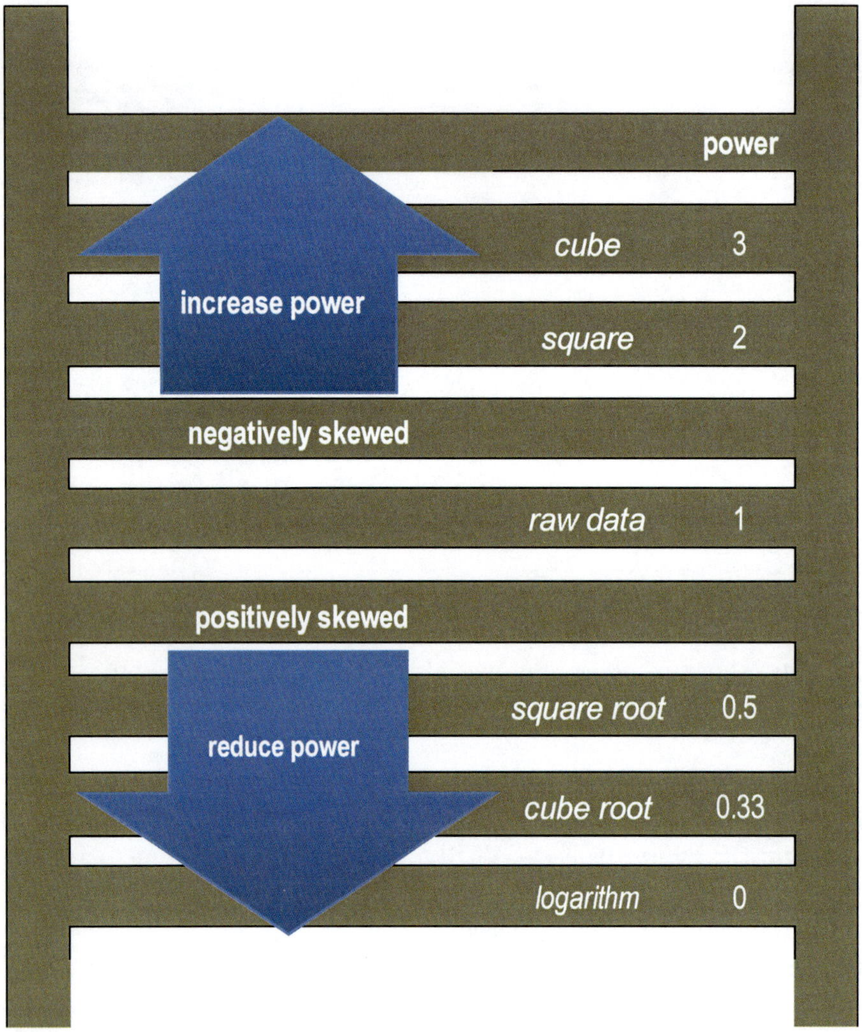

Figure 13.3 Tukey's Ladder of Powers

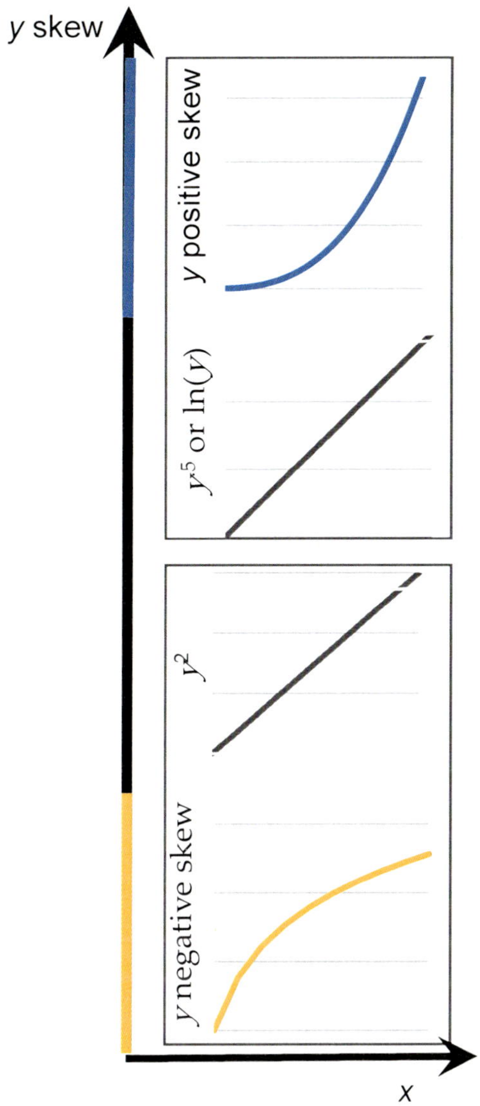

Figure 13.4 Rescaling to reduce skew linearizes

13.3 Rescaling y Builds in Interactions

Jointly, two drivers may make a larger difference than the sum of their individual influences. For example, advertising levels may be more effective when sales forces are larger. The impact of population growth in a country may influence imports more if growth in GDP has been relatively high. When the dependent variable is rescaled, the model becomes multiplicative, which produces interactions between predictors. With this potential benefit of improved fit and validity, comes the cost of transforming predictions in rescaled units back to the original units.

Example 13.1 LAN Airlines in 2011. LAN Airlines, a Chilean multinational, had achieved status as a major global carrier with innovative strategies. Two innovations distinguished LAN. With insightful route planning, the cargo and passenger businesses shared capacity. Passengers might be flown from Chile to Germany, where cargo was loaded for shipment to the United States. In the States, passengers boarded for flights to Chile, for example. After review of the success of low cost carriers, such as Ryan Air, LAN initiated a two tier fare system in 2007. International flights, which offered premium service, were sold at premium prices. Intra country flights were sold at discounted fares. The low cost intra country fares appealed to price conscious travelers and enabled purchase of more planes, including new, fuel efficient models. Critical to the success of the low cost program was passenger load, and the goal was to achieve passenger load of at least 75%.

Just as the low cost program was gaining ground, the global recession occurred, reducing air travel. It was not clear to LAN management whether the low cost program would continue to be effective in light of the altered world economy, or exactly the extent to which LAN business would suffer from the recession. To address these issues, forecasts of available capacity, passenger volumn, and load are desired.

Passenger load depended on

- available capacity, *ASKs,* available seat kilometers,
- the volume of passengers carried, *RPKs*, revenue passenger kilometers, and

Under the direction of Posie Holmes, the modeling team built naïve models of the two passenger load components using indicators for the *low cost program.* Model forecasts would be used integrated in a Monte Carlo simulation to forecast LAN passenger load in 2016.

After hiding the two most recent observations, Posie checked skewness of passenger *ASK* and *RPKs*. *ASKs* and *RPKs*, shown below in Figure 13.5, were mildly positively skewed, like the upper quadrant of Figure 13.1, reflecting increasing annual growth in both capacity and volume.

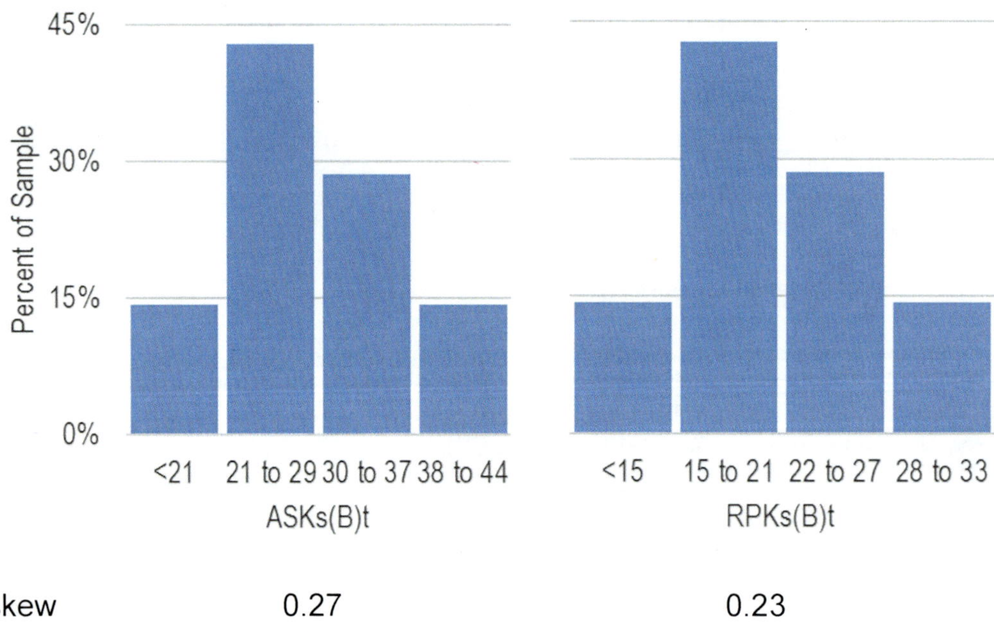

Figure 13.5 Skewness of passenger load components

The distributions of cube roots and natural logarithms of *ASKs* are shown in Figure 13.6. The natural logarithms of *ASKs* have skewness closer to zero, and would increase the chances that the naïve *ASKs* model would be valid for forecasting.

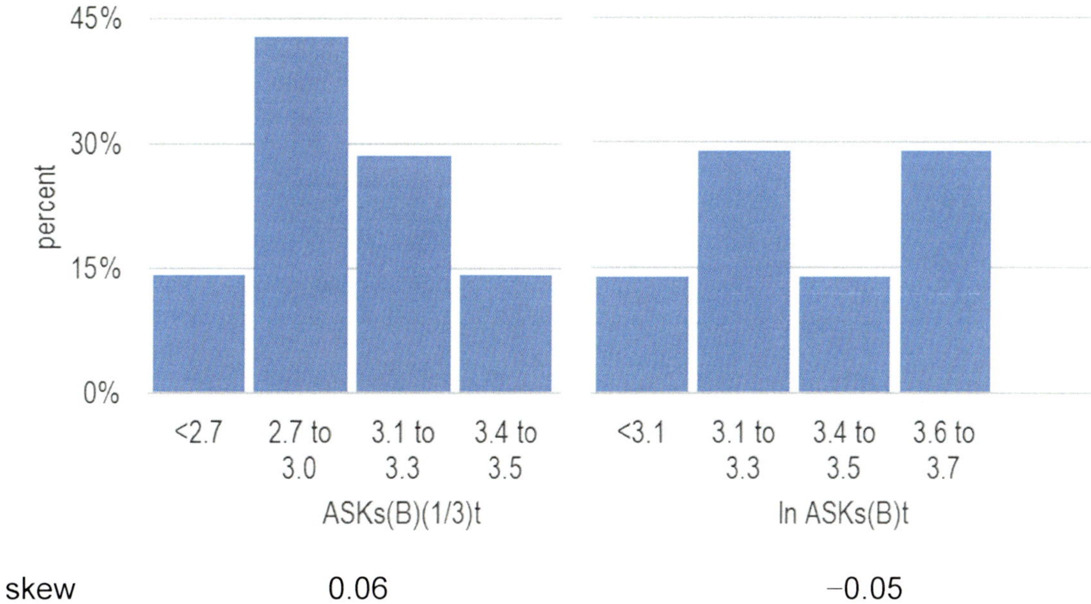

Figure 13.6 Skewness of ASKs rescaled to cube roots and natural logarithms

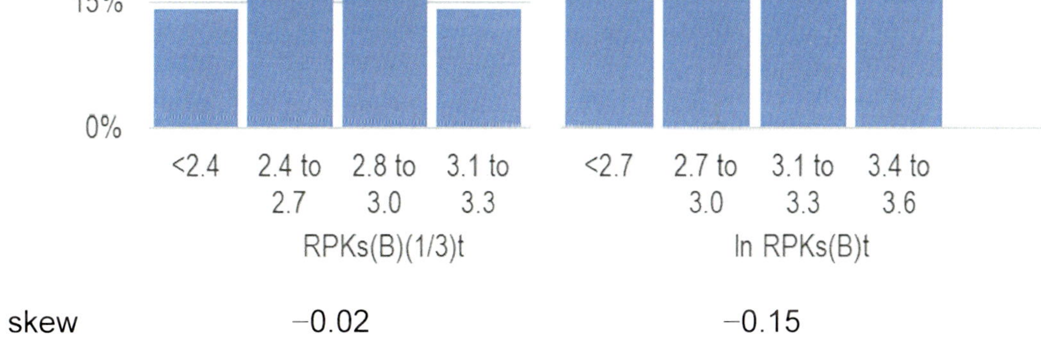

Figure 13.7 Skewness of RPKs rescaled to cube roots and natural logarithms

The cube roots and natural logarithms of RPKs are shown in Figure 13.7. The modeling team chose to use the cube roots of *RPKs* which have less skewness, closer to zero. Regression results are shown in Tables 13.1 and 13.2. For *Ln ASKs,* a short term low cost indicator improved DW, and reduced residuals in more recent years. For the cube roots of *RPKs*, a long term low cost indicator produced residuals free of positive autocorrelation which were relatively small in more recent years.

Table 13.1 Regression of ln ASKs

R Square	.999
Standard Error	.011
Observations	7

ANOVA

	df	*SS*	*MS*	*F*	*Significance F*
Regression	2	.452	.226	1886.2	.000
Residual	4	.0005	.0001		
Total	6	.453			

	Coefficients	*Standard Error*	*t Stat*	*p value*	*Lower 95%*	*Upper 95%*
Intercept	−243	4.7	−51.6	.0000	−257	−230
Year	.123	.002	52.3	.0000	.12	.13
Short term low cost	.037	.010	3.5	.02	.008	.07

DW	*T*	*k*	dL	dU
1.63	7	3	.47	1.90

Table 13.2 Regression of cube roots of RPKs

R Square	.999
Standard Error	.014
Observations	7

ANOVA

	df	*SS*	*MS*	*F*	*Significance F*
Regression	2	.477	.239	1263.4	.0000
Residual	4	.001	.0002		
Total	6	.478			

	Coefficients	*Standard Error*	*t Stat*	*p value*	*Lower 95%*	*Upper 95%*
Intercept	−232	10	−22.3	.0000	−261	−203
Year	.117	.005	22.5	.0000	.10	.131
low cost	.063	.021	3.0	.04	.004	.121

DW	*T*	*k*	*dL*	*dU*
1.97	7	3	.47	1.90

The models were significant, both the low cost indicator and trend were significant, and their influences were in the expected positive direction in both models.

The models made sense and accounted for nearly all of the annual variation in passenger load components. The modeling team moved on to validate.

Comparing lower and upper 95% prediction intervals for the two most recent years confirmed that both models had validity for forecasting. Actual *ASKs* and *RPKs* fell within the prediction intervals. The models were recalibrated using all available data:

$$\ln A\hat{S}Ks(B)_t = -240(lnB)^a + .047(lnB)^a \times Low\ Cost_t + .12(\frac{lnB}{year})^a \times t$$

RSquare: $.999^a$

$$R\hat{P}Ks(B)_t^{(\frac{1}{3})} = -240(B^{(\frac{1}{3})})^a + .053(B^{(\frac{1}{3})})^b \times Low\ Cost_t + .12(\frac{B^{(\frac{1}{3})}}{year})^a \times t$$

RSquare: $.998^a$

[a]Significant at .01 or better; [b]Significant at .05 or better.

The *ASKs* equation is in logarithms. To see the equation in the original scale of billion *ASKs*, use the exponential function to reverse the natural logarithms. With a model built from logarithms, the part worths are multiplicative:

$$e^{\ln A\hat{S}Ks(B)_t} = e^{-240(lnB)^a + .047(lnB)^b \times Low\ Cost_t + .11(\frac{lnB}{year})^a \times t}$$

$$A\hat{S}Ks(B)_t = e^{-240(lnB)^a + .047(lnB)^b \times Low\ Cost_t + .11(\frac{lnB}{year})^a \times t} \tag{13.1}$$

$$= 1.8\text{E}{-103}(\text{B}) \times e^{.047(lnB) \times Low\ Cost_t} \times e^{.11(\frac{lnB}{year})^a \times t}$$

$$= 1.8\text{E}{-103}(\text{B}) \times 1.05^{Low\ Cost_t} \times 1.13^t$$

The *RPKs* equation is in cube roots. To see the equation in the original scale of billion *RPKs*, cube both sides, rescaling back:

$$[R\hat{P}Ks(B)_t^{(\frac{1}{3})}]^3 = [-240(B^{(\frac{1}{3})})^a + .053(B^{(\frac{1}{3})})^b \times Low\ Cost_t + .12(\frac{B^{(\frac{1}{3})}}{year})^a \times t]^3$$

$$R\hat{P}Ks(B)_t = [-240(B^{(\frac{1}{3})})^a + .053(B^{(\frac{1}{3})})^b \times Low\ Cost_t + .12(\frac{B^{(\frac{1}{3})}}{year})^a \times t]^3 \tag{13.2}$$

The fits and forecasts are shown in Figure 13.8.

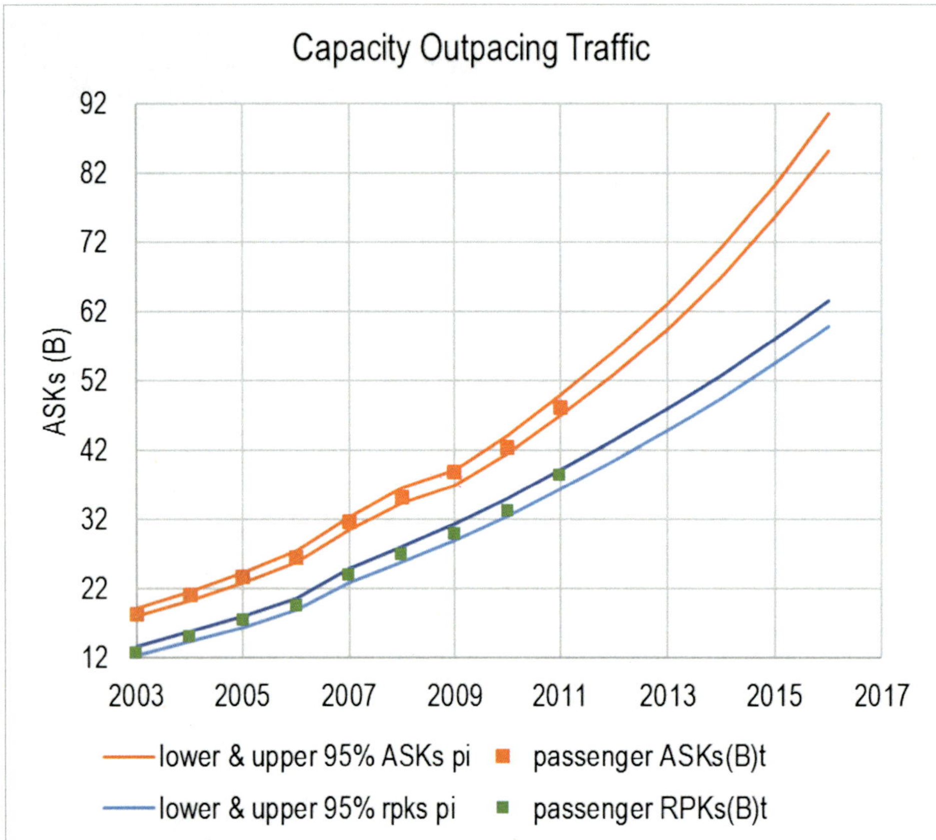

Figure 13.8 Fits and forecasts for ASKs and RPKs

13.4 The Margin of Error Is Not Constant with a Nonlinear Model

With a nonlinear model, the margin of error is not constant. The margins of error, half the distance between the lower and upper 95% prediction interval bounds, increase as the forecast date moves further from the present. The 2016 LAN forecast for passenger capacity can be expected to fall no further than 2.7 billion ASKs (= (90.6B − 85.2B) / 2) from actual, and the 2016 LAN forecast for passenger volume can be expected to fall no further than 1.9 billion RPKs (= (63.7 − 59.8) / 2) from actual.

13.5 Sensitivity Analysis Enables Scenario Comparisons

When a dependent variable is rescaled to build a nonlinear model, the model is multiplicative. The impact of each of the drivers depends on values of all of the other drivers. Expanding the right sides of the equations, the interactions are apparent. (Don't let this expansion scare you! The models would be presented in the forms shown in (13.1) and (13.2).)

The capacity equation features an exponential function on the right side, which can be written as the product of the three terms, the multiplicative part worths:

$$A\hat{S}Ks(B)_t = e^{-240^a + .047^b \times Low\ Cost_t + .12^a \times t}$$

$$= 1.8\text{E}{-}103(\text{B}) \times 1.05^{Low\ Cost_t} \times 1.13^t$$

For years before the Low Cost program was initiated, where the *Low Cost* indicator is set to zero, the equation becomes:

$$A\hat{S}Ks(B)_t = 1.8\text{E}{-}103 \times e^{.047 \times 0} \times e^{.12 \times t}$$

$$= 1.8\text{E}{-}103 \times 1.05^0 \times 1.13^t$$

$$= 1.8\text{E}{-}103 \times 1 \times 1.13^t$$

$$= 1.8\text{E}{-}103 \times 1.13^t$$

In years after the Low Cost program was initiated, where the *Low Cost* indicator is set to one, the equation becomes:

$$A\hat{S}Ks(B)_t = 1.8\text{E}{-}103 \times e^{.047 \times 1} \times e^{.12 \times t}$$

$$= 1.8\text{E}{-}103 \times 1.05^1 \quad \times 1.13^t$$

$$= 1.8\text{E}{-}103 \times 1.05 \quad \times 1.13^t e^{.12 \times t}$$

$$= 2.0\text{E}{-}103 \times 1.05 \ \times 1.13^{t^{.12 \times t}}$$

The *low cost* program, with a coefficient estimate of 1.05 multiplies capacity by 105%. While this multiplier is constant, the low cost impact is greater in years where capacity is greater.

Evaluating the impact of the Low Cost program can be accomplished by comparing capacity with and without the program, the distance between green and blue lines in Figure 13.9. In 2007 and 2008, the low cost program boosted capacity by 5%, 1.4B and 1.6B. In 2016, the Low Cost program is forecast to motivate no additional capacity.

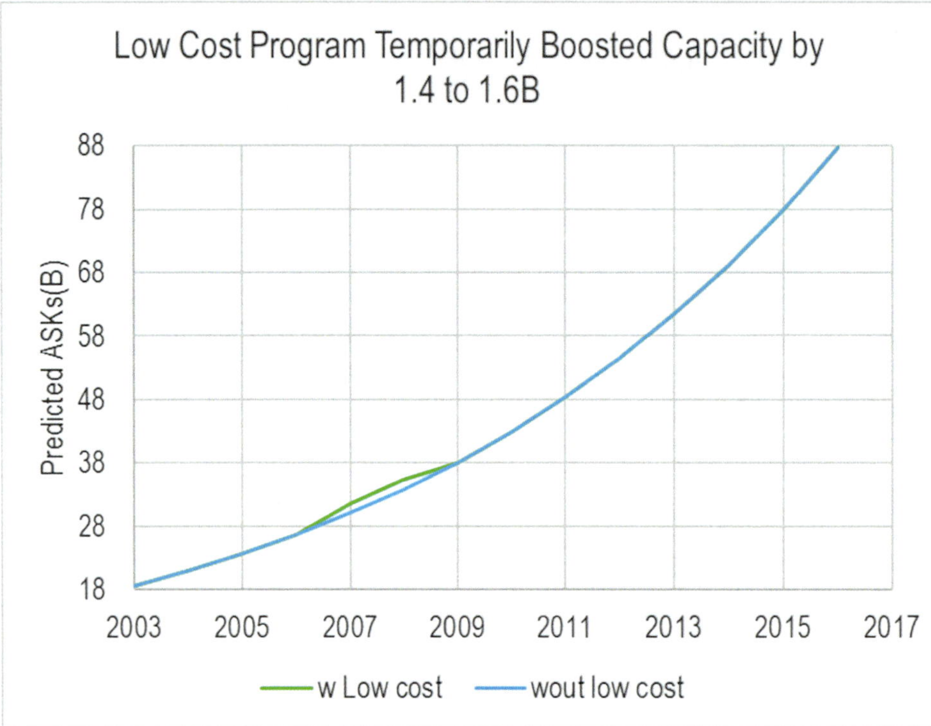

Figure 13.9 Impact of the low cost program on annual trend

The right side of the traffic equation features cubes. Expanding those:

$$R\hat{P}Ks(B)_t = [-238 + .053 \times Low\ Cost_t + .12 \times t]^3$$

$$= (-238 + .053 \times Low\ Cost_t + .12 \times t)$$

$$\times (-238 + .053 \times Low\ Cost_t + .12 \times t)$$

$$\times (-238 + .053 \times Low\ Cost_t + .12 \times t)$$

$$= -238^3$$

$$+3 \times (-238^2 \times .053 \times Low\ Cost_t - 238 \times .053^2 \times Low\ Cost_t^2)$$

$$+.053^3 \times Low\ Cost_t^3$$

$$+3 \times (-238^2 \times .12 \times t - 238 \times .12^2 \times t^2\ ) +.12^3 \times t^3$$

$$+3 \times (.053^2 \times .12 \times Low\ Cost_t^2 \times t + .053 \times .12^2 \times Low\ Cost_t \times t^2)$$

$$+ 6 \times -238 \times .053 \times .12 \times Low\ Cost_t \times t$$

The Low Cost program positively influences traffic, which exhibits a positive annual trend. The annual trend is greater in years following initiation of the Low Cost program.

The impact of this interaction can be seen by splitting the regression equation into two pieces:

- the baseline trend *t*:

$$R\hat{P}Ks(B)_t\ baseline\ trend = b_0^3$$
$$+3 \times (b_0^2 \times b_t \times t + b_0 \times b_t^2 \times t^2)$$
$$+ b_t^3 \times t^3$$
$$= -238^3$$
$$+3 \times (-238^2 \times .12 \times t - 238 \times .12^2 \times t^2)$$
$$+.12^3 \times t^3$$

- Terms with *Low Cost:*

$$R\hat{P}Ks(B)_t\ due\ \ to\ Low\ Cost = 3 \times (b_0^2 \times b_{lc} \times low\ cost_t + b_0 \times b_{lc}^2 \times low\ cost_t^2$$
$$+ b_t^2 \times b_{lc} \times t^2 \times low\ cost_t$$
$$+ b_t \times b_{lc}^2 \times t \times low\ cost_t^2)$$
$$+ 6 \times\ b_0 \times b_t \times b_{lc} \times t \times low\ cost_t$$
$$+ b_{lc}^3 \times low\ cost_t^3$$
$$= 3 \times (-238^2 \times .053 \times Low\ Cost_t$$
$$-238 \times .053^2 \times Low\ Cost_t^2$$
$$+ .053^2 \times .12 \times Low\ Cost_t^2 \times t$$
$$+.053 \times .12^2 \times Low\ Cost_t \times t^2)$$
$$+ 6 \times -238 \times .053 \times .12 \times Low\ Cost_t \times t$$
$$+ .053^3 \times Low\ Cost_t^3$$

Passenger volumes with and without the *Low Cost* program components are shown in Figure 13.10.

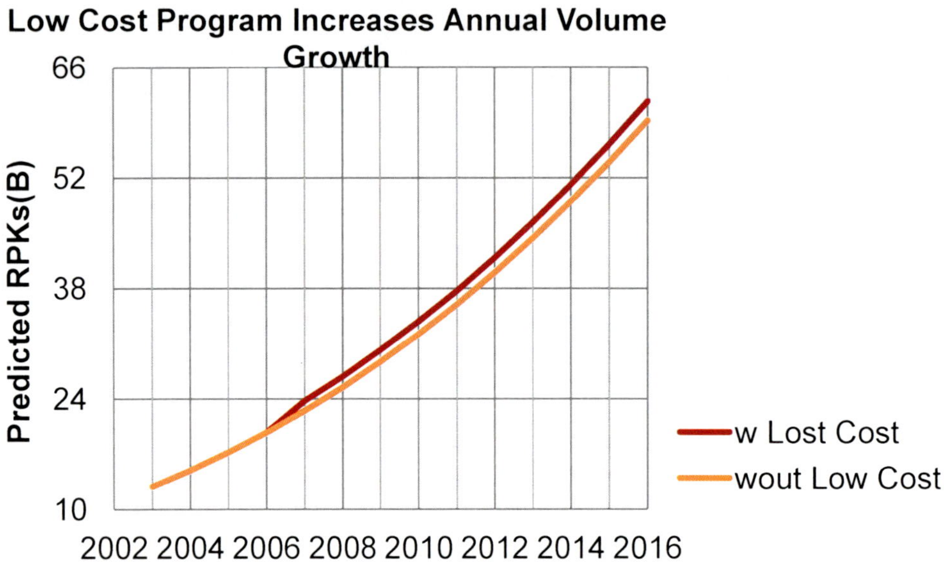

Figure 13.10 Volume trends with and without low cost program

Comparing 2016 forecasts with and without the Low Cost program, the distance between red and orange lines in Figure 13.10, the impact on passenger volume of the program is expected to add 2.4 billion RPKs.

Key to the success of LAN's profitable strategy was achieving at least 75% passenger load. If planes flew at less than 75% capacity, potential profits were not realized. The low cost program was spurring growth in both passenger traffic. Management needed forecasts of passenger load, as well. Passenger load could be as low as the ratio of the lower 95% prediction interval bound for volume to the higher 95% prediction interval bound for capacity, the "worst" case. The "best" case would be the ratio of the upper 95% prediction interval bound for volume to the lower 95% prediction interval bound for capacity. Table 13.3 compares these extreme possibilities, which are illustrated in Figure 13.11.

Table 13.3 "Best" and "worst" case passenger load forecasts

	RPKs (B)			*ASKs (B)*			*Load %*		
year	actual	lower 95%	upper 95%	actual	lower 95%	upper 95%	actual	“"worst”	“"best”
2003	12.7	12.2	13.6	18.3	18.0	19.1	69%	64%	75%
2004	15.1	14.2	15.7	21.1	20.3	21.6	72%	66%	77%
2005	17.5	16.4	18.1	23.7	22.9	24.3	74%	68%	79%
2006	19.5	18.8	20.7	26.4	25.8	27.4	74%	69%	80%
2007	24.0	22.7	24.8	31.6	30.4	32.4	76%	70%	81%
2008	27.0	25.8	28.0	35.2	34.3	36.5	77%	71%	82%
2009	29.8	29.0	31.4	38.8	36.9	39.2	77%	74%	85%
2010	33.1	32.5	35.1	42.4	41.6	44.2	78%	74%	84%
2011	38.4	36.3	39.1	48.2	46.9	49.8	80%	73%	83%
2012		40.4	43.4		52.8	56.1		72%	82%
2013		44.8	48.0		59.5	63.3		71%	81%
2014		49.5	52.9		67.1	71.3		69%	79%
2015		54.5	58.1		75.6	80.4		68%	77%
2016		59.8	63.7		85.2	90.6		66%	75%

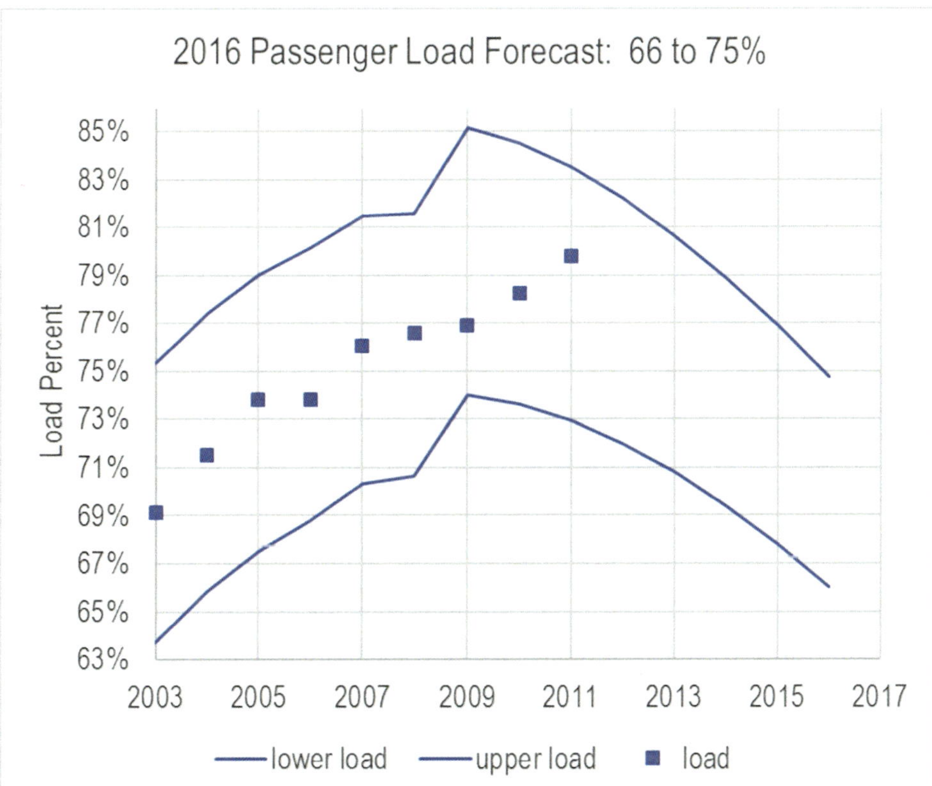

Figure 13.11 "Best" and "worst" passenger load forecasts

With the margin of error in load forecasts more than 4% (= (75% − 66%) / 2= 4.4%), executives could not effectively evaluate the passenger business. Both extremes were the ratio of extreme outcomes for both volume and capacity. The "best" and "worst" case load percentages would occur with less than one tenth of one percent chance (= 2.5% × 2.5% = .06%), making the "best" to "worst" load prediction interval a 99.9% confidence interval. A 95% confidence interval was needed, instead, which the modeling team could supply with Monte Carlo simulation using the regression model standard errors and predictions as assumptions for expected volume and capacity in 2016.

13.6 Nonlinear Models Inform Monte Carlo Simulation

To identify the 95% prediction interval for passenger load percentage in 2016, the modeling team set up a spreadsheet linking passenger capacity and volume to passenger load. Random samples of possible values for 2016 capacity and volume were generated, using the 2016 forecasts as most likely values and the regression standard errors as measures of dispersion. Since both capacity and volume were positively skewed, a sample of natural logarithms of 2016 ASKs was drawn, and a sample of cubed roots of 2016 RPKs was generated. Those were then rescaled to ASKs and RPKs in billions to find the sample of possible values for passenger load in 2016. The distribution of possible values in 2016 is shown in Figure 13.12, with 95% confidence interval 67 to 72%. There was a 95% chance that passenger load would be less than the 75% goal in 2016 if capacity and volume following existing trends. The margin of error in the 2016 forecast is half the 95% confidence interval: 2.5%.

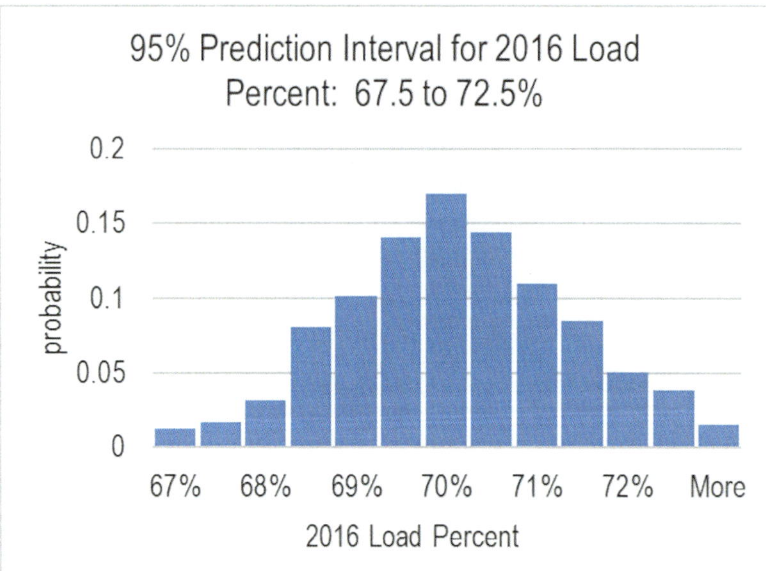

Figure 13.12 95% confidence interval for 2016 passenger load proportion

13.7 Gains from Nonlinear Rescaling Are Significant

To see the gain from building nonlinear models, compare results with those from simpler linear models. The naïve linear models of passenger capacity and volume, using the same sample, excluding the two most recent observations, are below and in Figure 13.13. Note that the low cost indicator was not significant in the capacity model of ASKs, and so it was removed.

$$A\hat{S}Ks(B) = -6940^{a} + 3.47^{a} \times t$$

$RSquare$: .989[a]

$$R\hat{P}Ks(B) = -4920^{a} + 2.11^{b} \times low\ cost + 2.46^{a} \times t$$

$RSquare$: .997[a]

[a]Significant at .01 or better; [b]significant at .05 or better.

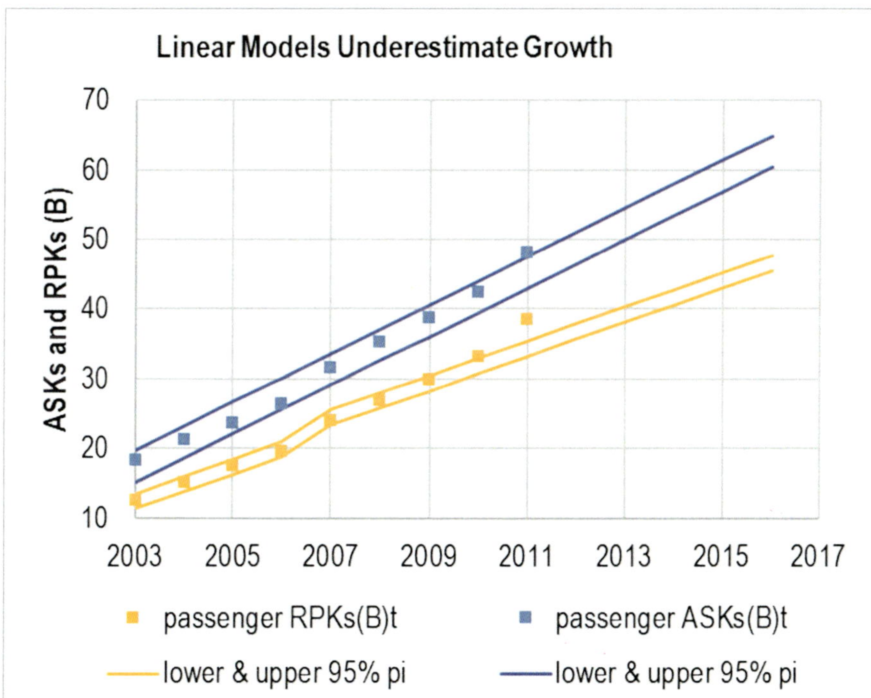

Figure 13.13 Linear model fits and forecasts are not valid

Neither linear model is valid. Both ASKs and RPKs are underestimated and neither model can be reliably used to forecast. Passenger capacity and volume is growing by increasingly larger increments annually, particularly since the Low Cost program was initiated. The linear models assume constant rates of growth, which do not fit historic data. In addition, the linear models ignore the interactions between the Low Cost program and annual trends, which have accelerated with the Low Cost program in place.

13.8 Nonlinear Models Offer the Promise of Better Fit and Better Behavior

It is a challenge to think of an example of truly linear (constant) response. Responses tend to be nonconstant and nonlinear. The fifth dip of ice cream is less appetizing than the first. Consumers become satiated at some point, and beyond that point, additional consumption is less valuable. Adding the twentieth stock to a portfolio makes less difference to diversification than adding the third. A second ad insertion in a magazine enhances recall more than a tenth ad insertion. As a consequence of nonconstant, changing marginal response, nonlinear models promise superior fit and better behaved models, with valid forecasts. Nonlinear models which feature a rescaled dependent variable incorporate interactions between drivers, adding another realistic and useful aspect. Nonlinear models do carry the cost of transformation to and back from logarithms, roots, or squares. In some cases, a linear model fits data quite well and is a reasonable approximation. Thinking logically about the response that you've set to explain and predict, and then looking at the distribution and skewness of your data and your residuals, will sometimes lead you toward the choice of a nonlinear alternative.

Skewness signals nonlinear response. Tukey's Ladder of Powers can help quickly determine the particular nonlinear model which will fit a dataset best. When a variable is positively skewed, rescaling to roots or natural logarithms reduces the positive skew. Negatively skewed variables are Normalized by squaring or cubing. The amount of difference corresponds to the power—square roots with power .5 are less radical than logarithms with power 0 and squares with power 2 are less extreme than cubes with power 3.

Excel 13.1 Rescale to Build and Fit Nonlinear Regression Models with Linear Regression

Passenger Load at LAN Airlines. Passenger load is a key measure of efficiency which captures the percent of capacity filled with passenger volume. A 2016 forecast of passenger load is needed to evaluate the Low Cost program, first introduced in 2007. Build naïve models to forecast passenger capacity, *ASKs(B)*, and passenger volume, *RPKs(B)*.

Historical annual data in **Excel 13 LAN Passenger Load** contains annual observations on *ASKs(B)* and *RPKs(B)*.

In order to validate the models for forecasting, hide the two most recent observations with data from 2010 and 2011.

Assess skewness and choose scales. Assess *skewness* of *ASKs(B)*and *RPKs(B),* using the function

=**SKEW(***array***)**.

	D	E	F
1	lty	*passenger* *ASKs(B)* $_t$	*passenger* *RPKs(B)* $_t$
14	2015		
15	2016		
16	skew	0.266	0.227

Add six columns of rescaled variables which are roots and natural logarithms of *ASKs* and *RPKs*, using

=*cell*^**.5**

=*cell*^**(1/3)**

=**LN(***cell***)**

G2 =SQRT(E2)

	G	H	I	J	K	L
1	sqrt asks (B) t	cubrt asks(B) t	ln asks (B) t	sqrt rpks (B) t	cubrt rpks (B) t	ln rpks (B) t
2	4.28	2.64	2.91	3.56	2.33	2.54
3	4.60	2.77	3.05	3.89	2.47	2.72
4	4.87	2.87	3.16	4.18	2.60	2.86

Find the skewness of the six rescaled variables:

	G	H	I	J	K	L
1	sqrt asks (B) t	cubrt asks(B) t	ln asks (B) t	sqrt rpks (B) t	cubrt rpks (B) t	ln rpks (B) t
15						
16	0.110	0.056	-0.054	0.046	-0.016	-0.145

Use the scales that produce skewness closest to zero, *ln ASKs*, and $RPKs^{(1/3)}$ and run regression with year *t*.

ASK model using natural logarithms

	A	B	C	D	E	F	G
1	SUMMARY OUTPUT						
2							
3	*Regression Statistics*						
4	Multiple R	0.99783					
5	R Square	0.99566					
6	Adjusted F	0.99479					
7	Standard E	0.01983					
8	Observatic	7					
9							
10	ANOVA						
11		*df*	*SS*	*MS*	*F*	*gnificance F*	
12	Regressior	1	0.45095	0.45095	1147.09	4.2E-07	
13	Residual	5	0.00197	0.00039			
14	Total	6	0.45292				
15							
16		*Coefficients*	*andard Err*	*t Stat*	*P-value*	*Lower 95%*	*Upper 95%*
17	Intercept	-251.281	7.51655	-33.4303	4.5E-07	-270.603	-231.959
18	t	0.12691	0.00375	33.8688	4.2E-07	0.11728	0.13654

Assess Durbin Watson, plot residuals, and set the major vertical units to the standard error.

	B	C	D
24	*Predicted ln asks (B) t*	*Residuals*	*DW*
25	2.91471	-0.0065	2.36387
26	3.04161	0.0099	
27	3.16852	-0.00358	
28	3.29543	-0.02206	
29	3.42233	0.02943	
30	3.54924	0.01113	
31	3.67615	-0.01832	

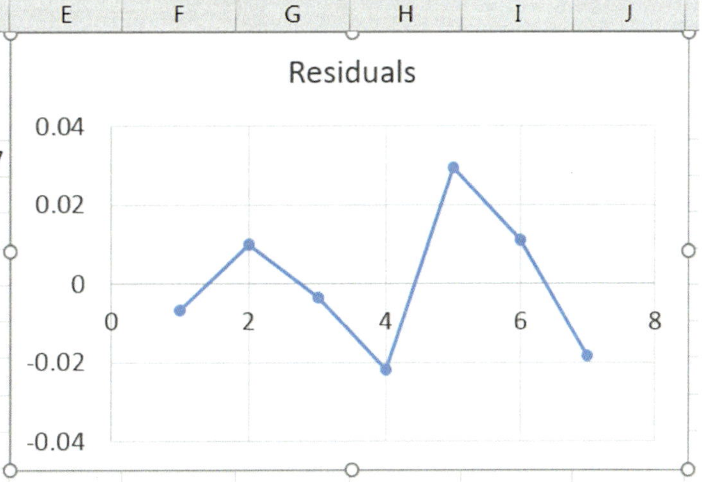

A shift, presumably due to the Low Cost program, is apparent in the *ln asks* residuals, beginning in the year 5, 2007.

Return to the data sheet and add a new column B, next to the year *t* for a *low cost* indicator, equal to 1 in years 2007 onward.

	A	B
1	*t*	*low cost t*
4	2005	0
5	2006	0
6	2007	1
7	2008	1
8	2009	1

Run regression including the *low cost* indicator, assess Durbin Watson, and plot residuals.

	A	B	C	D	E	F	G
1	SUMMARY OUTPUT						
2							
3	*Regression Statistics*						
4	Multiple R	0.9991					
5	R Square	0.99821					
6	Adjusted F	0.99731					
7	Standard E	0.01424					
8	Observatic	7					
9							
10	ANOVA						
11		*df*	*SS*	*MS*	*F*	*gnificance F*	
12	Regressior	2	0.45211	0.22605	1114.23	3.2E-06	
13	Residual	4	0.00081	0.0002			
14	Total	6	0.45292				
15							
16		*Coefficients*	*andard Err*	*t Stat*	*P-value*	*Lower 95%*	*Upper 95%*
17	Intercept	-228.996	10.7913	-21.2204	2.9E-05	-258.958	-199.034
18	t	0.11579	0.00538	21.5076	2.8E-05	0.10084	0.13073
19	low cost t	0.05189	0.02176	2.38511	0.07557	-0.00851	0.1123

	A	B	C	D	E
23	RESIDUAL OUTPUT			T k dL	dU
24				7. 3. 0.46723	1.89636
25	)bservatior	Predicted ln asks (B) t	Residuals	DW	
26	1	2.92583	-0.01762	1.62635	
27	2	3.04161	0.0099		
28	3	3.1574	0.00754		
29	4	3.27319	0.00018		
30	5	3.44087	0.0109		
31	6	3.55666	0.00371		
32	7	3.67244	-0.01461		

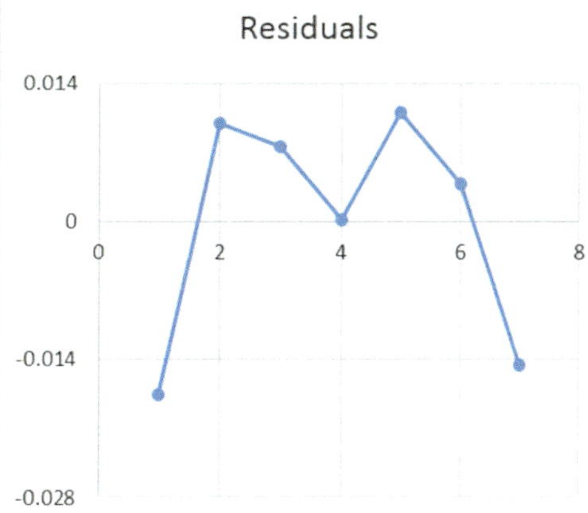

The residuals may exhibit positive autocorrelation. *Predicted ln asks* are too high in year 7, producing a relatively large, negative residual. The model fit in year 7 is worse than in years 2 through 6. Evidence suggests that the low cost program effectively drove capacity increases in 2007 and 2008, but the impact was short term, lasting only two years.

Return to the data sheet and add a *temporary low cost* indicator. Run regression with the *temporary low cost* indicator, assess Durbin Watson, and plot residuals.

	A	B	C	D	E	F	G
1	SUMMARY OUTPUT						
2							
3	*Regression Statistics*						
4	R Square	0.99894					
5	Standard Er	0.01095					
6	Observatio	7					
7							
8	ANOVA						
9		*df*	*SS*	*MS*	*F*	*gnificance F*	
10	Regression	2	0.45244	0.22622	1886.15	1.1E-06	
11	Residual	4	0.00048	0.00012			
12	Total	6	0.45292				
13							
14		*Coefficients*	*andard Err*	*t Stat*	*P-value*	*Lower 95%*	*Upper 95%*
15	Intercept	-243.417	4.71463	-51.6301	8.4E-07	-256.507	-230.327
16	t	0.12298	0.00235	52.3113	8E-07	0.11645	0.12951
17	temporary low cost t	0.03663	0.01041	3.51978	0.02445	0.00774	0.06553

	B	C	D	E
21	OUTPUT		T k dL	dU
22			7. 3. 0.46723	1.89636
23	*Predicted ln asks (B) t*	*Residuals*	*DW*	
24	2.91601	-0.00781	2.91764	
25	3.039	0.01252		
26	3.16198	0.00296		
27	3.28496	-0.0116		
28	3.44458	0.00719		
29	3.56756	-0.00719		
30	3.65391	0.00392		
31				
32				
33				

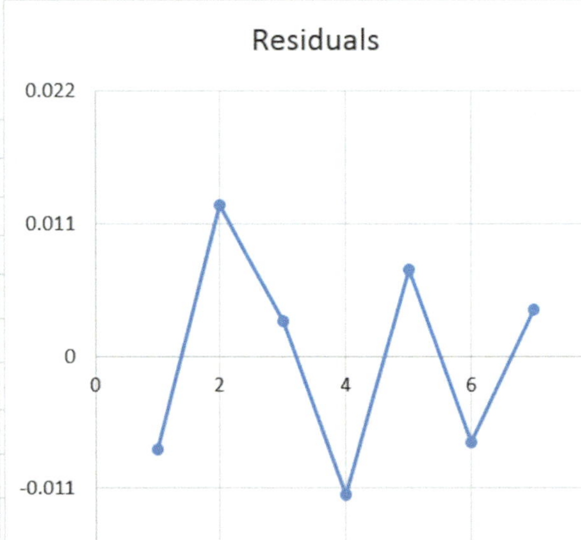

Residuals are now free of positive autocorrelation, and residuals in recents years are relatively small, indicating a good fit.

Assess the predictive validity of the *ln ASKs* model. Find the three multiplicative part worths to find predicted *ASKs (B)*.

For the intercept part worth, find the exponential function of the intercept in logarithms, $\exp(b_0)$.

M24 =L24-$I$5

	K	L	M	N
23	**ln asks (B) t**	predicted ln asks (B) t	lower (ln B)	upper (ln B)
24	2.91	2.92	2.89	2.95
25	3.05	3.04	3.01	3.07
26	3.16	3.16	3.13	3.19
27	3.27	3.28	3.25	3.32
28	3.45	3.44	3.41	3.47
29	3.56	3.57	3.54	3.60
30	3.66	3.65	3.62	3.68
31	**3.75**	3.78	**3.75**	**3.81**
32	**3.87**	3.90	**3.87**	**3.93**

The lower and upper prediction interval bounds capture actual *ln asks*, with the actuals lie on top of the lower prediction interval bound.

Recalibrate the model, running the regressions with all of the available data to update coefficients and forecasts. Copy and paste formulas for critical t and the margin of error from the validation sheet to the recalibration sheet.

	A	B	C	D	E	F	G
1	SUMMARY OUTPUT						
2							
3	*Regression Statistics*						
4	Multiple F	0.999465					
5	R Square	0.998929					
6	Adjusted I	0.998573	critical t	me (ln B)			
7	Standard E	0.012484	2.446912	0.030546			
8	Observati	9					
9							
10	ANOVA						
11		*df*	*SS*	*MS*	*F*	*gnificance F*	
12	Regressio	2	0.872511	0.436256	2799.421049	1.23E-09	
13	Residual	6	0.000935	0.000156			
14	Total	8	0.873446				
15							
16		*Coefficient*	*andard Err*	*t Stat*	*P-value*	*Lower 95%*	*Upper 95%*
17	Intercept	-236.587	3.251748	-72.757	4.53696E-10	-244.544	-228.631
18	t	0.119575	0.00162	73.7972	4.16686E-10	0.11561	0.12354
19	low cost t	0.046861	0.010063	4.656711	0.003479262	0.022237	0.071485

Update predictions: copy and paste data, part worth and predicted columns, and lower and upper columns from the validation regression sheet to the recalibrated regression sheet to reuse formulas.

	E	F	G	H	I	J	K	L	M	N	O	P	Q
25	*t*	*temporary low cost t*	*passenger ASKs(B)$_t$*	*passenger RPKs(B)$_t$*	*sqrt asks (B) t*	*cubrt asks(B) t*	*ln asks (B) t*	*intercept*	*trend pw (ln B)*	*temporary low cost pw*	*predicted ln asks (B) t*	*lower ln*	*upper ln*
26	2003	0	18.3	12.7	4.28	2.64	2.91	-237	240	0.000	2.92	2.89	2.95
27	2004	0	21.1	15.1	4.60	2.77	3.05	-237	240	0.000	3.04	3.01	3.07
28	2005	0	23.7	17.5	4.87	2.87	3.16	-237	240	0.000	3.16	3.13	3.19
29	2006	0	26.4	19.5	5.14	2.98	3.27	-237	240	0.000	3.28	3.25	3.31
30	2007	1	31.6	24.0	5.62	3.16	3.45	-237	240	0.047	3.45	3.42	3.48
31	2008	1	35.2	27.0	5.93	3.28	3.56	-237	240	0.047	3.57	3.54	3.60
32	2009	0	38.8	29.8	6.23	3.38	3.66	-237	240	0.000	3.64	3.61	3.67
33	2010	0	42.4	33.1	6.51	3.49	3.75	-237	240	0.000	3.76	3.73	3.79
34	2011	0	48.2	38.4	6.94	3.64	3.87	-237	240	0.000	3.88	3.85	3.91
35	2012	0						-237	241	0.000	4.00	3.97	4.03
36	2013	0						-237	241	0.000	4.12	4.09	4.15
37	2014	0						-237	241	0.000	4.24	4.21	4.27
38	2015	0						-237	241	0.000	4.36	4.33	4.39
39	2016	0						-237	241	0.000	4.48	4.45	4.51

Predictions are in natural logarithms for *ASKs*. To rescale back to *ASKs* in billions, create three new columns for *predicted, lower* and *upper*, using the exponential function

=**EXP(***cell***)**

T39 =EXP(Q39)

	O	P	Q	R	S	T
25	*predicted ln asks (B) t*	*lower ln*	*upper ln*	*predicted asks (B)*	*lower asks (B)*	*upper asks (B)*
26	2.92	2.89	2.95	18.6	18.0	19.1
27	3.04	3.01	3.07	20.9	20.3	21.6
28	3.16	3.13	3.19	23.6	22.9	24.3
29	3.28	3.25	3.31	26.6	25.8	27.4
30	3.45	3.42	3.48	31.4	30.4	32.4
31	3.57	3.54	3.60	35.4	34.3	36.5

To illustrate the fit and forecast, move lower and upper 95% prediction interval bounds for *ASKs* and actual *passenger ASKs* next to year *t*. Select *t, passenger ASKs(B)$_t$, lower,* and *upper,* and insert a scatterplot.

	E	F	G	H	I	N
25	*t*	*lower asks (B)*	*upper asks (B)*	*passenger ASKs(B)$_t$*		
26	2003	18.0	19.1	18.3		
27	2004	20.3	21.6	21.1		
28	2005	22.9	24.3	23.7		
29	2006	25.8	27.4	26.4		
30	2007	30.4	32.4	31.6		
31	2008	34.3	36.5	35.2		
32	2009	36.9	39.2	38.8		
33	2010	41.6	44.2	42.4		
34	2011	46.9	49.8	48.2		
35	2012	52.8	56.1			
36	2013	59.5	63.3		0	-237
37	2014	67.1	71.3		0	-237
38	2015	75.6	80.4		0	-237
39	2016	85.2	90.6		0	-237

Adjust axes, remove the decimals on the vertical axis, change *passenger asks(B)* from a line to markers, and change the color of one of the prediction interval bounds to match the other, add a vertical axis title and a stand alone chart title.

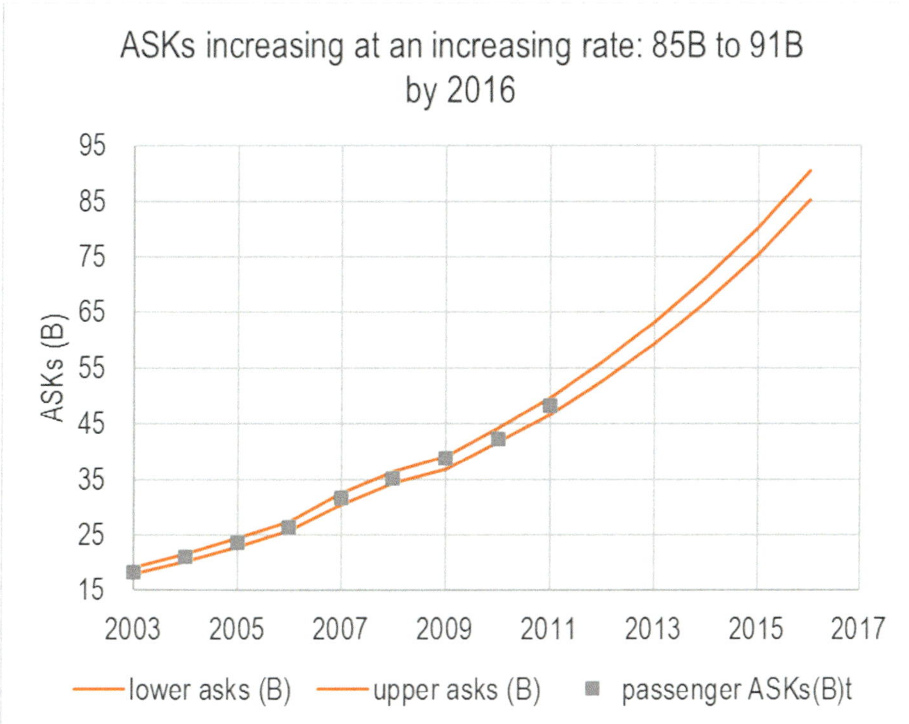

The RPK model using cube roots

Run regression of *cube roots rpks* with year *t* to quantify the trend. Assess Durbin Watson and plot residuals.

	A	B	C	D	E	F	G
1	SUMMARY OUTPUT						
2							
3	*Regression Statistics*						
4	Multiple F	0.997446					
5	R Square	0.994898					
6	Adjusted I	0.993878					
7	Standard I	0.022085					
8	Observati	7					
9							
10	ANOVA						
11		*df*	*SS*	*MS*	*F*	*gnificance F*	
12	Regressio	1	0.475576	0.475576	975.0812	6.32E-07	
13	Residual	5	0.002439	0.000488			
14	Total	6	0.478015				
15							
16		*Coefficient*	*andard Err*	*t Stat*	*P-value*	*Lower 95%*	*Upper 95%*
17	Intercept	-258.709	8.37224	-30.9008	6.66E-07	-280.23	-237.187
18	t	0.130326	0.004174	31.22629	6.32E-07	0.119597	0.141055

	B	C	D
24	*Predicted cubrt rpks (B) t*	*Residuals*	*DW*
25	2.334121	-0.00279	2.606398
26	2.464447	0.008613	
27	2.594773	0.001019	
28	2.725099	-0.0337	
29	2.855425	0.029123	
30	2.98575	0.012456	
31	3.116076	-0.01472	
32			
33			
34			
35			

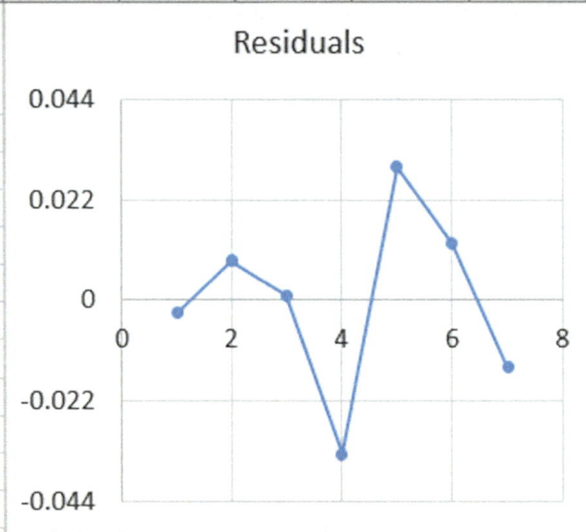

A shift, most likely due to the Low Cost program, is apparent in the *cube root rpks* regression, beginning in the year 5, 2007.

Run regression, including the ongoing *low cost* program indicator, assess Durbin Watson, and plot residuals.

	A	B	C	D	E	F	G
1	SUMMARY OUTPUT						
2							
3	*Regression Statistics*						
4	Multiple F	0.999209					
5	R Square	0.998419					
6	Adjusted I	0.997629	critical t	me			
7	Standard E	0.013743	2.776445	0.038158			
8	Observati	7					
9							
10	ANOVA						
11		*df*	*SS*	*MS*	*F*	*gnificance F*	
12	Regressio	2	0.477259	0.23863	1263.403	2.5E-06	
13	Residual	4	0.000756	0.000189			
14	Total	6	0.478015				
15							
16		*Coefficients*	*tandard Err*	*t Stat*	*P-value*	*Lower 95%*	*Upper 95%o*
17	Intercept	-231.797	10.41234	-22.2618	2.41E-05	-260.706	-202.888
18	t	0.116897	0.005194	22.50408	2.31E-05	0.102475	0.131319
19	long term	0.062668	0.020993	2.985164	0.040531	0.004382	0.120955

	B	C	D	E
23	OUTPUT		T k dL	dU
24			7. 3. 0.46723	1.89636
25	*Predicted cubrt rpks (B) t*	*Residuals*	*DW*	
26	2.34755	-0.01622	1.969504	
27	2.464447	0.008613		
28	2.581344	0.014448		
29	2.698241	-0.00684		
30	2.877806	0.006741		
31	2.994703	0.003503		
32	3.1116	-0.01024		
33				
34				

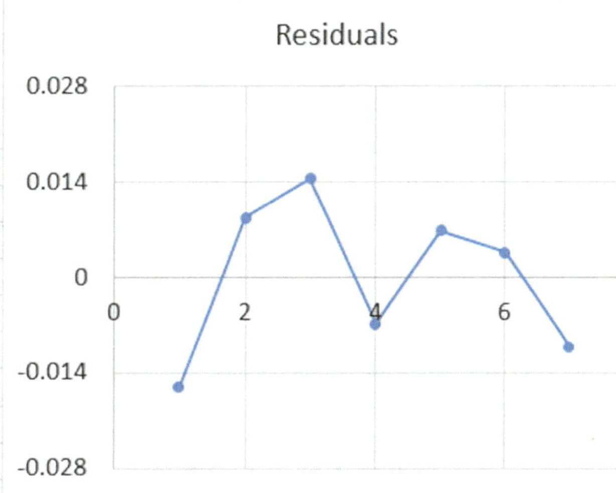

Residuals from the regression of *cube roots rpks* are free of positive autocorrelation, and residuals in recent years are all relatively small, indicating a good fit.

Assess the validity of the *cube root rpks* models for forecasting.

T26 =R26+$D$7

	N	O	P	Q	R	S	T
25	*cubrt rpks (B) t*	*intercept*	*trend pw*	*low cost pw*	*predicted cubrt rpks (B) t*	*lower (cubrt B)*	*upper (cubrt B)*
26	2.33	-232	234	0.000	2.35	2.31	2.39
27	2.47	-232	234	0.000	2.46	2.43	2.50
28	2.60	-232	234	0.000	2.58	2.54	2.62
29	2.69	-232	234	0.000	2.70	2.66	2.74
30	2.88	-232	235	0.063	2.88	2.84	2.92
31	3.00	-232	235	0.063	2.99	2.96	3.03
32	3.10	-232	235	0.063	3.11	3.07	3.15
33	3.21	-232	235	0.063	3.23	3.19	3.27
34	3.37	-232	235	0.063	3.35	3.31	3.38
35		-232	235	0.063	3.46	3.42	3.50
36		-232	235	0.063	3.58	3.54	3.62

The actual *cube root rpks* fall within the lower and upper prediction intervals in the two most recent years, providing evidence that the model is valid for forecasting.

Recalibrate the model and copy and paste the *critical t* and *margin of error* to reuse formulas.

	A	B	C	D	E	F	G
1	SUMMARY OUTPUT						
2							
3	*Regression Statistics*						
4	Multiple F	0.999139					
5	R Square	0.998279					
6	Adjusted I	0.997706	critical t	me			
7	Standard E	0.01688	2.446912	0.041304			
8	Observati	9					
9							
10	ANOVA						
11		*df*	*SS*	*MS*	*F*	*gnificance F*	
12	Regressio	2	0.991753	0.495876	1740.316	5.1E-09	
13	Residual	6	0.00171	0.000285			
14	Total	8	0.993462				
15							
16		*Coefficient*	*andard Err*	*t Stat*	*P-value*	*Lower 95%*	*Upper 95%ov*
17	Intercept	-237.386	8.736415	-27.172	1.64E-07	-258.763	-216.009
18	t	0.119685	0.004358	27.46083	1.54E-07	0.109021	0.13035
19	low cost t	0.052685	0.022647	2.326347	0.058935	-0.00273	0.1081

Copy and paste *t, low cost, passenger rpks (B), cube root rpks (B),* the past worths, *predicted cube root rpks (B), lower* and *upper cube root rpks* from the validation sheet to the recalibrated sheet to reuse formulas.

	E	F	G	H	I	J	K	L	M	N	O	P	Q	R	S	T
25	*t*	*low cost t*	*temporary low cost t*	*passenger ASKs(B)$_t$*	sqrt asks (B) t	cubrt asks(B) t	*ln asks (B) t*	*passenger RPKs(B)$_t$*	*sqrt rpks (B) t*	*cubrt rpks (B) t*	*intercept (cube rt B)*	*trend pw (cube rt B)*	*loc cost pw (cube rt B)*	*predicted cubrt rpks (B) t*	*lower cubrt*	*upper cubrt*
26	2003	0	0	18.3	4.28	2.64	2.91	12.7	3.56	2.33	-237	240	0.000	2.34	2.30	2.38
27	2004	0	0	21.1	4.60	2.77	3.05	15.1	3.89	2.47	-237	240	0.000	2.46	2.42	2.50
28	2005	0	0	23.7	4.87	2.87	3.16	17.5	4.18	2.60	-237	240	0.000	2.58	2.54	2.62
29	2006	0	0	26.4	5.14	2.98	3.27	19.5	4.42	2.69	-237	240	0.000	2.70	2.66	2.74
30	2007	1	1	31.6	5.62	3.16	3.45	24.0	4.90	2.88	-237	240	0.053	2.87	2.83	2.92
31	2008	1	1	35.2	5.93	3.28	3.56	27.0	5.19	3.00	-237	240	0.053	2.99	2.95	3.04
32	2009	1	0	38.8	6.23	3.38	3.66	29.8	5.46	3.10	-237	240	0.053	3.11	3.07	3.16
33	2010	1	0	42.4	6.51	3.49	3.75	33.1	5.76	3.21	-237	241	0.053	3.23	3.19	3.28
34	2011	1	0	48.2	6.94	3.64	3.87	38.4	6.20	3.37	-237	241	0.053	3.35	3.31	3.39
35	2012	1	0								-237	241	0.053	3.47	3.43	3.51
36	2013	1	0								-237	241	0.053	3.59	3.55	3.63
37	2014	1	0								-237	241	0.053	3.71	3.67	3.75
38	2015	1	0								-237	241	0.053	3.83	3.79	3.87
39	2016	1	0								-237	241	0.053	3.95	3.91	3.99

Predictions for *RPKs* are in cube roots. To rescale back to billions of *RPKs*, create three new columns, *predicted, lower* and *upper* using

=*cell*^**3**

U26 f_x 12.8682978420727

	R	S	T	U	V	W
25	*predicted cubrt rpks (B) t*	*lower cubrt*	*upper cubrt*	*predicted rpks (B) t*	*lower rpks (B)*	*upper rpks (B)*
26	2.34	2.30	2.38	12.9	12.2	13.6
27	2.46	2.42	2.50	14.9	14.2	15.7
28	2.58	2.54	2.62	17.2	16.4	18.1
29	2.70	2.66	2.74	19.7	18.8	20.7
30	2.87	2.83	2.92	23.8	22.7	24.8

Move *lower* and *upper RPKs(B)* next to years, *t,* select year *t, lower* and *upper RPKs(B)* and request a scatterplot.

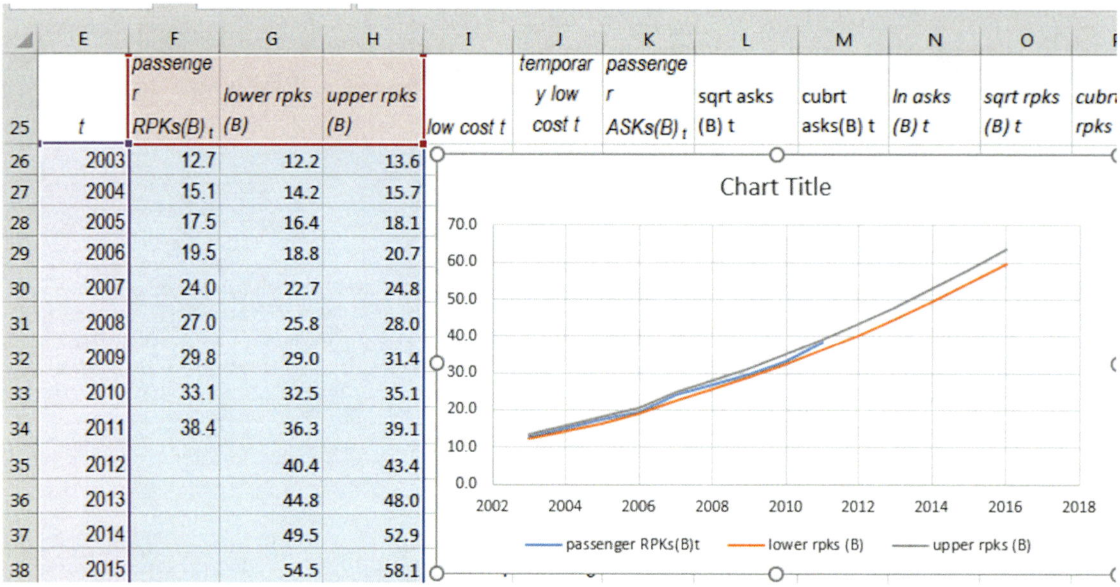

	E	F	G	H	I	J	K	L	M	N	O
25	t	passenger RPKs(B)$_t$	lower rpks (B)	upper rpks (B)	low cost t	temporary low cost t	passenger ASKs(B)$_t$	sqrt asks (B) t	cubrt asks(B) t	ln asks (B) t	sqrt rpks (B) t
26	2003	12.7	12.2	13.6							
27	2004	15.1	14.2	15.7							
28	2005	17.5	16.4	18.1							
29	2006	19.5	18.8	20.7							
30	2007	24.0	22.7	24.8							
31	2008	27.0	25.8	28.0							
32	2009	29.8	29.0	31.4							
33	2010	33.1	32.5	35.1							
34	2011	38.4	36.3	39.1							
35	2012		40.4	43.4							
36	2013		44.8	48.0							
37	2014		49.5	52.9							
38	2015		54.5	58.1							

Change actual *passenger RPKs(B)* from a line to markers, recolor one of the prediction interval bounds to match the other, rescale axes to make good use of space, remove the decimals from the vertical axis, add a vertical axis title and chart title, and set fontsize to 12.

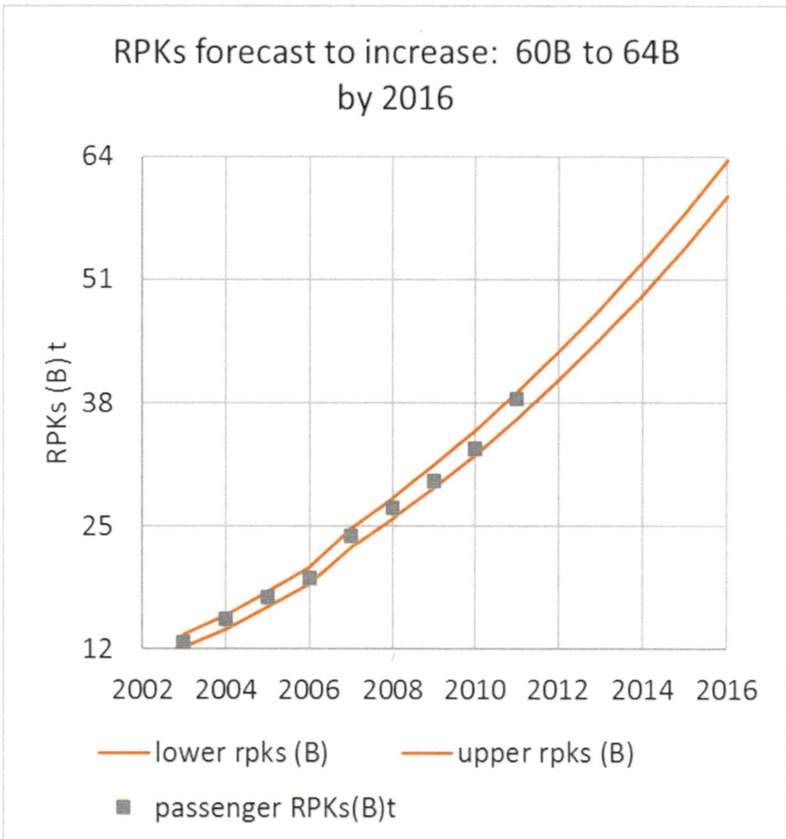

Since *load,* the ratio of *RPKs* to *ASKs*, is the desired forecast, plot fits and forecasts of both on the same graph. Copy and paste the *RPKs* graph into the *ASK* graph.

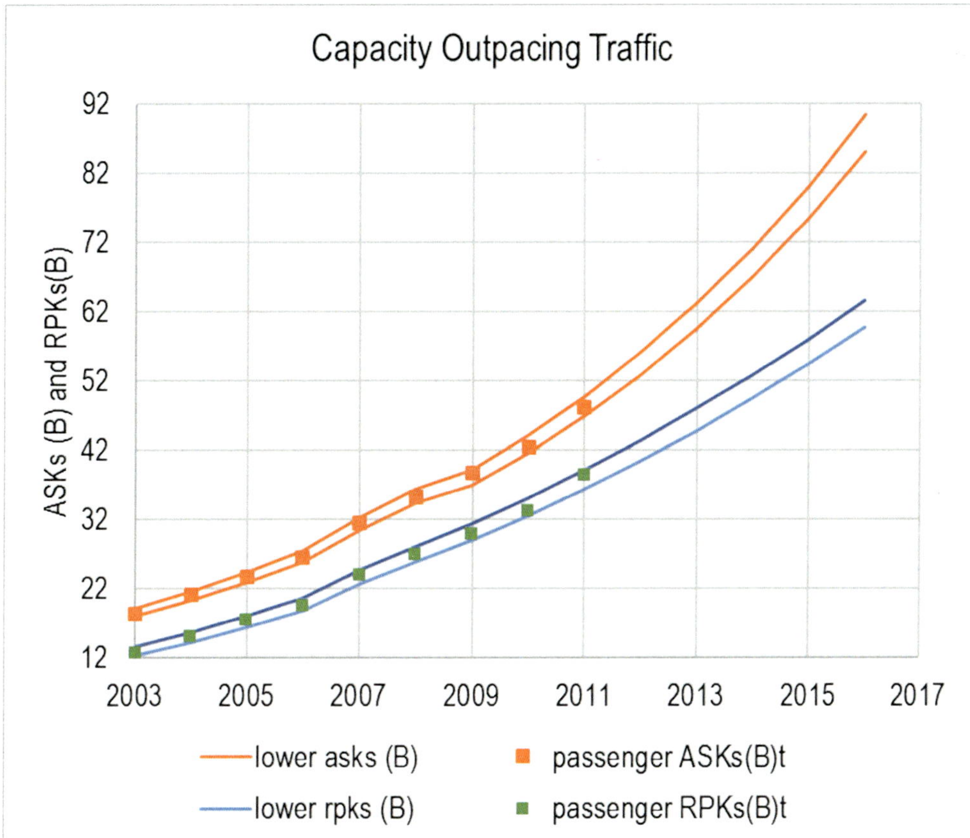

Excel 13.2 Compare Scenarios with Sensitivity Analysis

When a dependent variable has been rescaled, the model becomes multiplicative, and the impact of each driver depends on the values of other drivers. The *ASKs* equation is the product of three components, the intercept, the impact of the *low cost* program, and the trend:

$$A\hat{S}Ks(B)_t = e^{-240^a + .047^a \times Low\ Cost_t + .12^a \times t}$$

$$= 1.8\text{E}{-103} \times 1.05^{.Low\ Cost_t} \times 1.13^{.t}$$

To see the impacts of the annual *trend* and the *low cost* program, add four columns, the baseline, which is the exponential function of the intercept,

=exp($B$17),

the *low cost* multiplier, which is the exponential function of the *low cost* indicator coefficient, raised to the power of the *low cost* indicator (in column I), equal to one when the indicator is turned off, and equal to 1.05 when the indicator is turned on,

=exp($B$10)^I26,

the *trend,* which is the exponential function of the coefficient for *t,* the year, 1.13, raised to the power of the year, *t* (in column E),

=exp($B$18)^E26,

and their product, *predicted ASKs(B),*

X26 =U26*V26*W26

	U	V	W	X	Y
25	*baseline*	*low cost multipler*	*trend*	*predicted ASKs(B)*	ME (B)
26	1.78E-103	1.00	1.04E+104	18.6	0.57
27	1.78E-103	1.00	1.17E+104	20.9	0.64
28	1.78E-103	1.00	1.32E+104	23.6	0.72
29	1.78E-103	1.00	1.49E+104	26.6	0.81
30	1.78E-103	1.05	1.68E+104	31.4	0.96

Add one more column, *predicted ASKs(B) without low cost,* equal to the product of the *baseline* and the *trend.* Move *predicted ASKs(B)* and *predicted ASKs(B) without low cost* next to the year, select year *t, predicted ASKs(B)* and *Predicted ASKs(B) without low cost* and request a scatterplot to see the temporary impact of the low cost program.

	E	F	G	H	I	J	K	L	M
25	*t*	*predicted ASKs(B)*	*ASKs(B) without*	*lo (B*					*br ks*
26	2003	18.6	18.6						
27	2004	20.9	20.9						
28	2005	23.6	23.6						
29	2006	26.6	26.6						
30	2007	31.4	29.9						
31	2008	35.4	33.8						
32	2009	38.0	38.0						
33	2010	42.9	42.9						
34	2011	48.3	48.3						
35	2012	54.5	54.5						
36	2013	61.4	61.4	59.5	63.3		0		

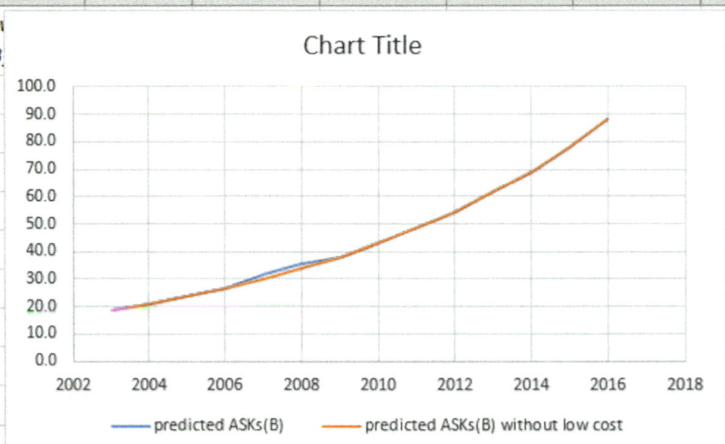

Adjust axes, add a vertical axis title, remove the decimals from the vertical axis, add a chart title, and adjust the fontsize to 12:

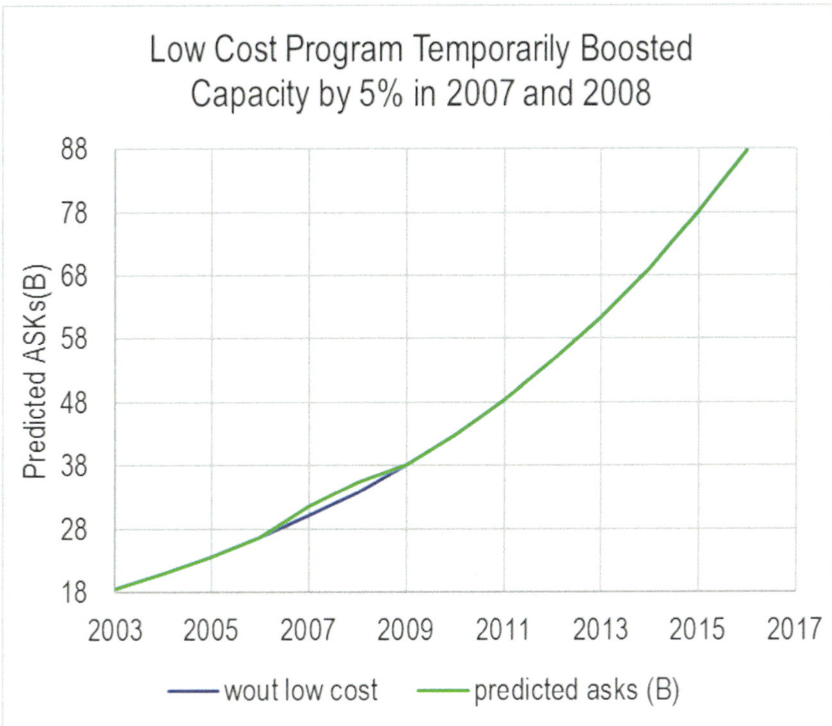

The *RPSs* equation can be split into two parts, that due to trend *t*, with four part worths, and that due to the impact of the *low cost* program, with six part worths:

- The trend *t*:

$$R\hat{P}Ks(B)_t\, trend = b_0^3 + 3 \times (b_0^2 \times b_t \times t + b_0 \times b_t^2 \times t^2) + b_t^3 \times t^3$$

- Impact of the *Low Cost* program:

$$Low\ C\hat{o}st\ impact(B) = 3 \times (b_0^2 \times b_{lc} \times low\ cost_t + b_0 \times b_{lc}^2 \times low\ cost_t^2 + b_t^2 \times b_{lc} \times t^2 \times low\ cost_t + b_t \times b_{lc}^2 \times t \times low\ cost_t^2) + 6 \times b_0 \times b_t \times b_{lc} \times t \times low\ cost_t + b_{lc}^3 \times low\ cost_t^3$$

Using the formulas shown above, add three columns in the *RPKs(1/3)* sheet, *trend without low cost program, low cost impact,* and their sum, *predicted RPKs(B) with low cost program*:

X26 =$B$17^3+3*($B$17^2*$B$18*E26+$B$17*$B$18^2*E26^2)+$B$18^3*E26^3

	X	Y	Z
25	trend (B) without Low Cost program	low cost impact (B)	rpks(B) with Low Cost Program
26	12.9	0.0	12.9
27	14.9	0.0	14.9
28	17.2	0.0	17.2
29	19.7	0.0	19.7
30	22.5	1.3	23.8
31	25.5	1.4	26.9
32	28.7	1.5	30.2
33	32.2	1.6	33.8
34	36.0	1.7	37.7
35	40.0	1.9	41.9
36	44.4	2.0	46.4
37	49.0	2.1	51.2
38	54.0	2.3	56.3
39	59.3	2.4	61.7

Notice that the impact of the *low cost* program increases each year, with the trend.

Move the *trend without low cost* and *predicted RPKs(B)t* columns next to year *t* then select data in the three columns and plot to see the impact of the *low cost* program:

	E	F	G
25	*t*	trend (B) without Low Cost program	rpks(B) with Low Cost Program
26	2003	12.9	12.9
27	2004	14.9	14.9
28	2005	17.2	17.2
29	2006	19.7	19.7
30	2007	22.5	23.8
31	2008	25.5	26.9
32	2009	28.7	30.2
33	2010	32.2	33.8
34	2011	36.0	37.7
35	2012	40.0	41.9
36	2013	44.4	46.4
37	2014	49.0	51.2
38	2015	54.0	56.3
39	2016	59.3	61.7
40			

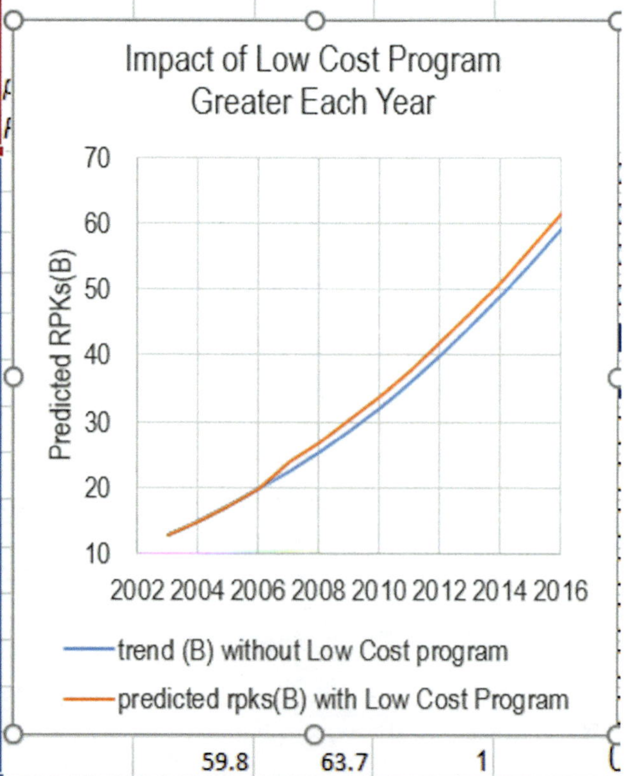

Adjust the vertical axis, add a vertical axis title, remove the decimals from the vertical axis, and add a chart title.

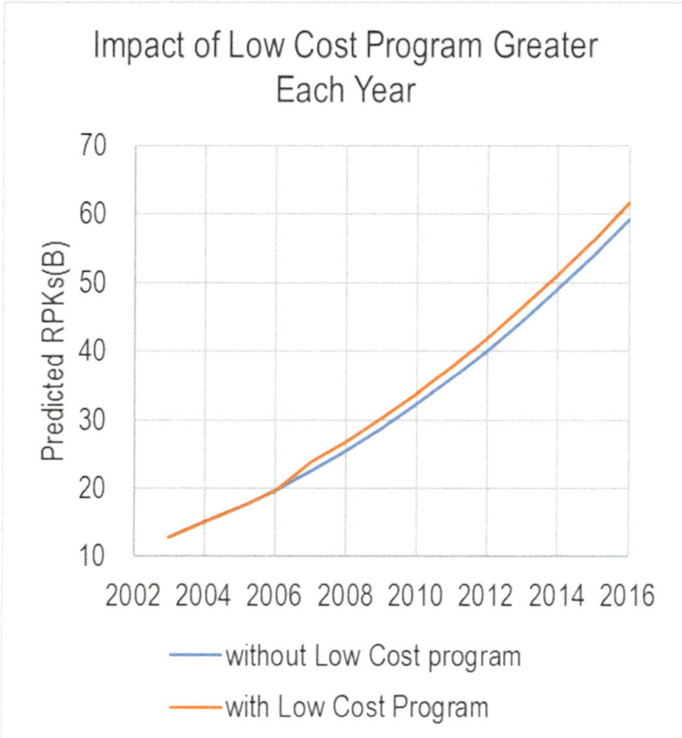

Excel 13.3 Use Nonlinear Regression Estimates with Monte Carlo Simulation

To estimate 2016 passenger load percents, samples of possible values for 2016 passenger capacity, ASKs, and passenger volume, RPKs, are needed. Both ASKs and RPKs were positively skewed in the historical data sample.

Generate Normal distributions of 1000 potential values for *ln ASKs* and $RPKs^{(1/3)}$, with means set at the predicted values for 2016 *ln ASKs* and *cube root RPKs* and standard deviations set at the standard errors from the *ln ASKs* and *cube root RPKs* regressions.

Rescale the simulated samples to *ASKs(B)* and *RPKs(B)*.

	AB	AC	AD	AE	AF	AG	AH
1		cube rt rpks	ln asks	cube rt rpks	rpks	ln asks	asks
2	M	3.95	4.48	3.97	62.6	4.46	86.7
3	SD	0.01688	0.01248	3.95	61.5	4.49	89.2
4				3.98	63.0	4.46	86.6
5				3.93	60.9	4.46	86.7
6				3.96	62.3	4.49	89.4

Find the sample of possible values for 2016 *passenger load percent* from the ratio of *RPKs(B)* to *ASKs(B)*:

AI2 =AF2/AH2

	AF	AG	AH	AI
1	rpks	ln asks	asks	load
2	62.6	4.46	86.7	0.722
3	61.5	4.49	89.2	0.690
4	63.0	4.46	86.6	0.728
5	60.9	4.46	86.7	0.702

Find the 95% prediction interval for 2016 *passenger load percent* using

=**PERCENTILE**(*array*,.975)

And

=**PERCENTILE**(*array*,.025)

AI1003 =PERCENTILE(AI2:AI1001,0.025)

	AE	AF	AG	AH	AI
1	cube rt rpks	rpks	ln asks	asks	load
1001	3.96	62.2	4.48	88.1	71%
1002				upper	72.3%
1003				lower	67.4%

Find the margin of error in the *2016 passenger load percent* forecast by dividing the 95% prediction interval by 2:

AI1004 =(AI1002-AI1003)/2

	AE	AF	AG	AH	AI
1	cube rt rpks	rpks	ln asks	asks	load
1001	3.96	62.2	4.48	88.1	71%
1002				upper	72.3%
1003				lower	67.4%
1004				me	2.5%

To see the distribution of possible 2016 passenger load percents, make a histogram of possible 2016 *passenger load percents*, using bin widths of .5% within the 95% prediction interval range, 67% to 73%.

	AI	AJ	AK
1	load	2016 load%	Frequency
2	72%	67.0%	13
3	69%	67.5%	17
4	73%	68.0%	32
5	70%	68.5%	81
6	70%	69.0%	101
7	68%	69.5%	141
8	68%	70.0%	170
9	71%	70.5%	144
10	71%	71.0%	110
11	70%	71.5%	85
12	69%	72.0%	51
13	72%	72.5%	39
14	73%	73.0%	11
15	71%	More	5

Add a column *probability,* equal to the frequency divided by the simulated sample size of 1000.

AL2 =AK2/1000

	AI	AJ	AK	AL
1	load	2016 load%	Frequency	probability
2	72%	67.0%	13	0.013
3	69%	67.5%	17	0.017
4	73%	68.0%	32	0.032
5	70%	68.5%	81	0.081
6	70%	69.0%	101	0.101
7	68%	69.5%	141	0.141
8	68%	70.0%	170	0.17
9	71%	70.5%	144	0.144
10	71%	71.0%	110	0.11
11	70%	71.5%	85	0.085
12	69%	72.0%	51	0.051
13	72%	72.5%	39	0.039
14	73%	73.0%	11	0.011
15	71%	More	5	0.005

Move *probability* next to *2016 load %* and request a column chart.

	AI	AJ	AK	AL
1	load	2016 load%	probability	Frequency
2	72%	67.0%	0.013	
3	69%	67.5%	0.017	
4	73%	68.0%	0.032	
5	70%	68.5%	0.081	
6	70%	69.0%	0.101	
7	68%	69.5%	0.141	
8	68%	70.0%	0.17	
9	71%	70.5%	0.144	
10	71%	71.0%	0.11	
11	70%	71.5%	0.085	
12	69%	72.0%	0.051	
13	72%	72.5%	0.039	
14	73%	73.0%	0.011	
15	71%	More	0.005	

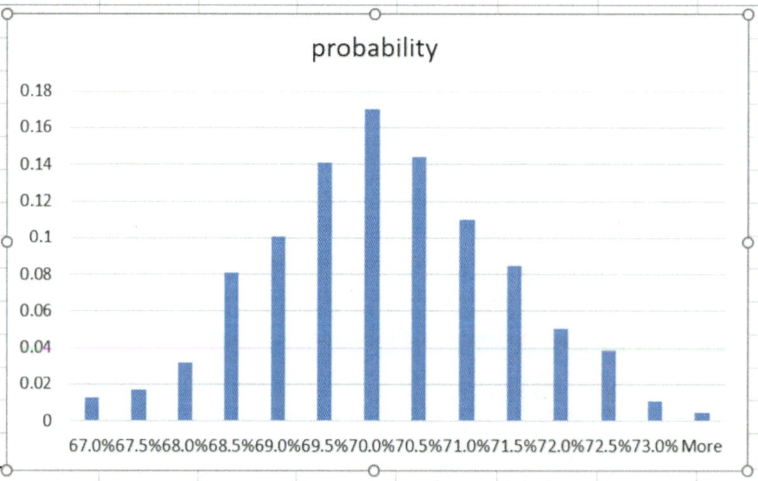

Add axes titles and a stand alone chart title and set font size to 12.

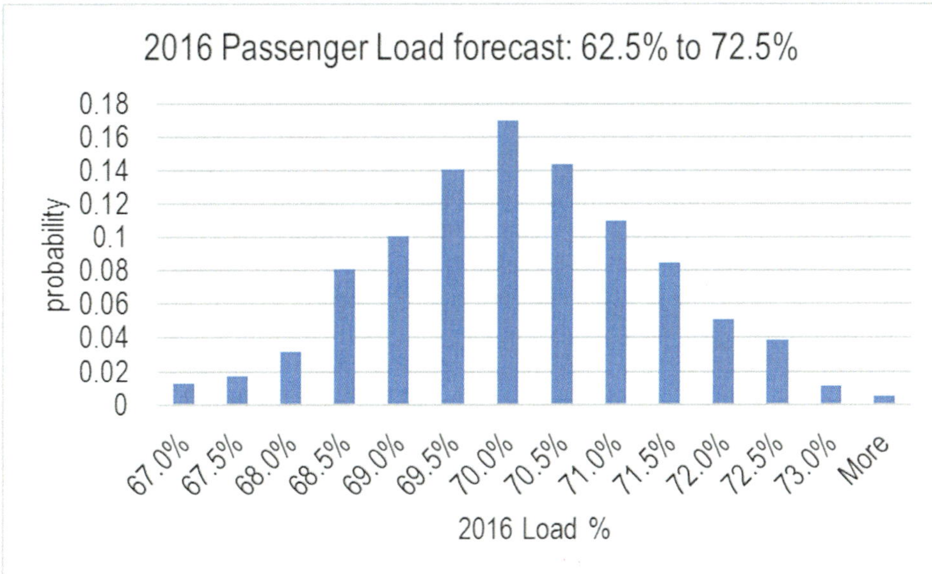

Lab 13.1 Nonlinear Forecasting LAN Airlines Passenger Revenues: Building the Model

The "low cost" model implemented in '07 included two goals: (1) increase passengers by 40%, and (2) acquire 20% more capacity.

Now in place five years, the low cost model appeared working:

> Passenger traffic had increased each year by an average of 21%, and
> capacity had increased each year by an average of 13%.

However, the lower prices on domestic flights were of some concern. Average prices (*revenue per ASK)* had temporarily fallen during the Lowcost Launch year, 2007, due to the appeal of cheaper domestic flights. Prices had increased the following year in 2008. However, the recession appeared to have discouraged international travel, bringing prices down again in 2009 and later years. Passenger revenues in 2009 had fallen below 2008 revenues. The long term impact of the recession on revenues could threaten the low cost program's success.

Information Needed. The Board is asking for evidence that revenue improvements from the low cost model are sustainable. You have been asked to

- determine the impacts of the low cost program launch on prices (*revenue per ASK*)
- determine the impact of the recession on *revenue per ASK*

Considerations. LAN's global business expansion occurred after the '01 industry disruption; data before '03 is not thought to represent well current businesses.
Management contends that all elements under their control will be managed according to the low cost plan presented. Recent growth in performance indicators is considered representative of future growth through '16.

Data are in **Lab 13.1 LAN Passenger Revenue.**

Assess skewness of passenger *Revenue per ASK*, <u>*using only years preceding the global recession*</u>, whose effects began in 2009. (The global recession shift could potentially mislead skewness assessment, since only a single year is available for model building.)

1. Skewness of passenger *Revenue per ASK (2003 through 2008)*: _____.

2. Are Revenues per ASK increasing at ___ an increasing rate? Or ___ a decreasing rate?

Rescale Revenue per ASK and choose the scale that best Normalizes Revenues per ASK, and then run regression with year t as the driver.

Durbin-Watson Critical Values: 5% Significance

T	K	dL	dU
7	2	0.7	1.36

3. Is your model free of positive autocorrelation (trend, cycles, shifts and shocks)?

   DW:_____ ___ yes ___ maybe ___no

4. Plot the residuals and describe what you see: ______________________

Add a *recession* indicator equal to 1 in years 2009 through 2016, 0 elsewhere, and a run a two driver regression.

Durbin-Watson Critical Values: 5% Significance

T	K	dL	dU
7	3	0.47	1.9

5. Is your model free of positive autocorrelation (trend, cycles, shifts and shocks)?

   DW:_____ ___ yes ___ maybe ___no

6. Plot the residuals and describe what you see: ______________________

Add a *temporary low cost* indicator, equal equal to 1 in '07, and equal to 0 in other years, and then run regressions to find average annual change in each performance indicator, accounting for temporary low cost and recession shifts.

Durbin-Watson Critical Values: 5% Significance

T	K	dL	dU
7	4	0.28	2.46

7. Is your model free of positive autocorrelation (trend, cycles, shifts and shocks)?

   DW:_____ ___ yes ___ maybe ___no

8. Plot the residuals and describe what you see: ______________________

9. Is your model valid? Y or N

Lab 13.2 Nonlinear Forecasting LAN Airlines Passenger Revenues: Describe the Model

1. Recalibrate your model and illustrate your fit and forecast for *Passenger revenue per ASK* in years '03 through '16.

2. Write your equation for *passenger revenue per ASK* in the original units (cents):

3. Estimate the impact of the recession on *passenger revenue per ASK* in '16:

   Impact of recession in '16: ________________________________

4. Illustrate the impact of the recession on passenger revenues per ASK in years '03 through '16.

5. What is the *margin of error* in your forecast of *revenue per ASK*? ____________

6. Passenger revenue depends on both capacity and prices. Find the "best" and "worst" outcomes in '16, with and without continuing global recession:

	'16 Capacity (ASKs)(B)	*'16 Revenue per ASK (cents)*	*'16 Revenue per ASK (cents) without recession*	*'16 Passenger revenue ($B)*	*'16 Passenger revenue ($B) without recession*
"Worst"	85.2				
"Best"	90.6				

7. How likely is the "best" case scenario? _____ or one in ________

Lab 13.3 Forecasting with Uncertain Drivers: LAN Passenger Revenues

After reviewing naïve models quantifying the impact of the low cost program and the global recession on passenger revenue, the Board is now pondering those impacts in the passenger business. Over the past five years, passenger revenues have grown at an average annual rate of 17.2%, due, in part of the successful low cost program. Is this level of growth sustainable, given the global recession?

According to the "best case" scenario, with continuing recession, revenues could grow at an average annual rate of 22.5%. However, in the "worst case," revenues would grow at an average annual rate of 21.6% if the recession continued.

If the global economy recovered, in the "best case," revenues could grow at an average annual rate of as much as 27.6%, and even in the "worst case," average annual revenue growth would be 26.7%.

The Board has asked you to provide a 95% prediction interval for passenger revenues in 2016, with and without continuing global recession, to narrow the possibilities to this more likely range.

Data are in **LAN forecast.xlsx.**

1. Use 2016 predictions for *ln ASKs* and *sqrt revenues per ASK* with standard errors from regression models with Monte Carlo simulation to forecast possible values for *passenger revenues* in 2016, with and without a continuing global recession:

	Passenger revenue ($B)	
	With continuing global recession	Without continuing global recession
Lower 95%		
Upper 95%		

2. What average annual growth rate from 2011 passenger revenues, $3.87B, is possible?

	Average annual growth in passenger revenue	
	With continuing global recession	Without continuing global recession
Lower 95%		
Upper 95%		

3. Illustrate the distribution of possible outcomes for revenue in the passenger business in '16, with and without continuing global recession with a column chart.

Assignment 13.1 Billionaires in 2020

Rolls-Royce believes that a major target market for supersonic business jets is the growing segment of billionaires. Forecast the trend in billionaires from data in **billionaires**.

1. Plot billionaires by year and identify patterns:

   ___ positive trend ___ shift ___ cycle

2. Find the skewness of billionaires: __________

3. The number of billionaires is increasing at: ____ an increasing ___ decreasing rate

4. Rescale to square roots and assess skewness:

	Square roots
Skew	

Build a naïve model of trend in billionaires, including a indicator of the global recession from 2009 through 2015.

5. Is your model free from unaccounted for trend, shifts, shocks and cycles? __ Y __ N

6. Plot the residuals and identify patterns:

___ positive trend ___ shift ___ cycle

7. Is your model valid? ___ Y ___ N

8. Recalibrate and report your forecast for 2020: ________________

9. What is the margin of error in the forecast for 2020? ____________

10. The global recession reduced the number of billionaires expected in 2020 by: ______

11. Present your equation for billionaires:

Assignment 13.2 Primary Aluminum Production in 2020

Alcoa executives believe that with recently added capacity in China, primary aluminum production will increase, consequently leading to lower prices. Forecast the trend in primary aluminum production from data in **primary aluminum**.

1. Plot primary aluminum production by year and identify patterns:

   ___ positive trend ___ shift ___ cycle

2. Find the skewness of primary aluminum production, before and after the global recession shift: '06 to '08 ___ '09 to '15____

3. Primary aluminum production is increasing at: ____ an increasing ___ decreasing rate

4. Rescale to squares and cubes and assess skewness, before and after the global recession shift:

	Squares	Cubes
Skew '06 to '08		
Skew '09 to '15		

Build a naïve model of trend in primary aluminum production, including an indicator of the global recession from 2009 through 2015.

5. Is your model free from unaccounted for trend, shifts, shocks and cycles? __ Y __ N

6. Plot the residuals and identify patterns: ___ positive trend ___ shift ___ cycle

7. Is your model valid? ___ Y ___ N

8. Recalibrate and report your forecast for 2020: _______________

9. What is the margin of error in the forecast for 2020? ___________

10. Plot your fit and forecast:

11. Present your equation for primary aluminum production:

12. The global recession reduced primary aluminum production expected in 2020 by: ___

Chapter 14
Nonlinear Explanatory Multiple Regression Models

In this chapter, the insights offered by nonlinear regression models built with multiple drivers are examined. In many cases, a dependent variable and its drivers are approximately Normal, with skewness between −1 and +1. Linear regression models often provide good fit for either cross sectional or time series, and are often valid for forecasting in time series. However, the choice of a nonlinear model enables acknowledgement of the interactions inherent in many cases. With nonlinear models, the impact of each driver depends on the values of other drivers. In a nonlinear model, driver influences are multiplicative, which adds an element of realism relative to linear regression models with constant response, and provides richer insights from sensitivity analysis.

Example 14.1 Marriott Hotel Pricing. Competition in the DC area has intensified among hotels in recent years. Marriott executives believe that the Marriott brand name commands a price premium, particularly for hotels that offer high quality accommodations and amenities. The modeling team is tasked with quantification of the Marriott brand premium. Available data comes from online reservation sites and includes *Guest rating, Star rating, starting room price,* and hotel name for 41 hotels in the DC area.

The sample distribution of starting room prices, shown in Figure 14.1, is positively skewed. Five percent appear to be *outliers*, more than three standard deviations above the sample mean, shown in red in Figure 14.1. Those relatively expensive hotels are of particular interest to Marriott, and only appear to be outliers because the distribution of prices is positively skewed. To Normalize the *Starting Room Prices* for use with linear regression, scales to reduce skewness, square roots and natural logarithms, were considered. The natural logarithms improve skewness and relatively expensive hotels no longer appear to be outliers.

C. Fraser, *Business Statistics for Competitive Advantage with Excel 2016*,
DOI 10.1007/978-3-319-32185-1_14

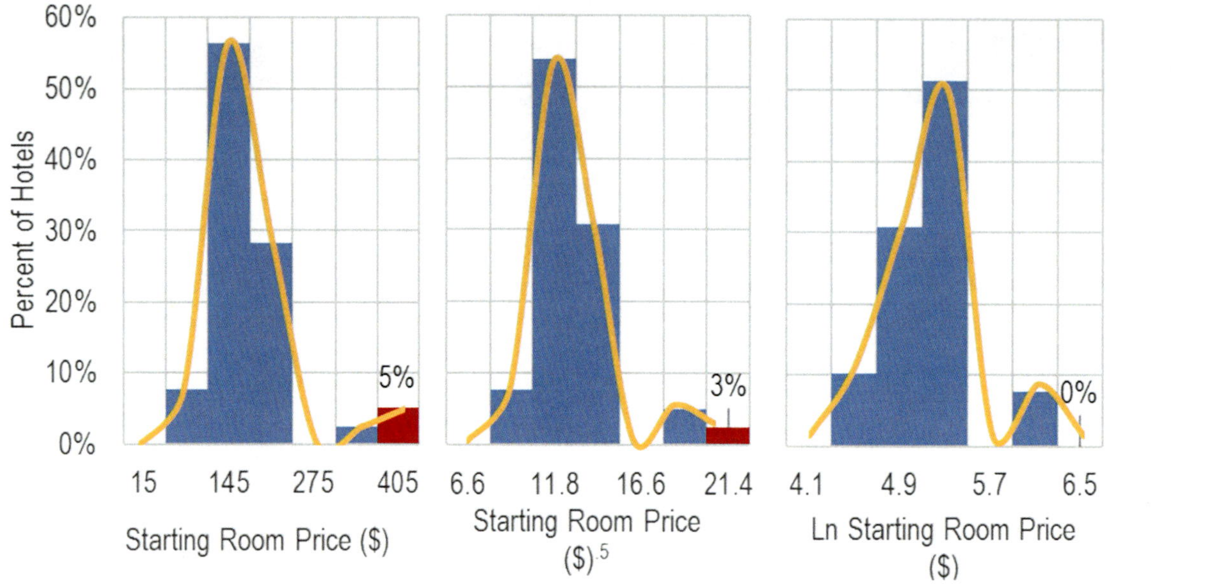

	Starting Room Price ($)	Starting Room Price ($)$^{.5}$	ln Starting Room Price ($)
skew	2.5	1.8	1.1

Figure 14.1 Distributions of starting room prices in dollars and rescaled to square roots and logarithms

The distribution of *Guest Ratings* is negatively skewed, with more than half above the sample mean, 4.2, but within one standard deviation of the sample mean. Increasing the power will improve skew. The squares and cubes are shown in Figure 14.2, and cubes produce a more Normal distribution:

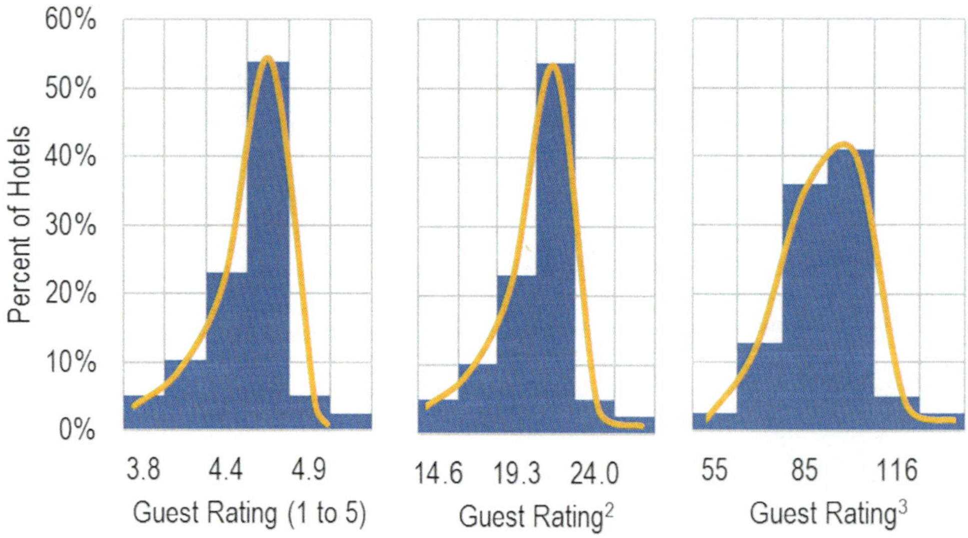

	Guest Rating	*Guest Rating*^{2}	*Guest Rating*^{3}
skew	−.62	−.38	−.16

Figure 14.2 Sample distribution of Guest Rating and Guest Rating squares and cubes

Stars are positively skewed. The distributions of *Stars,* their square roots and their natural logarithms are shown in Figure 14.3.

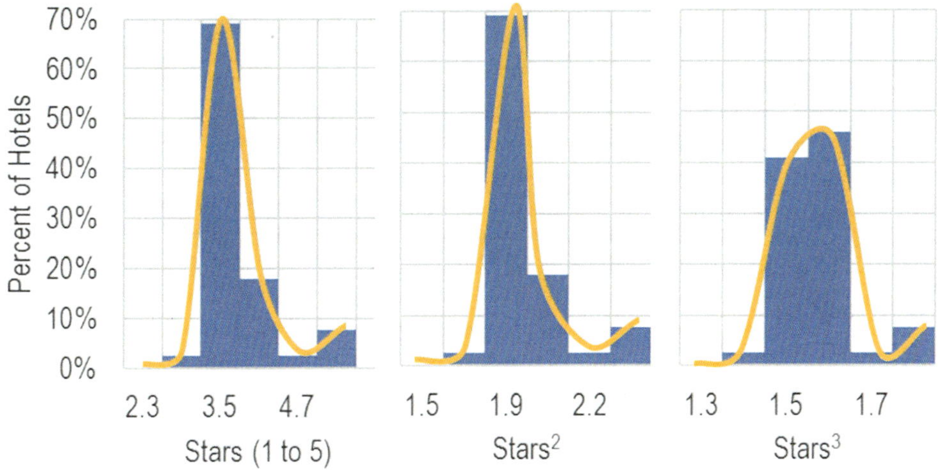

	Stars	*Stars*$^{.5}$	*Ln Stars*
skew	1.06	.90	.70

Figure 14.3 Sample distribution of Guest Rating and Guest Rating squares and cubes

The regression of *Ln Starting Room Prices* with a *Marriott* indicator, the cubes of *Guest Rating,* and *ln Stars* is shown in Table 14.1. The model is significant, and all three drivers are significant, with positive coefficient estimates. Together, quality ratings and the Marriott brand account for 69% of the variation in DC hotel starting room prices. The model equation is in natural logarithms:

$$Ln\,Starting\,\hat{R}oom\,Price(\$) = 2.61(ln\$)^{a} + .14(ln\$)^{.b} \times Marriott$$

$$+ .012\left(\frac{ln\$}{Rating^3}\right)^{a} \times Guest\,Rating^3$$

$$+1.01\left(\frac{\ln\$}{\ln rating}\right)^{a} \times \ln Stars$$

RSquare: .69[a]
[a]Significant at .01 or better; [b]Significant at .05.

Table 14.1 Regression of Ln starting room prices

Regression Statistics	
R Square	.692
Standard Error	.207
Observations	39

ANOVA

	df	*SS*	*MS*	*F*	*Significance F*
Regression	3	3.38	1.13	26.2	4.5E−9
Residual	35	1.50	.04		
Total	38	4.88			

	Coefficients	*Standard Error*	*t Stat*	*p value*	*One tail p value*
Intercept	2.6	.28	9.3	5E−11	
Marriott	.14	.080	1.7	.090	.045
lnStars	1.0	.21	4.8	3E−05	2E−05
Guest Rating^3	.012	.0023	5.2	9E−06	5E−06

To express the equation in dollars, the exponential function is used for both sides:

$$Ln\ Starting\ \hat{R}oom\ Price(\$) = 2.61(ln\$)^{a} + .14(ln\$)^{.b} \times Marriott$$

$$+ .012\left(\frac{ln\$}{Rating^3}\right)^{a} \times Guest\ Rating^3$$

$$+1.01\left(\frac{\ln\$}{\ln rating}\right)^{a} \times \ln Stars$$

$$\exp\left[Ln\ Starting\ \hat{R}oom\ Price(\$)\right] = \exp\left[2.61(ln\$)^{a} + .14(ln\$)^{.b} \times Marriott\right.$$

$$+ .012\left(\frac{ln\$}{Rating^3}\right)^{a} \times Guest\ Rating^3$$

$$\left.+1.01\left(\frac{\ln\$}{\ln rating}\right)^{a} \times \ln Stars\right]$$

$$Starting\ \hat{R}oom\ Price(\$) = \exp\left[2.61(ln\$)^{a} + .14(ln\$)^{.b} \times Marriott\right.$$

$$+ .012\left(\frac{ln\$}{Rating^3}\right)^{a} \times Guest\ Rating^3$$

$$\left.+1.01\left(\frac{\ln\$}{\ln rating}\right)^{a} \times \ln Stars\right]$$

The equation can be written to show the multiplicative influences of each of the drivers and the impact of the Marriott brand.

$$Starting\ \hat{R}oom\ Price\ (\$) = constant(\$) \times Marriott\ multiplier \times Guest\ Rating\ multiplier \times Stars\ multiplier$$

For Marriott hotels, with the *Marriott* indicator set to one:

$$Starting\ \hat{R}oom\ Price\ (\$) = 13.6(\$) \times 1.15 \times 1.01^{\text{Guest Rating}^3} \times Stars^{1.01}$$

And for other hotels, with the *Marriott* indicator set to zero:

$$Starting\ \hat{R}oom\ Price\ (\$) = 13.6(\$) \times 1.00 \times 1.01^{\text{Guest Rating}^3} \times Stars^{1.01}$$

The Marriott brand commands a price that is, on average, 115% (=exp($b_{Marriott}$) = exp (.14)) the price of other hotels of comparable quality. Improvements in either *Guest Ratings* or *Stars* will benefit Marriott hotels, commanding price increases that are 115% larger than price increases of competing hotels with the same improvement in ratings.

14.1 Sensitivity Analysis Reveals the Relative Strength of Drivers

When the dependent variable is rescaled to build a nonlinear model, the model is multiplicative. The impact of each of the drivers depends on values of all of the other drivers. Predicted *starting room prices* can be compared for contrasting scenarios, such as those which are linked to lower prices (lower quality hotels with lower *Star* and *Guest Ratings*) and those which are linked to higher prices (higher quality hotels with higher *Star* and *Guest Ratings*).

As an example, to identify the impact of *Guest* ratings on *Starting Room Prices* for Marriott hotels, the *Marriott* indicator is set to one. Since the *Stars* coefficient is positive, lower *Stars* would reduce *starting room price* and the impact of *Guest* rating will be lower than in the contrasting case in which *Stars* are higher. For both of these contrasting scenarios, find predicted *Starting Room Prices* at varying levels of *Guest Rating* to find its impact. Predicted prices by *Star* ratings for Marriott hotels with the best and worst *Guest Ratings* are illustrated in the left panel in Figure 14.3. Predicted prices by *Guest* ratings for Marriott hotels with the best and worst *Star Ratings* are illustrated in the right panel in Figure 14.4.

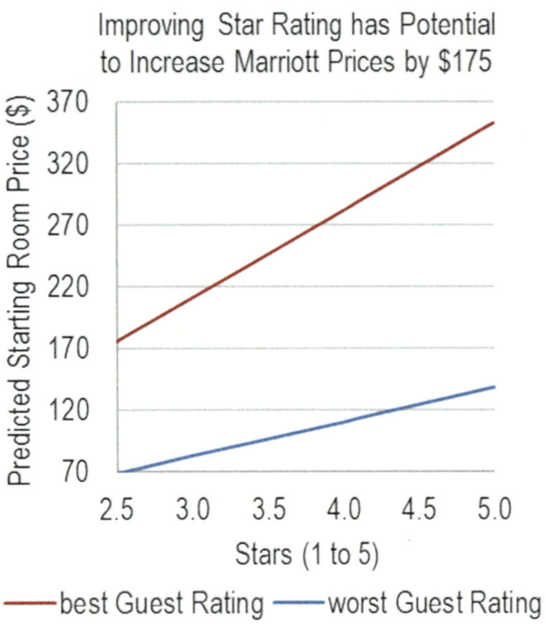

Figure 14.4 Predicted Marriott starting room prices by quality scenario

Marriott hotels rated high by guests benefit more from improvements in Star ratings. Starting Room Price response to Stars is increasing at a decreasing rate, which suggests that Star improvements of hotels with fewer Stars (and greater Guest Ratings) drive price increases more.

For Marriotts with a *Guest Rating* of 5, have a *Guest Rating* multiplier of 1.20:

$$Starting\ \hat{R}oom\ Price\ (\$) = 13.6(\$) \times 1.15 \times 1.01^{\text{Guest Rating}^3} \times Stars^{1.01}$$

$$= 13.6(\$) \times 1.15 \times 1.01^{5^3} \times Stars^{1.01}$$

$$= 13.6(\$) \times 1.15 \times 1.20 \times Stars^{1.01}$$

In contrast, for Marriotts with the sample minimum *Guest Rating* of 3.6, have a *Guest Rating* multipler of 1.13:

$$Starting\ \hat{R}oom\ Price\ (\$) = 13.6(\$) \times 1.15 \times 1.01^{\text{Guest Rating}^3} \times Stars^{1.01}$$

$$= 13.6(\$) \times 1.15 \times 1.01^{3.6^3} \times Stars^{1.01}$$

$$= 13.6(\$) \times 1.15 \times 1.14 \times Stars^{1.01}$$

Star improvements will be 1.05 (=1.20/1.14) times more influential for hotels with *Guest Ratings* of 5, relative to those with *Guest Ratings* of 3.6. Improvements to higher rated hotels will be more productive than improvements to lower rated hotels.

Five star Marriott hotels benefit more than hotels with fewer Stars from Guest Rating improvements. Starting Room Price response to Guest Rating is increasing at an increasing rate, particularly for higher Star hotels.

The impact of the scenario is apparent from the equation. Marriotts with five *Stars* have a *Star* multiplier of 5.05:

$$Starting\ \hat{R}oom\ Price\ (\$) = 13.6(\$) \times 1.15 \times 1.01^{\text{Guest Rating}^3} \times Stars^{1.01}$$

$$= 13.6(\$) \times 1.15 \times 1.01^{\text{Guest Rating}^3} \times 5^{1.01}$$

$$= 13.6(\$) \times 1.15 \times 1.01^{\text{Guest Rating}^3} \times 5.05$$

In contrast, Marriotts with 2.5 *Stars* have a *Star* multiplier half as large, of 2.51.

$$Starting\ \hat{R}oom\ Price\ (\$) = 13.6(\$) \times 1.15 \times 1.01^{\text{Guest Rating}^3} \times Stars^{1.01}$$

$$= 13.6(\$) \times 1.15 \times 1.01^{\text{Guest Rating}^3} \times 2.5^{1.01}$$

$$= 13.6(\$) \times 1.15 \times 1.01^{\text{Guest Rating}^3} \times 2.51$$

14.2 Sensitivity Analysis with Nonlinear Models Reveals Interactions

Nonlinear models which feature a rescaled dependent variable feature built in interactions. The impact of one driver depends upon the values of other drivers. Comparing predicted values under hypothetical scenarios allows quantification of the importance of each driver and more accurate inference for alternative scenarios under consideration.

Logarithmic models are multiplicative, with each driver multiplying the impact of others. Models built with roots are additive, but incorporate pairwise and higher order interactions. Either will often add realism when modeling performance in business. Rarely do drivers impact performance in isolation. Had the Marriott modeling team been content to use a linear model, they would not have discovered the importance of the brand name, the *Marriott* advantage from quality improvements relative to competing brands' improvements, or the joint impact of Star ratings and Guest ratings, which reinforce each other.

Excel 14.1 Build a Nonlinear Model with Cross Sectional Data

Pricing of Marriott Hotels. Hotel pricing reflects the quality of accommodations and amenities offered by competing hotels, as well as the value of a chain's brand name. Quantify the impact of hotel quality, reflected in *Star* and *Guest ratings,* as well as the Marriott brand name, with a model of Starting Room Prices in the DC area. Data for 39 hotels, including nine Marriott hotels are in Excel 14 Pricing Marriott hotels, and include *starting room price, Stars,* and *Guest Rating.*

Use Excel's **SKEW(*array*)** function to assess *skewness* of *Starting Room Prices ($), Stars,* and *Guest Rating.*

B41 =SKEW(B2:B40)

	A	B	C	D
1	*Hotel*	*Starting Room Price*	*Guest Rating (1 to 5)*	*Stars (1 to 5)*
39	Westin	76	4.2	3.5
40	Westin Grand	139	4.4	4
41	skew	2.46	-0.62	1.06

Starting room price is positively skewed, *Guest Rating* is negatively skewed, and *Stars* is slightly negatively skewd. Make three new columns to consider shrinking *Starting Room Prices* with square roots, cube roots and natural logarithms. Add two new columns to consider the squares and cubes of *Guest Rating,* and add three new columns to shrink *Stars* to square roots, cube roots and natural logarithms. Assess skewness of the rescaled variables:

B41 =SKEW(B2:B40)

	A	B	C	D	E	F	G	H	I	J	K	L
1	*Hotel*	*Starting Room Price*	*Guest Rating (1 to 5)*	*Stars (1 to 5)*	*sqrt price*	*cube rt price*	*ln price*	*guest rating sqr*	*guest rating cube*	*sqrt stars*	*cube rt stars*	*ln stars*
40	Westin Grand	139	4.4	4	11.8	5.18	4.9	19.4	85.2	2	1.6	1.4
41	skew	2.46	-0.62	1.06	1.8	1.6	1.1	-0.38	-0.16	0.88	0.82	0.70

The natural logarithm of *Starting Room Price* reduces skewness to 1.1. The squares of *Guest Rating* improve skew to −.16, close to Normal skewness of zero. The natural logarithms of *Stars* reduces skewness to approximately Normal at .70. Use these, with the Marriott indicator in regression:

	A	B	C	D	E	F	G
1	SUMMARY OUTPUT						
2							
3	*Regression Statistics*						
4	Multiple F	0.831942					
5	R Square	0.692127					
6	Adjusted I	0.665738					
7	Standard E	0.207241					
8	Observati	39					
9							
10	ANOVA						
11		*df*	*SS*	*MS*	*F*	*nificance F*	
12	Regressio	3	3.379364	1.126455	26.2278	5E-09	
13	Residual	35	1.503211	0.042949			
14	Total	38	4.882575				
15							
16		*Coefficients*	*andard Err*	*t Stat*	*P-value*	*ower 95%*	*Upper 95%*
17	Intercept	2.609044	0.279116	9.34751	4.81E-11	2.0424	3.17568
18	Marriott	0.139283	0.079754	1.746401	0.089514	-0.023	0.30119
19	guest ratii	0.011991	0.002318	5.173949	9.48E-06	0.0073	0.0167
20	ln stars	1.006209	0.211761	4.75163	3.39E-05	0.5763	1.43611

To assess the *Normality* of the residuals, find the residual skewness:

	A	B	C
63	37	5.866647	0.112238
64	38	4.757983	-0.42725
65	39	5.025397	-0.09092
66		skew	-0.46903

The residuals are approximately Normal.

To see *predicted Starting Room Price ($)* values, find *predicted starting room price ($)* from the product of the multiplicative part worths.

Copy *Guest Rating, Stars, Marriott, and Guest Rating cube* and paste next to residuals.

Make the multiplicative part worths. Find the multiplicative constant in dollars from the exponential function of the intercept, exp(b_0).

I27 =EXP($B$17)

	D	E	F	G	H	I
26	Hotel	Guest Rating (1 to 5)	Stars (1 to 5)	Marriott	guest rating cube	constant ($)
27	Best West	4.6	3	0	97.34	13.6
28	Capital Hil	4.6	3.5	0	97.34	13.6

Find the *Marriott* part worth from the exponential function of the *Marriott* coefficient to the power of the *Marriott* indicator, $\exp(b_{Marriott})^{Marriott}$.

J27 =EXP($B$18)^G27

	E	F	G	H	I	J
26	Guest Rating (1 to 5)	Stars (1 to 5)	Marriott	guest rating cube	constant ($)	Marriott pw
27	4.6	3	0	97.34	13.6	1.00
28	4.6	3.5	0	97.34	13.6	1.00
29	4	3	1	64.00	13.6	1.15
30	4.3	3	1	79.51	13.6	1.15

Find the *Guest Rating* part worth from the exponential function of the *Guest Rating* coefficient raised to the *Guest Rating cube* power.

K27 =EXP($B$19)^H27

	F	G	H	I	J	K
26	Stars (1 to 5)	Marriott	guest rating cube	constant ($)	Marriott pw	Guest Rating pw
27	3	0	97.34	13.6	1.00	3.2
28	3.5	0	97.34	13.6	1.00	3.2
29	3	1	64.00	13.6	1.15	2.2
30	3	1	79.51	13.6	1.15	2.6

Find the *Stars* part worth from *Stars* to the power of the *Stars* coefficient.

L27 =F27^$B$20

	H	I	J	K	L
26	guest rating cube	constant ($)	Marriott pw	Guest Rating pw	Stars pw
27	97.34	13.6	1.00	3.2	3.0
28	97.34	13.6	1.00	3.2	3.5
29	64.00	13.6	1.15	2.2	3.0
30	79.51	13.6	1.15	2.6	3.0

Find *predicted Starting Room price ($)* from the product of the multiplicative part worths.

M27 =I27*J27*K27*L27

	I	J	K	L	M
26	constant ($)	Marriott pw	Guest Rating pw	Stars pw	predicted price ($)
27	13.6	1.00	3.2	3.0	132
28	13.6	1.00	3.2	3.5	154
29	13.6	1.15	2.2	3.0	102

Excel 14.2 Sensitivity Analysis of Scenarios and Driver Influence

In a multiplicative model, driver influence depends upon the values of other drivers. To isolate the importance of a driver, compare predicted *Starting Room Prices* of hypothetical Marriott hotels, with the other driver first set at the sample minimum, and then at the sample maximum.

Determine the difference in Marriott *starting room prices* driven by differences in *Guest Rating* by adding sixteen new rows to the dataset which describe two sets of eight hypothetical Marriott hotels

- eight with *Star* ratings (in column F, below) of 2.5, the sample minimum, and
- eight with Five *Star* ratings, the sample maximum.

Within each set of eight, hypothetical hotels are identical, except that they differ only with respect to *Guest Rating* (in column E, below), from 3.6, the sample minimum, to 5, the sample maximum. Since the minimum to the maximum is 1.4 (=5−3.6), use *Guest Rating* increments of .2.

	D	E	F	G	H	I	J	K	L	M
66	Stars 2.5, GR 3.6	3.6	2.5	1	46.66	13.6	1.15	1.7	2.5	69
67	Stars 2.5, GR 3.8	3.8	2.5	1	54.87	13.6	1.15	1.9	2.5	76
68	Stars 2.5,GR 4	4	2.5	1	64.00	13.6	1.15	2.2	2.5	85
69	Stars 2.5, GR 4.2	4.2	2.5	1	74.09	13.6	1.15	2.4	2.5	95
70	Stars 2.5, GR 4.4	4.4	2.5	1	85.18	13.6	1.15	2.8	2.5	109
71	Stars 2.5, GR 4.6	4.6	2.5	1	97.34	13.6	1.15	3.2	2.5	126
72	Stars 2.5, GR 4.8	4.8	2.5	1	110.59	13.6	1.15	3.8	2.5	148
73	Stars 2.5, GR 5	5	2.5	1	125.00	13.6	1.15	4.5	2.5	176
74	Stars 5, GR 3.6	3.6	5	1	46.66	13.6	1.15	1.7	5.1	138
75	Stars 5, GR 3.8	3.8	5	1	54.87	13.6	1.15	1.9	5.1	152
76	Stars 5, GR 4	4	5	1	64.00	13.6	1.15	2.2	5.1	170
77	Stars 5, GR 4.2	4.2	5	1	74.09	13.6	1.15	2.4	5.1	192
78	Stars 5, GR 4.4	4.4	5	1	85.18	13.6	1.15	2.8	5.1	219
79	Stars 5, GR 4.6	4.6	5	1	97.34	13.6	1.15	3.2	5.1	253
80	Stars 5, GR 4.8	4.8	5	1	110.59	13.6	1.15	3.8	5.1	297
81	Stars 5, GR 5	5	5	1	125.00	13.6	1.15	4.5	5.1	353

Move *predicted starting room price ($)* to the right of *Guest Rating* and plot *predicted Starting Room price ($)* by *Guest Rating* for hypothetical hotels with the minimum *Star* rating of 2.5.

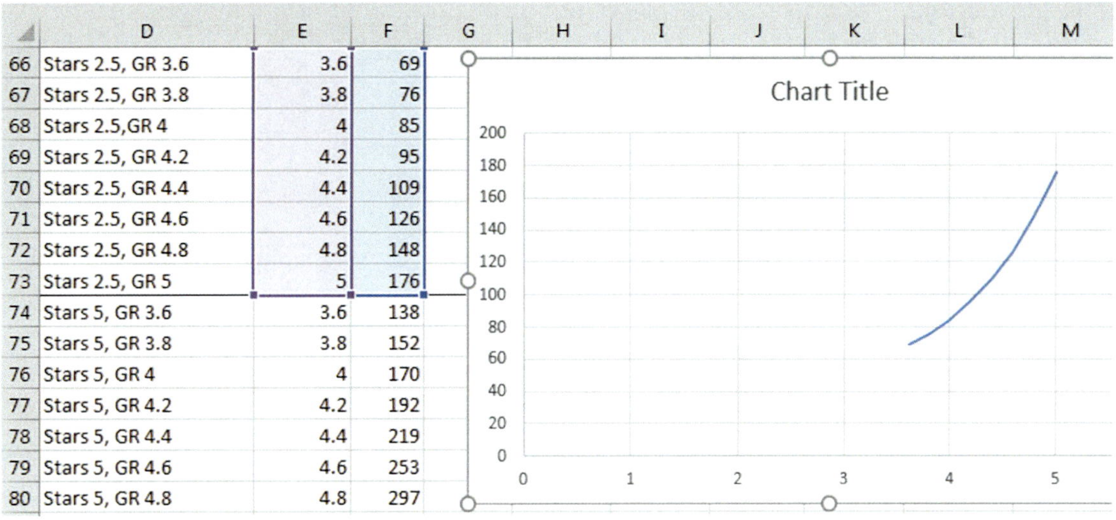

Add the response to *Guest Rating* for hotels with 5 *Stars*. Right click the scatterplot and choose Select Data, then Add.

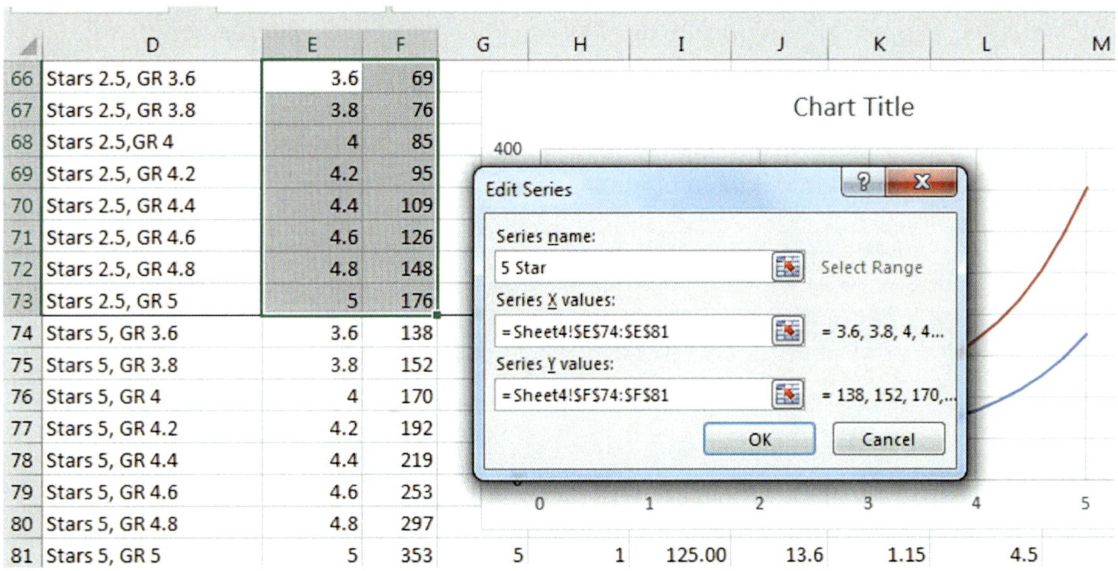

Edit Series One to name as *2.5 Stars*.

Add axes labels, rescale axes, adjust font size and add a chart title.

Repeat sensitivity analysis, now focusing on price response to *Stars*. Determine the difference in Marriott *starting room prices* driven by differences in *Stars* by adding twelve new rows to the dataset which describe two sets of six hypothetical Marriott hotels

- six with *Gest Rating* (now in column E) of 3.6, the sample minimum, and
- six with *Guest Rating* of 5.0, the sample maximum.

Within each set of six, hypothetical hotels are identical, except that they differ only with respect to *Stars* (now in column G), from 2.5, the sample minimum, to 5, the sample maximum. Since the minimum to the maximum is 2.5 (=5−2.5), use *Guest Rating* increments of .5.

	D	E	F	G	H	I	J	K	L	M
82	GR 3.6,Stars 2.5	3.6	69	2.5	1	46.7	13.6	1.15	1.7	2.5
83	GR 3.6, Stars 3	3.6	83	3.0	1	46.7	13.6	1.15	1.7	3.0
84	GR 3.6, Stars 3.5	3.6	96	3.5	1	46.7	13.6	1.15	1.7	3.5
85	GR 3.6, Stars 4	3.6	110	4.0	1	46.7	13.6	1.15	1.7	4.0
86	GR 3.6, Stars 4.5	3.6	124	4.5	1	46.7	13.6	1.15	1.7	4.5
87	GR 3.6, Stars 6	3.6	138	5.0	1	46.7	13.6	1.15	1.7	5.1
88	GR 5,Stars 2.5	5	176	2.5	1	125.0	13.6	1.15	4.5	2.5
89	GR 5, Stars 3	5	211	3.0	1	125.0	13.6	1.15	4.5	3.0
90	GR 5, Stars 3.5	5	247	3.5	1	125.0	13.6	1.15	4.5	3.5
91	GR 5, Stars 4	5	282	4.0	1	125.0	13.6	1.15	4.5	4.0
92	GR 5, Stars 4.5	5	318	4.5	1	125.0	13.6	1.15	4.5	4.5
93	GR 5, Stars 6	5	353	5.0	1	125.0	13.6	1.15	4.5	5.1

Request a scatterplot of the six lowest guest rated hypothetical hotels *predicted starting room prices* by *Stars,* and then add the second series of six highest guest rated hotel prices.

Since both scatterplots share a common vertical *y* axis, they can share the vertical axis and be shown side by side for comparison.

Lab 14 Mattel's Acquisition of Radica

Mattel revenues recovered quickly from a brief downturn in '09 due to the global recession. Management credits much of this resilience to the acquisition of Radica late in '06. Radica added electronic games aimed at the preadolescent market of children ages 9 through 13. Mattel targets both younger children, under age 9, as well as middle school aged children, ages 9 through 13. Mattel management is counting on demand from the growing segment of middle school aged children, 9 through 13, to fuel revenues.

Build a *valid* model of Mattel revenues to forecast revenues in '14 and '15 from data in **Mattel Radica**. The dataset contains *Mattel Revenues* (B$) in billion dollars and *U.S. population (M) of under 5* and *5 to 9 year olds* in millions for selected years '98 through '13. Use years '98 through '11 to build your model, holding out revenues in '12 and '13 to validate.

Create lags for the populations of under 5 and 5 to 9 year olds which will enable forecasts for '14 and '15. (Note that lags of two years produce populations under 7 and 7 to 11 year olds; lags of four years produce populations under 9 and 9 to 13 year olds; lags of six years produce populations under 11 and 11 to 15 year olds; In other words, children who were 9 in '12 are eleven in '14.)

1. Find skewness:

	Skewness
Revenues	
Under 5 lag	
5 to 9 lag	

2. Rescale to reduce skewness:

	sqrt	Cube rt	ln	Sqr	cube
Revenues					
Under 5 lag					
5 to 9 lag					

3. Plot Mattel revenues and describe what you see:

___ trend ___ shift ___ shock ___ other:________________________

Create an indicator variable *Radica* equal to 0 in years '98 through '05 and equal to 1 in years 2005 through 2015, and build a one driver model.

4. Assess Durbin Watson. dL:____ dU: ____ DW: ____

Are residuals pattern free? ___ N ___ Maybe ___ Y

Copy residuals, paste into the data page, and then mark decreases in '08 through '11.

Mark decreases in lags of populations under 5 and 5 to 9 in '08 through '11.

5. Which is the better choice to add to your regression, matching pattern in the four most recent years?

___ rescaled under 5 lag ___ rescaled 5 to 9 lag

Build a two driver model.

6. Assess Durbin Watson. dL:____ dU: ____ DW: ____

   Are residuals pattern free? ___ N ___ Maybe ___ Y

Residuals indicate that the global recession did reduce revenues temporarily in '09. Add an indicator *recession* equal to 0 in all years except '09, and equal to 1 in '09. Build a three driver model.

7. Assess Durbin Watson. dL:____ dU: ____ DW: ____

   Are residuals pattern free? ___ N ___ Maybe ___ Y

8. Is your model valid? ___N ___ Y

9. Recalibrate and present your forecast for 2015: _____

10. The margin of error in your forecast for 2015: _____

11. Illustrate your model fit and forecast:

12. Present your regression equation in the original scale of billion dollars:

13. Find the Radica and recession multipliers:

    Radica multiplier: _____ recession multiplier: _____

14. Which of these age groups drive revenues? ___ under 5 ___ 5 to 9 years

Assignment 14 Identifying Promising Global Markets

Harley-Davidson would like to identify the most promising global markets for motorcycle sales.

Some managers believe that motorcycle sales potential is greater in developed countries with higher GDP. Others believe that per capita GDP may a better indicator, with wealther drivers choosing cars instead of motorcycles.

Management believes that motorcycle sales potential will necessarily be greater in more populated countries. Some believe that population density may matter more, since motorcycles may be preferred to cars for parking and commuting in larger cities.

Build a model to identify the drivers of motorcycle market potential. **14 Global Moto** contains measures of

Motorcycle sales, GDP, per capita GDP, population, and *population density*

for 20 countries with the highest motorcycle sales. A number of these variables are skewed and ought to be rescaled.

1. Identify outliers and list. Which countries have unusually high motorcycle sales?

2. Which economic and population variables drive motorcycle sales?

___ GDP ___ per capita GDP ___ population ___ population density

Explain how you reached your conclusions, including statistics that you used to decide:

3. Present your regression equation in the original scale of motorcycles sold:

4. Illustrate the impact of the two most influential drivers, using the same scale for the y axis on both plots.

5. Compare predicted sales with actual sales in all countries in the sample to identify two markets with the greatest unrealized potential that Harley-Davidson should target:

Case 14.1 Promising Global Markets for EVs

Shiso Motors has designed an inexpensive Electric Vehicle (EV) which targets global segments where air pollution is severe. Shiso executives believe that the level of carbon emissions is a good surrogate for potential demand.

It is generally believed that economic productivity, population, and carbon based fuels drive emissions, and Shiso managers would like quantify these impacts, in order to prioritize targeted global segments.

There is disagreement whether GDP or GDP per capita is the stronger driver of emissions.

Some believe that population is a key driver, while others argue that GDP per capita matters more.

It is thought that emissions are highest in the BRIC countries, though rapidly emerging markets that have not yet reached the level of economic development of the BRICs may be attractive.

A. Build a Model to Explain Emissions Differences across Countries

Shiso has asked you to build a model to explain differences in emissions across countries.

Shiso contains data on *emissions, GDP, population, urban population, GDP per capita, oil production*, and fuel sources used to generate electricity in 32 countries. The 32 include developed nations, the *BRIC* nations, and two emerging market segments, "*emerging*" and "*fast emerging*."

Use natural logarithms when rescaling variables would improve skewness.

1. Which potential continuous drivers influence emissions?

___ GDP ___ GDP per capita ___ Population ___ Urban population

___ Oil Production ___ Electricity generated from oil

___ Electricity generated from coal

2. Write your model of emissions (in the original scale of *kt*) in a form which shows the separate impacts of the intercept (first) and each continuous driver.

3. Find the multipliers for each of the drivers. (Leave multipliers blank for potential drivers not in your model.)

   Emerging multiplier: ____ Fast emerging multiplier: ____ BRIC multiplier: ___

   GDP multiplier: ____ GDP per capita multiplier: ____

   Population multiplier: ____ Urban population multiplier: ____

Oil production multiplier: ____ Electricity generated from oil multiplier: ____

Electricity generated from coal multiplier: ____

4. One of the Shiso managers is convinced that emissions (and potential demand for an EV) are higher in countries where more electricity is produced from oil. Is this manager correct? ___ Y ___ N

B. Sensitivity Analysis of Driver Impacts

1. Illustrate the impacts of two continuous drivers on expected emissions in three global segments:

   a. Make three graphs, one for emerging, one for fast emerging, and one for BRICs, to illustrate the differences in expected emissions in response to differences in six to eight levels of a continuous driver, *in the original scale*. Copy and paste two of the graphs into the third graph.

   b. Make a second graph from graphs for emerging, fast emerging and BRICs, to illustrate the differences in expected emissions in response to differences in six to eight levels of a second continuous driver *in the original scale.* Rescale the vertical axis to match the vertical axis that you used in b. Show this graph to the right of the graph in b.

2. Are differences in emissions across global markets increasing at an *increasing* or *decreasing* rate with increasing values of these two continuous drivers?

   _____increasing at an increasing rate _____increasing at a diminishing rate

3. Emissions in South Africa are noticeably higher than emissions in Thailand. Explain why this is the case:

4. If GDP grows by 10% in both China and Columbia, how much will expected emissions increase in each of the two countries?

   ______% increase in China ______% increase in Columbia

Case 14.2 Chasing Whole Foods' Success

Trader Vics management is considering an attempt to replicate Whole Foods' success by implementing a process for selecting new store locations that mirrors Whole Foods' selection process, detailed in their website:

> If you have a retail location you think would make a good site for Whole Foods Market, Inc., please review the following guidelines carefully for consideration:
>
> **200,000 people or more in a 20-minute drive time**
> 25,000–50,000 Square Feet
> **Large number of college-educated residents**
> Abundant parking available for our exclusive use
> Stand alone preferred, would consider complementary
> Easy access from roadways, lighted intersection
> Excellent visibility, directly off of the street
> Must be located in a high traffic area (foot and/or vehicle)

James Sud, executive vice president of growth and development at WFM, describes WFM's store location selection success:

> "We're still very proud of the fact that in our 30-plus year history, we've never had a store that we opened ourselves ever fail. So we're really determined to keep that track record alive,"

Some executives at Trader Vic's believe that States in the Heartland may be the most promising locations for new stores, since residents of Heartland States are thought earn more college *degrees per person.*

Data in **14 WFM Stores** contains these data for each State:

Heartland State or not
WFM stores in 2011
colleges
2010 population
college *degrees* conferred *per person* in 2010
2010 people per sq mi

Help Trader Vics' strategists by identifying drivers of WFM's store selection (*stores* in a State).

The variables are skewed. Rescale to reduce skewness. (Note that logarithms cannot be used with *WFM stores* since some states have zero stores.)

1. What drives number of *stores* in a State? ___ *colleges* ___*degrees per person*

   ___*population* ___*people per sq mi*

2. Write the equation for your final model of *stores* in a State.

3. Embed a graph illustrating the impact on *stores* of the two most important drivers.

4. Test managers' suspicion that *Heartland* State residents earn more college *degrees per person* than residents of other States. Present and interpret statistics that you used.

5. Using the data, identify differences between *Heartland* States and other States. Present and interpret the statistics that you used:

6. Based on the WFM store selection process revealed by your model results, which two States ought Trader Vics consider first for new store locations?

7. Based on your results, explain why Trader Vics should or should not focus on Heartland States for new store locations:

Case 14.3 Promising Global Markets for Water Purification

Alcoa has developed a process to remove pollutants from water contaminated by coal mining. Alcoa executives would like to know if the coal intensities (for electricity generation and for export sales) of countries drive water pollutants.

Data are available for 37 coal producing countries on annual water pollutants by country, as well as use of coal to generate electricity, and coal rents from export sales. In addition to coal intensity, it is thought that level of development, reflected in a country's GDP, and mineral rents, the value of exported minerals, may also drive the level of annual water pollution.

It is thought that water pollution is greater among the rapidly industrializing BRIC and emerging countries.

I. Build a Model to Explain Water Pollution Differences Across Countries

Alcoa has asked you to build a model to explain differences in annual water pollution across countries. **Water pollution segmentation** contains data on *water pollution, coal use to generate electricity, coal rents from export, mineral rents from export,* and *GDP* in 37 countries. The 37 include developed nations, the *BRIC*s, emerging countries, and underdeveloped countries.

For variables which are zero for some countries, limit your choices of scales to square roots, cube roots, squares, or cubes. (One cannot rescale zero values with logarithms.)

1. Which potential continuous drivers do influence emissions?

___ GDP ___ coal to generate electricity ___ coal rents ___ mineral rents

2. Write your model of water pollution (in the original scale of K BODs), specifying units of all continuous variables in your model:

3. One of the Alcoa managers is convinced that water pollution (and potential demand for the Alcoa process) is higher in countries where mineral rents are higher. Is this manager correct?

4. In order to use linear regression, the residuals must be approximately Normally distributed. Are your model residuals approximately Normal? ___Y or ___N

   Cite the statistic (and its value) that you used to reach your conclusion: _________

I. Sensitivity Analysis of Driver Impacts

1. Find expected water pollution in a hypothetical baseline *emerging market* country in which all continuous drivers are at their medians. Next, find expected water pollution in a hypothetical *emerging market* country with the most influential driver at the 5%, 10%, 25%, 50%, 75%, 90% and 95%. Plot expected water pollution by values of the most influential driver to illustrate response. Adjust axes to make good use of space, and include axes labels with units.

   Make three more graphs to illustrate the differences in expected water pollution in an *emerging, a developing,* and *a developed* country, in response to differences in levels of the most influential driver, using the same baseline "median polluted" country. Show driver levels at the 5%, 10%, 25%, 50%, 75%, 90% and 95%, illustrate response.

Copy and paste three of your four graphs into the first, for comparison by level of development.

2. Water pollution levels in BRIC countries are ___% higher than those in developed countries.
3. Water pollution in Ukraine is nearly four times that in Hungary. Provide one reason based on your results to explain this difference.
4. In which countries do differences in GDP make the greatest differences in water pollution?

___ Emerging ___ BRIC ___ developing ___ developed

Index

C. Fraser, *Business Statistics for Competitive Advantage with Excel 2016*,
DOI 10.1007/978-3-319-32185-1